Ash Fox

MAIA SZALAVITZ is one of the premier American journalists covering addiction and drugs. Her book *Help at Any Cost* is the first book-length exposé of the "tough love" business that dominates addiction treatment, particularly for teens. She is coauthor of *Born for Love* and *The Boy Who Was Raised as a Dog*, both with Dr. Bruce D. Perry, and her work often appears in publications such as TIME.com, *The New York Times*, *New York* magazine, *VICE*, *Scientific American*, *Elle*, *Psychology Today*, and *The Guardian* among others. She is a 2015–2016 Soros Justice Fellow and has received major awards for her coverage of neuroscience and addiction from the American Psychological Association, the Drug Policy Alliance, and the American College of Neuropsychopharmacology.

she exposes her own fears and pain. . . . A dense blending of self-exposure, surprising statistics, and solid science reporting that presents addiction as a misunderstood coping mechanism, a problem whose true nature is not yet recognized by policymakers or the public."   —*Kirkus Reviews*

"Journalist Szalavitz offers a multifaceted, ground-up renovation of the concept of addiction—both its causes and its cures."   —*Publishers Weekly*

Unbroken Brain

Unbroken Brain

A Revolutionary
New Way of
Understanding
Addiction

MAIA SZALAVITZ

PICADOR ST. MARTIN'S PRESS NEW YORK

picadorusa.com • picadorbookroom.tumblr.com
twitter.com/picadorusa • facebook.com/picadorusa

Picador® is a U.S. registered trademark and is used by Macmillan Publishing Group, LLC, under license from Pan Books Limited.

For book club information, please visit facebook.com/picadorbookclub or e-mail marketing@picadorusa.com.

Portions of chapter 18 first appeared in slightly different form in *Pacific Standard* magazine in M. Szalavitz, "The Drug Lord with a Social Mission," *Pacific Standard*, March 2015.

Portions of chapter 19 first appeared in slightly different form in *Pacific Standard* magazine in M. Szalavitz, "How America Overdosed on Drug Courts," *Pacific Standard*, May 18, 2015.

Designed by Richard Oriolo

The Library of Congress has cataloged the St. Martin's Press edition as follows:

Names: Szalavitz, Maia, author.
Title: Unbroken brain : a revolutionary new way of understanding addiction /
    Maia Szalavitz.
Description: New York : St. Martin's Press, [2016]
Identifiers: LCCN 2015044722 | ISBN 9781250055828 (hardcover) |
    ISBN 9781466859562 (e- book)
Subjects: LCSH: Substance abuse—Treatment. | Compulsive behavior. |
    BISAC: SELF-HELP/Substance Abuse & Addictions/General.
Classification: LCC RC564.S975 2016 | DDC 362.29—dc23
LC record available at http://lccn.loc.gov/2015044722

Picador Paperback ISBN 978-1-250-11644-4

Our books may be purchased in bulk for promotional, educational, or business use. Please contact your local bookseller or the Macmillan Corporate and Premium Sales Department at 1-800-221-7945, extension 5442, or by e-mail at MacmillanSpecialMarkets@macmillan.com.

First published by St. Martin's Press

First Picador Edition: May 2017

10   9   8   7   6   5   4

To Ted

Special acknowledgment to the Soros Justice Fellowship Program,

which supported the completion of this book.

CONTENTS

AUTHOR'S NOTE

Writing about addiction presents many challenges. One of the most difficult is the omnipresence of stigmatizing and inaccurate language. Advocates have successfully advanced "person first" language to describe people with other types of conditions like schizophrenia and bipolar disorder, observing that characterizing individuals entirely by their illness is dehumanizing. But this advance has yet to reach most media coverage of addiction, which also spews degrading terms like "junkie," "drunk," and "crackhead" in ways that are inconceivable in other health reporting.

Consequently, I have tried throughout to use "person with addiction" or "person with alcoholism" in place of "addict" or "alcoholic." When I do use the shorthand, it is only because the alternative would have been unduly clunky or repetitive or because I am referring to a stereotype. Further, I have only included the demeaning term "abuse" to describe drug problems when citing unfortunately named government agencies or the now-obsolete psychiatric diagnosis of "substance abuse," which formerly labeled problems milder than addiction. The preferred term for less severe drug problems is "substance misuse," which doesn't automatically associate drug users with child abusers, molestation, and domestic violence. ("Drug use" is the term for substance use that is not associated with harm or addiction.)

Moreover, when writing about autism, I have used the term "autistic person" rather than "person with autism" because that is the preferred language of many autistic advocates. These activists see autism as intrinsic to who they are and view saying "person with autism" as being similar to describing a woman as a "person with femaleness." I have also used "addicted people"—in this case, to minimize repetition and awkwardness.

Finally, I want to note that in the memoir sections of this book, names have been changed.

Unbroken Brain

Introduction

There is often a struggle, and sometimes, even more interestingly, a collusion between the powers of pathology and creation.

—OLIVER SACKS

I AM LYING ON MY BACK inside the thin metal tube of the brain scanner at the Semel Institute for Neuroscience and Human Behavior at UCLA, trying not to think about coffins and earthquakes. On my thigh is a rubber ball to squeeze in case of panic, which can immediately liberate me from the giant white donut-shaped machine; my head is now centered deep in the hole. Earlier, as I was propelled inside on sliding rails, I couldn't help but be reminded of the drawers in which corpses are kept in morgues. Although I'm wearing earplugs, the machine's metallic roar—complete with occasional shaking and shrill beeping—seems deafening. Since I am claustrophobic and abhor loud noise, I try to focus on my breathing. One task I will perform here is supposed to measure impulse control, but it is taking nearly all of mine not to immediately crush the squeeze ball and escape.

I'm not being scanned because a doctor has ordered it. I've actually chosen to put myself into this tight spot as part of an experiment. I want to understand more about addiction: my own history of it and what it means more generally. How did I go from being a "gifted" child and Ivy League scholarship student to injecting cocaine and heroin up to 40 times a day? Why did I recover at 23, when many others take much longer or succumb? More important, what determines who gets hooked, who recovers, and who does not? And how can we as a society do better at addressing addiction? As I wait in the scanner, I recall the last days of my drug use, a distressing period in 1988 when I spent my time either shooting up, selling drugs, or trying to buy. I consider what has changed—and what hasn't.

Sadly, had I nodded off in the '80s and somehow been revived in 2015, I wouldn't find much different about how we frame and deal with addiction. Sure, at least four states and Washington, D.C., have legalized recreational marijuana sales. That would be shocking to anyone whose last memories were of the "Just Say No" years. And yes, addictive behavior is back in the media spotlight, though these days it's not crack but Internet addiction, sex addiction, food addiction, gaming addiction, and the tragic drumbeat of prescription overdose deaths (celebrity and otherwise) that get the most attention. Overdoses are now, in fact, the number one cause of accidental death, surpassing even auto fatalities.

Indeed, today, more people than ever before see themselves as addicted or recovering from substance addiction: 1 in 10 American adults—more than 23 million people—said they'd kicked some type of drug or alcohol addiction in their lifetime, in a large national survey conducted in 2012. At least another 23 million currently suffer from some type of substance use disorder. That doesn't even count the millions who consider themselves addicted to or recovering from behaviors like sex, gambling, or online activities—nor does it include food-related disorders. With the 2013 declaration by the American Medical Association that obesity, like addiction, is a disease, up to one in three Americans may now qualify due to their body weight.

At the same time, Big Pharma, Big Food, Big Tobacco, Big Alcohol, and Big Business in general all seem to intimately understand addiction and how to manipulate it. However, most of the American public—including most people with drug problems and their families—do not. Trapped in outdated ideas—many unchanged since the flapper days of Prohibition—we continue to recycle the same tired debates and enforce counterproductive criminalization strategies. But it doesn't have to be this way.

I propose here a new perspective, one that could help end this stagnation and suggest a way forward in treating, preventing, and otherwise managing addictive behavior. As this book will demonstrate, addiction is not a sin or a choice. But it's not a chronic, progressive brain disease like Alzheimer's, either. Instead, addiction is a developmental disorder—a problem involving timing and learning, more similar to autism, attention deficit hyperactivity disorder (ADHD), and dyslexia than it is to mumps or cancer. This is clear both from abundant data and from the lived experience of people with addictions.

Like autism, addiction involves difficulties in connecting with others; like ADHD, it can also be outgrown in a surprisingly large number of cases. Moreover, like other developmental disorders, addiction can be associated with talents and benefits—not just deficits. For example, people with ADHD often thrive as entrepreneurs or explorers, while autistic people can excel at detail-oriented tasks and many are highly talented musicians, artists, mathematicians, and programmers. Dyslexia can improve visual processing and pattern finding, which is also helpful in science and math careers. Addiction is frequently linked with intense drive and obsessiveness, which can fuel all types of success if channeled appropriately—and some believe that the "outsider" perspective of people with illegal drug addictions is linked with creativity. In all of these conditions, the boundaries between normal and problem behavior are fuzzy.

Of course, in some ways addiction appears extremely unlike other developmental disorders, most prominently because it involves apparently deliberate and repeated choices, some of which, like taking illegal drugs, are considered inherently immoral. Early-life trauma also can play an important role in addiction, whereas it plays no role in autism. These differences mask important similarities, however. In both autism and addictions, for example, repetitive coping behaviors are frequently misinterpreted as the source of the problem, rather than being seen as attempts at solutions. In fact, severely neglected children often develop autistic-like behavior such as constantly rocking as a way to soothe or stimulate themselves—and maltreated children often appear to have ADHD because they are hypervigilant to "distractions" like the sound of a door slamming.

In all of these conditions—including autism itself—repetitive, vigilant, or destructive behaviors are not usually the primary problem. Instead, they are typically a coping mechanism, a way to try to manage an environment that frequently feels threatening and overwhelming. Similarly, addictive behavior is often a search for safety rather than an attempt to rebel or a selfish turn inward (a charge previously made against autistic children as well). We'll see throughout this book how misinterpreting understandable attempts at self-protection as hedonistic, selfish, or "crazy" has needlessly stigmatized people with developmental disorders including addiction—and, as a result, has increased associated disability rather than helping.

Critically, addiction is not created simply by exposure to drugs, nor is it the inevitable outcome of having a certain personality type or genetic background, though these factors play a role. Instead, addiction is a learned relationship between the timing and pattern of the exposure to substances or other potentially addictive experiences and a person's predispositions, cultural and physical environment, and social and emotional needs. Brain maturation stage is also important: Addiction is far less common in people who use drugs for the first time after age 25, and it often remits with or without treatment among people in their mid-20s, just as the brain becomes fully adult. In fact, 90% of all substance addictions start in adolescence, and most illegal drug addictions end by age 30.

The implications of the developmental perspective are far-reaching. For one, if addiction is a learning disorder, fighting a "war on drugs" is useless. Surprisingly, only 10–20% of those who try even the most stigmatized drugs like heroin, crack, and methamphetamine become addicted. And that group, which tends to have a significant history of childhood trauma and/or preexisting mental illness, will usually find some way of compulsively self-medicating, no matter how much we crack down on one substance or another. In this context, trying to end addiction by attempting to eliminate particular drugs is like trying to cure compulsive hand washing by banning one soap after another. Although you might get people to use more or less harmful substances while in the grips of their compulsions, you aren't addressing the real problem.

Second, given that addiction is a learning disorder, it isn't necessarily a lifelong problem that demands chronic treatment and the acceptance of a stigmatized identity: studies find that the majority of cocaine, alcohol, prescription drug, and cannabis addictions end before people are in their mid-30s and most do so without treatment. Similarly, between one third and one half of children diagnosed with ADHD no longer meet criteria for it as adults, and treatment doesn't seem to affect whether they outgrow the disorder or not, although it certainly can affect their ability to thrive. Finally, the learning perspective offers insight into other conditions—from anxiety disorders to schizophrenia, bipolar disorder to depression—that often precede addiction and could benefit from similar approaches.

Challenging both the idea of the addict's "broken brain" and the notion of a simple "addictive personality," *Unbroken Brain* offers a new way of thinking about drugs, craving, and compulsions, whether seen in behavior as extreme as shooting drugs or as ordinary as dieting.

AS I WAIT for the scanner to measure the soft structures under my skull, I can't help but think about the organ that scientists consider to be the most complex object in the known universe. I know that all of our experiences are somehow written into our brains. Somewhere in the sinuous curves and pulsing surfaces of my own must be the echoes of everything I've ever learned, whether I can now recollect it or not, and every choice I've ever made, consciously or otherwise.

And somewhere in my gray and white matter are the neural structures that put me at high risk for addiction even before I'd taken drugs; here, too, is any lingering residue of chemical changes made by the substances themselves. Everything I am and everything I've ever been has at some point been represented here chemically, structurally, or electrically: not just addiction but now over 25 years of recovery and decades of other life experiences.

I hope that my scans can help me illustrate why learning matters in this disorder. After decades of reading, reporting, and writing on addiction, hundreds of interviews with experts, and even more with drug users and former users—many of whom have experienced addiction—I have come to believe that learning is the key to better treatment, prevention, and policy. While scientists have long recognized that learning is critical to addiction, most of the public does not—or is not aware of the implications of seeing it this way. However, trying to understand addiction without recognizing the role of learning is like trying to analyze songs and symphonies without knowing music theory: you can intuitively identify discord and beauty, but you miss the deep structure that shapes and predicts harmony.

Failing to recognize the true nature of addiction has also come at a catastrophic price. It prevents us from effectively tackling all types of drug problems, whether in terms of prevention, treatment, or policy. It buries the need for individualized approaches. It also keeps debates on these issues stalled in sterile arguments over whether addiction should be seen as a crime or a disease.

Further, misunderstanding addiction allows drug policy to continue to be used as a political and racial football, since our ongoing use of ineffective tactics has produced widespread despair about affected people and families. In fact, however, research shows that, overall, addiction is the psychiatric disorder with the *highest* odds of recovery, not the worst prognosis—as many have been led to believe.

Addiction doesn't just happen to people because they come across a particular chemical and begin taking it regularly. It is learned and has a history rooted in their individual, social, and cultural development. We think it's a simple brain disease or a matter of criminal behavior because we don't understand its developmental history and how that plays out in wildly varied ways to create a suite of problems that only look the same superficially. Understanding the role of learning illuminates what's really going on and what to do about it.

Properly understood, the addicted brain isn't broken—it's simply undergone a different course of development. Like ADHD or autism, addiction is what you might call a wiring difference, not necessarily a destruction of tissue, although some doses of some drugs can indeed injure brain cells. While, like anything else that is learned, addiction may get more engrained with time, people actually have *increased* odds of recovery as they age, not reduced chances. This apparent paradox makes much more sense if seen as part of a developmental disorder that can change with life stages.

Moreover, as parents and teachers everywhere know well, it's almost impossible to force or coerce learning—especially to alter behavior that has already become habitual. As B. F. Skinner himself observed, "A person who has been punished is not less inclined to behave in a given way; at best, he learns how to avoid punishment." Fear and threat also literally shunt energy away from the areas of the brain involved in self-control and abstract reasoning—the exact opposite of what you want when you are trying to teach someone new ways of thinking and acting. Changing behavior is far easier if you use social support, empathy, and positive incentives, as a great deal of psychology research—though often ignored in addiction treatment and policy—demonstrates. This has obvious implications for the prospects of altering addiction via the criminal justice system.

Finally, the role of learning and development in addiction means that unlike in most physical diseases, cultural, social, and psychological factors are inextricably woven into its biological fabric. Pull any thread alone and the entire idea unravels into an incomprehensible tangle. Label addiction as merely biological, psychological, social, or cultural and it cannot be understood. Incorporate the importance of learning, context, and development, however, and it all becomes much more explicable and tractable.

Seeing addiction as a learning disorder allows us to answer many previously perplexing questions, such as why addicted people can make apparently free choices like hiding their drug use and planning to ensure an ongoing supply while failing to change their habits when they result in more harm than good. Learning helps explain why cultural trends and genetics can both have big influences and why addictive behavior is so varied. Further, learning and development elucidate why factors like employment and social support affect recovery in a far greater way than they do with physical illness. Sadly, cancer rarely disappears when someone falls in love and marries—but alcoholism and other addictions can and often do remit.

In the rest of this book, we'll see just how intimately learning is involved in every aspect of addiction: from the molecular brain changes that result from certain patterns of drug use and experience to the associations between drugs and particular cues and memories that are mediated by individual, familial, cultural, and historical circumstances. Using my own experience as a case study, I'll show how one addiction played out through one pathway and why others take different routes. Though my specific story is undoubtedly unusual, its particulars illustrate the universality of learning in the addiction process and why its singular nature in all cases is critical to understanding the larger problem.

Here, we'll see how addiction affects a very specific type of learning, involving ancient brain pathways that evolved to promote survival and reproduction. Since those are the fundamental tasks of any biological organism, they produce highly motivated behavior. When starving, when in love, and when parenting, being able to persist despite negative consequences—the essence of addictive behavior—is not a bug, but a feature, as programmers say. It can be the difference between life and death, between success and failure. However, when brain pathways intended to promote eating, social connection, reproduction, and parenting are diverted into addiction, their blessings can become curses. Love and addiction are alterations of the same brain circuits, which is why caring and connection are essential to recovery, too.

The world is finally recognizing that the punitive American approach to addiction, which has dominated drug policy for the last century, is failing. In order to move beyond it, a new understanding of the disorder and its relationship to drugs and other behaviors is needed. Only by learning what addiction is—and is not—can we begin to find better ways of overcoming it. And only by understanding addicted people as individuals and treating them with compassion can we learn better and far more effective ways to reduce the harm associated with drugs.

As I lie in the scanner, a snippet of a Talking Heads song pops into my head. It goes, "Well, how did I get here?" That's the mystery of every addiction—and to solve it, we need to look at it from an addict's perspective, elucidating general principles by examining the specific and particular.

ONE

Needle Point

Heroin was the only thing that really worked, the only thing that stopped him scampering around in a hamster's wheel of unanswerable questions. Heroin was the cavalry . . . [it] landed purring at the base of his skull, and wrapped itself darkly around his nervous system, like a black cat curling up on its favourite cushion.

—EDWARD ST. AUBYN, *BAD NEWS*

BY JULY OF 1988, MY LIFE had narrowed to the point of a needle. I was living with my boyfriend, Matt, and selling cocaine. My only daily goals were, first, to slog to a methadone program, and then, somehow, to ensure we made sufficient money to get high and pay for rent and cat supplies. That summer was simultaneously the best and the worst time of my life. It was the best because in August, I would successfully kick the cocaine and heroin addiction that had left me weighing 85 pounds, with angry tracks dotting all four limbs, my hair a thin, overbleached Madonna-trying-to-be-Marilyn blonde and my eyes distant and blank. It was also the worst because, well, I wouldn't exactly *recommend* active addiction and early recovery.

I was 23. I was out on bail, facing a mandatory minimum 15-to-life sentence on a 1986 cocaine charge, under New York State's Rockefeller drug laws. I had been busted with 2.5 kilos of cocaine, which made me look like a high-level dealer, but the truth was that most of it belonged to Matt's supplier, who had asked him to stash it.

Few would have predicted such a future for a child who read at three, who tried to channel her social awkwardness and became "most likely to succeed" in eighth grade, and who excelled academically enough to be admitted to Columbia in its first class of women in 1983. But Columbia was now in my past. I couldn't study while facing the stress of a felony case; in fact, I couldn't do much at all, not even tasks of basic self-care like household cleaning, bathing, and laundry.

About here's where I'm supposed to tell you that I'm different, that I wasn't your "typical addict." The American media repeatedly assures us that such an addict certainly isn't white, female, educated, or middle class. But I'm not going to do that. History demonstrates that the idea of the "typical addict" is itself a cruel stereotype forged in a period of intense racism, which has a great deal to do with why our drug treatment system and drug policies are both draconian and ineffective. The whole notion is one of the hidden obstacles that keeps us from truly understanding drug issues. To do better, we need to understand what addiction really is—and how our misguided attempts to define it have actually caused great harm.

In the 1980s, when I was addicted, great emphasis was placed on the distinction between "physical" and "psychological" addiction, and folk belief in the importance of this difference remains surprisingly common. Physical addiction was seen as medical: it was primarily a problem of dependence, of biologically coming to need a drug to function without being physically ill. Indeed, the official term for the problem in psychiatry's diagnostic manual, in the '80s and until 2013, was "substance dependence."

"Psychological" addiction, however, was seen as moral: it meant you had lost control over your mind and were weak willed, selfish, and bad. Physical addiction was real; psychological addiction was all in your head. Unfortunately, as people like me learned the hard way, the physical need for the drug to avoid withdrawal symptoms is not the core of the problem. Instead, psychology, and the learning that influences it, matters much, much more. In the summer of 1988, that psychology dominated my life.

*   *   *

ONE OF MATT'S favorite words was "fetid," and that aptly described our living conditions that summer. Our $750/month rental in Astoria, not far from the Triborough Bridge, was essentially a square divided into four rooms, sparsely furnished, with a bare, stained futon mattress on the floor in one bedroom, many books, comic books, records, CDs, a high-end stereo system, and a few tables and chairs.

Scattered around were the detritus of drug habits: bent, blackened spoons and bulbous glass crack pipes, some of which were broken and had charred metal screens in their bowls. A few neon-orange syringe tops could be seen atop piles of dirty laundry, mine almost entirely black. In the corner of one bedroom was a desk with an early PC and dot matrix printer, which I used to file articles I wrote for the stoner magazine *High Times*. (My first national column, written under the pseudonym Maura Less, was called "Piss Patrol" and covered urine testing.)

A litter box stood in another corner, and our long-haired gray tiger cat, Smeek, padded around, showing off his massive puffy tail. Smeek, at least, was well loved and perhaps a bit too well fed. But otherwise we lived in filth and disarray—and the litter certainly wasn't always clean, a situation Smeek would sometimes protest by thinking outside the box, often on the scattered papers and clothes.

Meanwhile, Matt had become grotesquely obsessed with his bodily functions and terrified of being arrested by firefighters. He thought that the men in the red trucks were somehow able to monitor and detect the fumes from the cocaine he smoked. He always kept the shades down, cautiously peeking out occasionally to see if the firefighters were on to him. This once drily witty and artistic Jewish boy from Long Island now sat indoors most days, wearing only tighty-whities and surrounded by garbage, convinced that freebasing was destroying his digestive tract but unable to stop himself.

Every morning, I'd tell myself that I wasn't going to shoot coke, knowing that it would only make me anxious, obsessive, and paranoid (though, at least not about firemen!). I'd drag myself to the fortress-like methadone program near the elevated subway lines at the base of the 59th Street Bridge. I had chosen to get this treatment; I knew being physically dependent on heroin was a problem and I wanted help detoxing from it. I thought, in fact, that doing this would be all that was needed to get me back on track.

I was thoroughly steeped in America's paradoxical view of addiction: I thought it was simultaneously a moral and a medical problem. I couldn't accept that I had the moral problem; I thought that would mean that my intelligence—the only aspect of myself that I valued—was weak and corrupted. So, I told myself I was "just physically addicted" and that methadone would fix that.

The idea was to "wean" me from an illegal opiate by providing a safe, clean, noninjectable legal one in progressively lower amounts over the course of six months. The Bridge Plaza Methadone Maintenance Treatment Program started by first "stabilizing" me on what I now know is far too low a dose. An effective dose of methadone varies from patient to patient but is typically over 60 milligrams (I was given 30) and ultimately reduces craving for heroin without producing euphoria. At such a dose, methadone also blocks your high if you relapse. I never experienced that.

But even if the program had gotten the original dose right it wouldn't have mattered much. They began "tapering" me almost immediately, decreasing the methadone in what research had already shown by that time to be a highly ineffective way of using the drug. Consequently, as that data could have predicted, I titrated up my heroin use as they ran the methadone down, rendering the entire exercise useless while sustaining my physical dependence. I felt hopeless, trapped.

So, I changed tactics. First, I convinced myself that the methadone actually made the detox process worse. The word on the street was that kicking heroin was "easier" than methadone detox because the worst of heroin withdrawal lasts around two harsh weeks, while methadone withdrawal is more protracted, lasting months (though if done right, I later learned, it should be less severe). My new plan to quit drugs became this one: I would complete the methadone detox and afterward, just do heroin for a few weeks to get the methadone out of my system. *Then* I'd stop for good. And yeah, I'd stop the cocaine at that point, too. Today, however, I'd just have one more shot.

Since Matt and I were selling cocaine and virtually always had it around, that one injection would soon lead to dozens. I'd dig for one of my few remaining accessible veins, awaiting the moment when I'd strike it rich, watching the blood blossom up into the barrel like a gusher of oil. But even when that happened easily, the euphoria no longer sparkled. It was contaminated with paranoia, shadowed by a looming overhang of objectless dread. What had started as a burst of excitement that opened for me a sense of endless opportunity and capacity was now fraught with fear and a feeling of being stuck, not liberated.

Desire curdled into dread that only prompted more fruitless and frustrating desire for more.

Wired to the gills, shaking, unable to relax, my heart seeming to pound louder than it should, I'd then realize that the only thing that would help was heroin. That would produce an epic quest to cop in what were then dire neighborhoods in Bushwick, Brooklyn, or Manhattan's Lower East Side.

I was terrified of getting arrested while buying—not only for the usual reasons, but because I was afraid of what effect it would have on the conditions of my bail and, therefore, on both of my divorced parents' houses, which served together as collateral instead of $50,000 in cash. The high bail was set because the 2.5 kilos of cocaine I'd been charged with possessing when I got busted legally qualified me as a high-level dealer, although that was hardly the reality. It was actually my first arrest.

Consequently, to minimize rearrest risk, I wouldn't buy for myself, but I would instead ride along with friends who would score on the street for me in return for a share of the drugs. We'd drive in someone's clunker of a car to Bushwick on the Brooklyn-Queens Expressway, shivering past the acres of graveyards that divide the two boroughs. When we got close, I'd slump out of sight in the passenger seat or backseat. My race and ragged looks made it obvious why we were in the neighborhood. I'd wait anxiously as whoever had agreed to drive darted into decrepit, graffitied buildings, always taking what seemed like years to return.

The heroin, if we managed to procure decent stuff—not ineffective, adulterated dross—would win me a few blessed hours of blissful calm. When I got home, I'd heat the spoon to prepare the heroin, which I dissolved in water, adding a dash of coke when it cooled, then injecting the mixture. If the drugs were good and my tolerance wasn't too high (a rare concurrence at this stage of my addiction), the first hit would be heavenly. Like a flourish from a brass section, the cocaine would trumpet a burst of exhilaration as I pressed the plunger in; I could taste its icy flavor at the back of my throat. A few moments later, the warmer, soothing harmony of the heroin would take over. Every atom in my body felt calm, safe, fed, content, and, most of all, loved.

Unfortunately, however, before long, I'd decide that another shot of cocaine would be nice. That would start a compulsive cycle of "just one more," until the cocaine's anxious alertness blotted out the heroin's sedative effect entirely. After a sleepless night, the next day would be exactly the same, starting with the humiliations of the methadone program.

Low-slung, located in a forbidding light industrial area in the shadow of the elevated tracks of the N and 7 trains, surrounded by businesses like auto parts suppliers, the place looked like a prison. Every feature bespoke an emphasis on security and the siege mentality that comes with trying to retain valuables when you see all of your customers as criminals. Rain, snow, sleet, or hail—there was no weather extreme enough to keep a line from forming outside before the clinic's early-morning opening time so we could be medicated to stave off withdrawal. At that minute—sometimes a bit later, but never a second before—the heavy metal door would creak open and we'd be herded into a mantrap and monitored by video camera. After the first door closed, a second, equally armored and imposing portal would open to admit us.

Next, we'd line up inside. Often, you would have to provide a urine sample before you'd get your bitter cocktail of methadone and orangeade from a nurse using a precisely calibrated and heavily defended machine. This process posed a problem for me that summer: I was virtually always dehydrated from shooting drugs the night before. And if I had to give an "observed urine," the woman charged with that lovely job would usually have to stand there and wait until I was able to eke out enough pee. They could have just asked if I'd taken drugs: I don't think I ever gave the place a "clean" urine, and my using should have been a signal that I needed additional assistance. But that approach would have required them to see me as a fellow human being who was ill, not as just another junkie—and it would have required genuinely individualized therapy, not just bureaucratic rules.

This was my first personal encounter with supposedly professional "help" for addiction: a system that calls you "dirty" if you relapse; one that assumes you are a liar, a thief, or worse and responds to increasing symptoms of addiction not by offering more help but by punishment or expulsion. Indeed, as it became clear that my "detox" was failing, I asked my counselor if I could stay on methadone longer to see if I could stabilize and improve, but I was told that I hadn't used heroin for long enough to be given long-term methadone treatment. And besides, I took too much cocaine.

In other words, while it was clear that I wouldn't successfully detox, my problem was both "too bad" (the coke) and not "bad enough" (too few years on heroin) for me to get more help. The fact that I exhibited symptoms of addiction was basically why I was expelled from treatment for it. I wasn't even offered a referral for any type of additional rehab or medical care—despite the fact that it was the peak of the HIV epidemic in intravenous drug users in the United

States and I was in New York City, the epicenter of that epidemic. At least half of the injection drug users in the city were already infected; among these were many of my friends, with whom I could have shared needles. In any other area of medicine during a global pandemic, such "care" would be considered malpractice.

Still, that's what I received—and what sadly is still all too common today, with at least one third of all methadone programs still failing to provide an adequate dose. But as unbelievable as it now seems even to me, despite shooting up dozens of times a day and facing felony drug charges, despite being on a methadone program for heroin addicts and having dropped out of college following my arrest, I didn't yet see myself as a real drug addict.

That would change on August 4. That was the day I recognized that I was about to cross a line and meet my own carefully designed criteria for addiction (specially created, mainly, to try to exclude myself, I must admit). While recovery stories are often told as though they result from sudden insight that prompts life-altering action, in reality, studies find that psychological breakthroughs are not the typical path to change and rarely lead directly or in any linear way to alterations in behavior. Indeed, research suggests that having an intention to do something only predicts engaging in the desired behavior about 33% of the time, even for people *without* drug problems. Learning a new behavior typically takes time.

My experience, however, was somewhat different. My story may be an example of a recovery path that researcher William Miller has labeled "quantum change," in which a minority of people do suddenly completely shift course— as opposed to the more typical gradual process of fits and starts. It could also be the case that my brain's natural maturation process finally reached the point where my "executive function" could begin to put the brakes on the regions that create desire—and this change allowed my epiphany to save my life.

Either way, it had been one of those days dominated by the flat sense of terror that occupied my consciousness whenever I wasn't sufficiently drugged. I was no longer on methadone, which meant that I was back to heroin and coke. I felt the insistent premonitory angst of withdrawal. To elude it, that afternoon, I went to the Lower East Side with Heather, the girlfriend of one of Matt's friends. It was hot on the streets, literally and figuratively. No one seemed to be selling. Finally, however, she spotted a tout—a person who introduces buyers to sellers—and went off to make the transaction while I stood and waited.

I thought my heart would jump out of my chest every time I heard a loud noise or saw a car that might be the police. It was a few years after Operation Pressure Point, the first of several major police crackdowns that constantly swept users and low-level dealers into the justice system. I was on a street that led through a series of low, squat housing projects, which somehow seemed menacing in their red brick uniformity. I wanted to hide. Passersby seemed to look right past me, as though I was an unfortunate part of the street furniture, like an overflowing garbage can. I hung my head and looked away.

A dream I'd had the previous week popped into my mind. In it, I'd been struggling to fend off some type of space alien parasite that wanted to embed itself in my brain. I watched it merge with others, giving them unspeakable bliss but destroying their individuality and controlling their every move. Once it got you, you were part of it and couldn't be yourself anymore. I tried to flee, but nowhere was safe, and as a friend betrayed my hiding place to it, I started to be engulfed. I woke up, panicked and shaken. In retrospect, it seems like something in me was preparing for change.

I started to worry that Heather had gotten arrested and that I would be stuck on this corner in the hot sun for hours, getting sicker and sicker. But suddenly, she appeared from around a corner, walking with the quick, purposeful pace that I knew meant that she had scored.

When we got back to Queens, though, everything went wrong. There were several different brands in the two "bundles" (ten bags per $100 bundle: one for her and her partner, one for me) that she had purchased. Street heroin suppliers stamp their products with often ghoulishly humorous brand names—evergreens like D.O.A., Seven to Life, and Poison alternating with ripped-from-the-headlines snarks like today's Obamacare. The variety of brands, however, wasn't a good sign: if it's not a rip-off, you generally get one or at most two stamps from any particular street seller.

I dumped the contents of one bag into a spoon as soon as I got into the kitchen. I was definitely starting to feel ill now—withdrawal is actually much worse when you get the drug in your possession and know you can have it soon, but not yet. It didn't smell right as it hit the water. Still, I shot it anyway, to no effect.

And that was when my mind—no thanks, I suspect, to the pharmacology of the apparently inert substance in the packet—really did change. I found myself begging Heather's boyfriend, a man I didn't especially like, for some of the differently branded bags she had given him. His street name was Beaver. Brown-haired and bearded with a distinct overbite, he rather resembled one of those dam-building creatures, though not because of their reputed industriousness. He was usually extremely laid back.

I found myself pleading with him for another brand of heroin in the hopes that at least one of the bags might contain real dope. I couldn't afford to buy more. I felt desperate, with withdrawal clearly starting to close in. And, I argued, since he wasn't physically dependent at the time, the only consequence for him would be that he just wouldn't be able to get high that day. I, on the other hand, would be sick. Besides, I knew I had to appear in court the next day, one in an endless string of continuances that my lawyer sought to delay the day of reckoning on my terrifying Rockefeller case. I'd already "completed" the methadone program; I was supposed to be well.

But Beaver wouldn't budge. And in my panic, the thought crossed my mind that I might try to seduce him, just to get the drugs. It didn't matter that both my boyfriend and his girlfriend were in the same room with us. It didn't matter that I had a really snobbish attitude toward him and was not attracted to him. It didn't matter that the idea was basically completely insane and unlikely to succeed. My brain was searching for any strategy, any way of obtaining heroin, however absurd.

When that thought occurred to me, however, it shocked me. I don't mean to judge those who exchange sex for drugs or engage in prostitution to support their habits: if harm occurs, they themselves are the most likely victims and many women who do this have lengthy histories of childhood sexual abuse. However, I personally couldn't fathom the idea. For one, I'm seriously sensitive to touch, shame, and rejection. Second, I had this bizarre misconception that you had to be exceptionally beautiful in order to get men to pay for it, and, well, my self-esteem was, comically, not high enough to believe I'd pass as a credible prostitute.

Consequently, the idea of seducing Beaver was so beyond what I'd ever normally consider that it immediately forced me to contemplate what had previously been unthinkable: that I was a debased addict. Physically, psychologically, whatever you wanted to call it: the line I was about to cross was, for me, a sign that I was clearly in what I'd formerly seen as alien territory, where "those" weak-willed addicts dwelled.

Now, something in me began to shift. I considered my early drug use— marijuana and hash in high school, LSD at Grateful Dead shows, cocaine in chic nightclubs. I pictured myself back in my dorm at Columbia, weighing out quarter grams of what I then saw as a harmless drug that brought joy and aided study. I remembered the parties, especially at one of New York's most exclusive addresses, a Fifth Avenue apartment high above 86th Street, overlooking Central Park—somewhere I'd never have been invited if I hadn't been selling drugs. Then, I flashed on the late nights unable to stop, unable to sleep, with a bleeding nose and all that glamour drained away. I thought about being suspended from school for dealing and later, after being readmitted, my arrest. I saw myself snorting heroin, then shooting it, and now here in this filthy apartment, begging a man I didn't like for drugs.

I looked around and it was as if I was seeing my home, and myself in it, for the first time. The cat litter hadn't been changed in far too long, and it gave off an evil stench. Everything was covered in a layer of grime—cat hair clotted with dust atop soiled laundry; burnt, shattered crack pipes on yellowing newspapers. The atmosphere was nauseating. I was suddenly amazed at my surroundings and unable to comprehend how I had lived in such a state for years.

It was as if a camera had pulled back, and instantly, rather than seeing an incomprehensible mass of lines and colors, I could see my life for what it was. As I begged for heroin, the lens through which I saw the world shattered, and suddenly, everything I thought was certain, everything I thought I knew about drugs and my life, was no longer sure.

That was when I decided that I needed help. My father would be meeting me in the courtroom the next day; he always showed up, no matter how wretched and hopeless I seemed. My mother, in contrast, had found attending my court appearances so upsetting that she stayed away. For at least six months, she'd begun limiting even our phone conversations to brief talks, in which she'd always suggest rehab. This tactic had been recommended to her by her therapist and in the drug counseling classes she'd been taking to train for a second career. I resolved to ask my dad to take me upstate to her house, so my mom could help me get into treatment. I remember my final night of using—that bad shot turned out to be my last one—as a blur of unpleasant sensations that heralded my ongoing descent into withdrawal.

I went to court sick and sweaty. I wore a black, but at least brightly flowered, "business" dress, my bones jutting out. I was so weak, I could barely open the wooden half door to the defendant's well; my lawyer, Donald Vogelman, a prominent defense attorney, gallantly did it for me. He was tall and solidly built, in his late 30s or early 40s, with dark hair and a strong Brooklyn accent.

Judge Leslie Crocker Snyder, a former prosecutor famed as the "Dragon

Lady" for her harsh sentencing policies, later said she had been considering locking me up "for my own good" if I hadn't sought help that day. She had straight, perfectly coiffed blond hair. Her voice and body language visibly projected power and authority, not just her robes. She was so well known for her harsh sentencing that she needed round-the-clock police protection from hit men sent by drug lords angered by sentences stretching up to 120 years.

As I stood before her in court, I was so painfully thin and pale that I could have been mistaken for a cancer patient and was in fact soon suspected of having anorexia. I hadn't been trying to lose weight; it was just that cocaine took away my appetite. When I did bother to eat, my diet consisted almost exclusively of Boston cream pie and other sweets from Astoria's excellent and plentiful bakeries.

My hair also testified to my poor health: once an over-the-top Jew-fro, it was now brittle and thin and hung in clumps, as if I'd had chemotherapy. Not only had I damaged it with too much bleach, I'd also developed trichotillomania, a habit of compulsively pulling strands out, which can be associated with cocaine addiction. Added all together, I looked more than twice my age. With my pupils dilated by withdrawal, I looked haunted. My mother described me at this stage of my addiction as looking as though there was "no one home behind my eyes."

I started inpatient detox later that day. August 4, 1988, is the day I now see as the beginning of my recovery. What changed my mind wasn't just the weight of all that legal pressure or my worsening health—it was my sudden realization that I met my own personal criteria for addiction. I hadn't stopped at the "worst" moment in my using or when I "hit bottom"—that would have been either the day I got suspended from Columbia or the night I got arrested and taken out of my apartment in handcuffs, or when my father bailed me out after that arrest.

Instead, I stopped when I diagnosed myself or, basically, learned that I was an addict.

TWO

A History of Addiction

I have absolutely no pleasure in the stimulants in which I sometimes so madly indulge. It has not been in the pursuit of pleasure that I have periled life and reputation and reason. It has been the desperate attempt to escape from torturing memories, from a sense of insupportable loneliness and a dread of some strange impending doom.

—EDGAR ALLAN POE

THE FIRST EMERGENCY ROOM WE TRIED wouldn't admit me, claiming they didn't treat "junkies." Although my mother had been trying for months to get me to accept help, she had been caught by surprise when I finally agreed to it. She didn't know where to take me, immediately ruling out one of the local hospitals. My youngest sister worked there as a candy striper—and my mom didn't want her to be embarrassed by my condition. I wound up at what was then known as Community General Hospital of Sullivan County, which accepted me into a 7-day detox program.

I lay on a gurney, shaking and crying, holding my mother's hand. At some point, I was given a shot that seemed to have no effect. The nurse wouldn't say what it was. I later learned that I'd been injected with naloxone, an opioid antagonist that is merciful when used to revive overdose victims but slightly sadistic for someone already in withdrawal, as it can intensify the symptoms. It works as an antidote, removing any remaining opioid drugs from their receptors. (It is the same drug now being made available to addicts, their loved ones, and police officers to reverse overdose, and this way of using it undoubtedly saves lives.)

I guess the "therapeutic" rationale for administering naloxone in detox was the theory that this would more rapidly remove the drugs from my system; the punitive one is that it increases your distress. Naloxone worsens withdrawal symptoms and it was given without my consent—another sign of the paradoxical way addiction is viewed as both a sin and a disease. The idea that addicted people should have the right to informed consent like any other patient wasn't even considered. Though treatment and punishment were supposed to be opposite approaches, in fact, harsh moralistic tactics were the rule, not the exception, when I sought help, and they are still part of the treatment experience for most people with addictions.

Before I sought help, I'd read about rehabs where addicts were "put on the hot seat" and dressed down, with everyone present ripping into their every flaw and defect, getting in their face and shouting, often spitting on them. I'd heard about such sessions lasting hours without breaks for the bathroom or sleep, about complete lack of privacy, and about relentless attacks aimed at shattering your identity. I'd seen these kinds of programs honored with celebrity charity dinners and featured uncritically in the mainstream media. I knew that they were widely accepted. In fact, fear of the brutal and deliberately humiliating tactics I'd heard about in inpatient addiction treatment had been a big part of why I hadn't tried it earlier.

I was, in general, terrified of other people; I used drugs for emotional protection and social comfort. The idea of being made even more vulnerable and "broken down" as a way to heal addiction seemed like the exact opposite of what I needed. Being the center of inescapable public condemnation with no control over my environment was my idea of hell, not help. I suffered no shortage of shame, self-loathing, and guilt; I did not inject cocaine dozens of times a day because I was proud of myself.

And for people who are introverted or oversensitive, having no private space or time and being forced into lengthy periods of group activity, even if it is friendly, can feel like torture. If hostile and unremitting personal attacks are focused on you, it's downright traumatic—particularly for those who have been bullied or abused in childhood, which includes a large proportion of people with addictions. One study found lasting psychological damage in nine percent of even normal college students who participated in an "attack therapy" encounter group modeled on one used commonly in drug treatment. The effects on those who are not mentally healthy, of course, are even worse. However, I didn't feel like I had any choice at that point other than to try to get whatever kind of treatment was on offer.

I didn't know how history had twisted both my own view of addiction and the way I would be treated for it.

THE USE OF mood-altering substances probably predates the rise of humans and other primates. Many species deliberately seek intoxicating plants or the alcohol produced in rotting fruit. Cats, big and small, for instance, ingest and roll in catnip for no other discernible reason than pleasure (though it may have the convenient side effect of killing parasites). Horses will seek out locoweed, which, as the name suggests, makes them behave bizarrely—and they'll eat it again, even after they have been sickened by it. The stimulating properties of coffee are said to have been revealed when people observed an extra spring in the steps of goats that had eaten the plant's beans. Some archaeologists even claim that civilization itself began when humans settled down to grow grain— not because they wanted it for food, but to make beer.

Whatever the evolutionary precursors of drug use are, a permanently "drug-free" human culture has yet to be discovered. Like music, language, art, and tool use, the pursuit of altered states of consciousness is a human universal. With access to few alternatives, Siberian shamans imbibe reindeer and human urine to maximize the psychedelic yield of *Amanita muscaria* mushrooms (the metabolite that is excreted may be stronger than the substance initially ingested); on nearly the opposite side of the world, New Zealanders party with untested "research chemicals" synthesized by Chinese chemists. Drug use spans time and culture. It is a rare human who has never taken a drug to alter her mood; statistically, it is non-users who are abnormal.

Indeed, today, around two thirds of Americans over 12 have had at least one drink in the last year, and 1 in 5 are current smokers. (In the 1940s and '50s, a whopping 67% of men smoked.) Among people ages 21 to 25, 60% have taken an illegal drug at least once—overwhelmingly marijuana—and 20% have taken one in the past month. Moreover, around half of us could suffer from physical withdrawal symptoms if denied our daily coffee. While Americans are relatively prodigious drug users—topping the charts in the use of many substances—we are far from alone in our psychoactive predilections.

Drug *addiction,* however, is far more rare—typically affecting 10–20% of users of the most common substances, with the exception of nicotine in the form of cigarettes, which addicts about one third of those who try smoking. Again, with the exception of tobacco, these statistics are typically consistent across time in national surveys, which compare the number of people who have ever tried a drug with the number who currently use it and meet diagnostic criteria for addiction. They are also consistent in studies that follow drug users for long periods and in research on the prevalence of various psychiatric disorders, including addictions, in large international populations. Though drug education programs tend to avoid publicizing these statistics, the expert consensus is that serious addiction only affects a minority of those who try even the most highly addictive drugs, and even among this group, recovery without treatment is the rule rather than the exception.

Importantly, the idea of addiction itself is also a relatively modern concept. To understand what it really is and how our thinking about it has become so distorted, it's crucial to know a bit about the history of this idea and see how politics and cultural biases have predominated over science in addressing it. My recognition of my own problems with cocaine and heroin was shaped by this troubling history—and only after I learned about the origins of the stereotypes about addiction and drug users was I able to see how much damage they have done. Without taking this history into account, it's hard to understand why we treat addiction the way we do—and why, to change it, we need to recognize the role of learning.

Historically, the word *addiction* referred to a social relationship of bondage: its Latin root means "enslaved by" or "bound to." At first, what we now call addiction was seen as a voluntary, if inadvisable, choice. The Bible, for example, describes a "drunkard" as a "lover of wine"; addiction here is an overindulgence in pleasure, not a joyless compulsion. "God sends many sore judgments on a people that addict themselves to intemperance in drinking," preached the Puritan Samuel Danforth, reflecting the idea that addiction was characterized as habitual, but deliberate, engagement in frequent excess. Those who chose to act on their addictions were seen as loving their intoxication too much, but otherwise, they were no different from any other kind of sinner. This link between unhealthy love and addiction has since been repeatedly rediscovered, but it tends to be misunderstood, as I'll explore further in chapter 11.

"Opium eaters," like Thomas De Quincey, who famously wrote his "confessions" about his experience in 1821, also spoke of the "extensive power" and "marvelous agency" of this drug. He described the high as a "panacea for all human woes" and "the abyss of divine enjoyment." De Quincey gave equally lavish descriptions of the agony of withdrawal but still concluded that "my case is at least a proof that opium, after a seventeen years' use and an eight years' abuse of its powers, may still be renounced." However, like many who followed him, the author of what is believed to be the world's first addiction memoir later relapsed. And as his successors would soon do, De Quincey paradoxically both exalted and demonized the drug, warning of its fatally seductive powers while presenting a story of enjoying and then overcoming them.

The idea that addiction was a form of chemical slavery became increasingly popular a few decades later in the middle of the nineteenth century. It may not be a coincidence that during the same period, the United States was also consumed by debates over race and the actual slave trade. Blatant racism and ideas about bondage have played a role in concepts of addiction and drug policy right from the start. As a result, confronting the role of race in our concepts of addiction is critical to developing better definitions of the problem, treatment, and policy.

The American physician and Declaration of Independence signer Benjamin Rush was among the first to call alcoholism a "disease of the will." He first wrote about this notion in the late 1700s, publishing the first edition of his pamphlet, *An Enquiry into the Effects of Spirituous Liquors upon the Human Body, and Their Influence upon the Happiness of Society,* in 1784. In the book, he blamed the physical effects of alcohol itself for ultimately overwhelming all attempts at controlled drinking in those who drank heavily (although, oddly, he thought the problem was only hard liquor, not beer or wine). But this disease concept did not really catch on until decades later, ultimately fueling the Prohibition movement toward the turn of the century and thereafter.

Intriguingly, Rush was also a leading abolitionist: he founded the first antislavery society in the United States. However, as with addiction, he saw being black as a disease. He called this illness "negritude" and thought that it could only be cured by becoming white. (He believed this was possible because he'd seen cases of vitiligo, a disease that produces lighter-colored patches on darker skin.) Rush himself doesn't seem to have linked the disease of "negritude" with racist stereotypes about the character of African Americans—and the fact that he named both blackness and addiction as diseases does seem to have been a genuine coincidence.

It is a telling one, however, because of the way drug policy has always been entwined with race and with ideas about who is one of "us" and who is one of "them." Indeed, it is not at all coincidental that the negative personality traits that characterize racist stereotypes are virtually identical to the depraved characteristics said to define people with addiction—from criminal propensities, laziness, promiscuity, violence, and childishness to deviousness and an inability to tell the truth. These twisted stereotypes have long been used both to promote harsh drug laws and to try to discredit people of the races and cultures with which prohibited substances are associated.

Moreover, the selective enforcement of such laws then creates further associations between crime, race, culture, and drugs, producing a vicious cycle. Consequently, the historical roots of addiction in the period of so-called scientific racism still influence our perception today. If you've viewed media coverage of crack or heroin addiction "spreading to the middle class" or read stories written by a white person who claims to be "not your typical addict," you are essentially hearing the echoes of the racist origins of contemporary ideas about addictions. These echoes run deep, even among those who think they are not being influenced by them.

Our concepts of addiction, in fact, have been enmeshed with fears about ethnicity, class, and foreigners since the idea was first applied to drug problems and used as a reason to ban substances. Intoxicants preferred by "us" have always been seen as nondrugs, medicines and tonics—while substances taken by "them" are tagged as dangerous drugs without legitimate uses. Consequently, addicts cannot be nice people "like us"—they are scary, bad, crazy people who can only be stopped by extreme measures. Or rather, if they are seen as members of stigmatized minorities, we advocate harsh, punitive treatment; and if someone "like us" does turn out to be affected, we see that person as exceptional and deserving of a different and much kinder approach.

The first American state laws against cocaine, for example, were passed in the South in the depths of the Jim Crow era. According to historian David Musto, "The fear of the cocainized black coincided with the peak of lynchings, legal segregation and voting laws all designed to remove political and social power from him." Southern sheriffs claimed that the drug not only made black men into better marksmen, but it also rendered them practically bulletproof, requiring police to use higher-caliber weapons. Even worse, it induced black men to rape white women—and could be used to lure "innocent" white women, who otherwise would never dream of having interracial affairs.

Similarly, state and local laws in California and other western states banning opium in the 1880s were first passed after supporters stoked racist fears about Chinese laborers who built the transcontinental railroad. In the South, cocaine was said to allow black men to lay claim to white women. But in the west, virtually identical charges were made against Chinese men and their lascivious "opium dens," even though pharmacologically, the drugs typically have nearly opposite effects. (Cocaine is a stimulant and opium a depressant.) For example, one law enforcement report from California at the time says that the police "have found white women and Chinamen side by side under the effects of the drug—a humiliating sight to anyone who has anything left of manhood."

Explicitly racist journalism and political campaigning helped drive the passage of the 1914 Harrison Narcotics Act. Under the guise of a tax law, it made cocaine and opium and its derivatives de facto illegal, with certain exceptions for medical use. While there were other factors like battles over labeling and fights for control over sales by professional groups like doctors, pharmacists, and industry, racism was evident throughout the political debate and the media messaging. Even supposedly straitlaced publications like the *New York Times* were not immune: a 1905 *New York Times* news story was titled "Negro Cocaine Evil," and a 1914 op-ed proclaimed, "Negro Cocaine Fiends Are a New Southern Menace." Testifying before Congress in favor of the antidrug law, one "expert" stated, "Most of the attacks upon white women of the South are the direct result of a cocaine-crazed Negro brain."

Marijuana prohibition, too, was deeply influenced by racism. Its main proponent, Harry Anslinger, used racist rhetoric to push through the federal law that effectively banned the drug in 1937. Anslinger, a former alcohol Prohibition agent who headed the Federal Bureau of Narcotics from its founding until 1962, said plainly that the main reason to ban cannabis was "its effect on the degenerate races." He claimed that "reefer makes darkies think they're as good as white men" and warned, "[t]here are 100,000 total marijuana smokers in the U.S., and most are Negroes, Hispanics, Filipinos and entertainers. Their Satanic music, jazz and swing result from marijuana use. This marijuana causes white women to seek sexual relations with Negroes, entertainers and any others." Although such rhetoric—along with films like 1936's *Reefer Madness*—may seem laughable now it helped create the drug laws we have today.

When particular drugs weren't associated with "the dangerous classes," however, addiction tended to be seen as a medical problem. For example, before the 1906 Pure Food and Drug Act required makers of "patent medicines" to list their ingredients, heroin, cocaine, and marijuana were frequently present in "tonics" and pills that could simply be purchased at pharmacies. The "typical opiate addict" of the time was actually a mother and housewife who had become inadvertently dependent after buying widely marketed remedies that came with no warnings. It's not clear how much of this dependence was simply medical use for pain and how much was actual addictive behavior, but historical accounts show that some women absolutely were using the drugs to get high or escape. (The Eugene O'Neill play *Long Day's Journey into Night* features one such woman, Mary, as a character.)

These addicts weren't seen as a menace to society, however. Instead, they were patients to be pitied, educated, protected from unscrupulous drug companies, and cared for by doctors. In fact, simply labeling opiate-containing medicines cut their use by between 25% and 50% in the years after the 1906 law passed, showing clearly that measures short of criminalization can affect the use of even the most addictive drugs and that education is a powerful part of prevention.

As the twentieth century continued, however, ideas about addiction began to change. Earlier, Rush and others had described alcoholism as a disease characterized by loss of control—one that was caused by exposure to "ardent spirits" that were destructive to human freedom. The problem was in the drug itself, in this case, alcohol: anyone exposed could become addicted if they continued drinking heavily for long enough, through no fault of their own, they argued. Not surprisingly, this was the ideology that ultimately helped lead to Prohibition, which lasted from 1920 to 1933. After all, if alcohol caused alcoholism—and all of the domestic violence, bar fights, poverty, and degradation that were associated with it—banning it would end, or at least significantly reduce, these social problems.

Notably, as with the passage of other drug laws, racism and prejudice against foreigners and immigrants were critical in ensuring the enactment of Prohibition. One of its biggest supporters was the Ku Klux Klan, which was revitalized in the 1920s in part through its embrace of this cause. One historian wrote that "support for Prohibition represented the single most important bond between Klansmen throughout the nation," and others have described the group's attacks on bootleggers and an overlap in membership between the Klan and the

Anti-Saloon League, which was a key group that pushed the legislation that banned alcohol. The racial animus here wasn't just against blacks, however; in fact, it primarily focused on immigrant groups that Prohibition supporters linked with excessive drinking like Germans, Irish people, Jews, and Italians.

Of course, the disastrous results of Prohibition are now notorious. Although it appeared to have some positive effects like reducing alcohol-related hospitalizations and cirrhosis deaths initially, the same declines were also seen at the time in countries with temperance movements that did not enact prohibitions. Some insurers estimated that alcoholism rates rose by 300% as Prohibition continued. Meanwhile, the murder rate went from 6.5 per 100,000 in 1918 before Prohibition to 9.7 per 100,000 in 1933, the year of repeal, nearly a 50% rise. Suggesting that the relationship was likely to be causal, this rate then fell back under 6 per 100,000 by 1942.

Perhaps worse, in a less well-known episode that shows just how much stigma Americans can attach to types of drug use they have decided to hate, thousands of people—research suggests as many as 10,000—were killed by government attempts to keep people from drinking alcohol used in manufacturing processes during Prohibition. No one has ever been held to account for these preventable deaths. But in 1926, the Coolidge administration began ordering manufacturers to add poisons like methyl alcohol, gasoline, chloroform, carbolic acid, and acetone to industrial alcohol in an attempt to prevent it from being diverted to bootleggers. That year, in New York City alone, 1,200 people were sickened and 400 died as a result. It soon became clear that the law was both unenforceable and counterproductive.

With Prohibition's fall, the idea that alcoholism is simply a result of exposure to large amounts of booze began to be discredited as well. Both the repeal movement and Alcoholics Anonymous, which was founded in 1935, began to promote a slightly different disease model. Instead of seeing alcohol as the sole cause of alcoholism, they started to view drinking as a symptom. Now the problem wasn't merely the substance: it was the user's relationship to the drug. While most people could safely handle liquor, alcoholics couldn't. They had an "allergy" to alcohol. Making it legal, then, would allow this group to be treated medically, while leaving normal drinkers unmolested and wresting control of the industry and much-needed tax dollars away from the mob.

These ideas about the failure of Prohibition to arrest addiction are now widely accepted—at least in the case of alcohol. However, they aren't nearly as well established in relation to illegal drugs, even though addiction rates for users of heroin, methamphetamine, and cocaine are comparable to those seen with alcohol, and addiction rates for marijuana are lower.

TO ME, AS I lay detoxing, it was obvious that addiction was not seen the way other illnesses—mental or physical—were. The image I had of an addict was essentially the standard stereotype, one that I now see as harmful and deceptive: a person who would lie, cheat, steal, have sex for money, perhaps even kill just for a fix. Of course, I'd seen some drug users around me who behaved that way—at least in terms of the lying, prostitution, and stealing. At the time, however, I had no idea where the idea that all people with addiction behave like this came from. I'd simply accepted this moral model.

And since *I* hadn't been doing things that I thought were wrong to get drugs—I naively saw dealing as helpful because I wanted other people to provide good drugs to me, too!—I didn't think I fit the picture, even as I injected drugs multiple times, every day. That is, until I found myself considering trying to seduce Beaver.

My conception of addiction was also, it must be said, shaped by the fact that when I'd started snorting cocaine, it was generally not viewed as particularly addictive. Although this, too, is hard to believe in the wake of crack—and the literally thousands of media stories in the '80s and '90s that described it as the most addictive drug ever discovered—it reveals the devilish difficulty of accurately defining this problem. At that time what this scientific misunderstanding did, at least in my case, was inadvertently encourage my drug use.

In 1982, *Scientific American* had described cocaine as being not addictive in the classical sense; I'd read the article either at home where my father subscribed or for a class at Columbia. The authors, Craig Van Dyke and Robert Byck, had argued that the behavioral pattern of those who snort cocaine is "comparable to that experienced by many people with peanuts or potato chips. It may interfere with other activities of the individual, but it may be a source of enjoyment as well." In other words, snorting coke was no more addictive than junk food. That assertion was met soon afterward with ridicule and outrage, when crack was blamed for skyrocketing crime rates and what looked like addicted skeletons began appearing nightly on the news. (Today, many experts argue that, in fact, junk food is at least as addictive as coke and the comparison is no longer shocking in light of the high prevalence of obesity—then, however, it was genuinely seen as both absurd and beyond the pale.)

But what the authors were trying to get at was a contradiction that has frustrated thinking about addiction since the term was first applied to drug problems in the 1800s. Depressant drugs like alcohol, opium, and heroin produce physical dependence—first, a requirement of more of the drug simply to achieve the same high (tolerance), and then, if heavy use continues for long enough, a physiological need for it to stave off unpleasant withdrawal symptoms like nausea and shakiness.

In contrast, stimulants like cocaine and methamphetamine don't cause complete tolerance. In fact, they can also have the opposite result, known as sensitization, where some effects actually are greater with lower doses as time goes on. (Sadly for people with addiction, the effects that get larger with a reduced dose are the unpleasant ones like anxiety and paranoia, not the fun ones.) Stimulant withdrawal also doesn't make you physically ill like heroin or alcohol withdrawal does; nearly all of its signs can be dismissed as "psychological" rather than "physical" and include things like irritability, craving, depression, and sleep disturbances.

The issues around tolerance and the lack of physical signs of dependence like vomiting and diarrhea in stimulant addiction made scientists see stimulant problems as less severe. You might *want* cocaine or speed, the thinking went, but you didn't *need* it like a real heroin junkie. Addiction was generally depicted at that time as an ongoing fight against withdrawal—so if it didn't lead to visible withdrawal sickness, a drug shouldn't be too addictive. Hence marijuana, amphetamines, and cocaine were seen as nonaddictive, while heroin and alcohol could create real addicts. Physical symptoms were seen as real and measurable; psychological symptoms were minimized and not to be taken seriously. The fact that both kinds of symptoms ultimately had to be expressed via chemical or structural changes in the brain in order to affect the body was ignored.

I took this perspective, too—that's why I'd entered the methadone program. I thought that simply relieving opioid withdrawal symptoms by slowly tapering the drugs would cure me and that quitting the cocaine, which had no physical withdrawal syndrome, would be easy. I found out for myself every day in 1988 how wrong and misdirected these ideas were.

And so, by the time I started detox, I wasn't yet clear on what addiction was, but at least I knew what it was not. That is, addiction isn't just a physical need for a substance to avoid withdrawal, though that doesn't help matters any. Physical withdrawal symptoms are nothing compared to psychological desires: what matters in addiction is what you want or, yes, believe you need, not whether you feel sick or even how sick you feel. During the five previous times I'd made it through physical withdrawal over the course of my opioid addiction, I found that when I was the most sick was not when I had the most craving—nearly the opposite was true. I was at greatest risk of taking heroin again when I felt better and decided that withdrawal hadn't been so bad and that I could use occasionally and avoid becoming physically dependent.

I'd soon find that the question of whether body or mind matters more— and where the boundary between mind and brain, mental and physical is—runs through most of the key issues and fault lines in addiction.

THREE

The Nature of Addiction

You don't wake up one morning and decide to be a drug addict. It takes at least three months' shooting twice a day to get any habit at all. . . . I think it no exaggeration to say it takes about a year and several hundred injections to make an addict.

—WILLIAM S. BURROUGHS, *JUNKY*

AFTER MY MOTHER LEFT ME AT the hospital, I was taken upstairs to a semi-private room. I was provided with a thin cotton bathrobe and a pair of green Styrofoam slippers with smiley faces on the toes. That first day of detox, I vomited incessantly. I was given a drug called clonidine to help ease some of the withdrawal symptoms. Clonidine is typically used to treat blood pressure. It doesn't relieve anxiety or cause a high—but it does dampen the activity of the stress system, which, not surprisingly, becomes hyperactive during withdrawal. It helped a little—perhaps mainly by making me tired and woozy—but I still couldn't sleep.

The first night, in fact, I didn't get any rest at all, just tossed and turned and threw up. My legs ached terribly and I banged them against the bed in a vain effort to knock the pain out. This common response is probably why they call heroin withdrawal "kicking." When I ran to the bathroom, I couldn't bear seeing my hugely dilated pupils in the mirror over the sink. Their gaping vacancy reminded me that my eyes were now in the opposite state of the blissfully tiny pinpoint pupils that you exhibit while high on opioids. Everything was painful and loud and bright and discomfiting. And the temperature was never right. I was either freezing or sweating profusely.

I had three emotional states in detox: depression, euphoria, and boredom. I swung rapidly between them. One moment, I'd be manic and grandiose: it was all going to work out and I would soon be back on my rightful path to fame, fortune, and greatness! The next, I'd be crushed by physical discomfort and plunged into utter despair about my legal situation: 15 years in prison, mandatory minimum, no parole, no chance at a better life. Then, I'd be overcome with desire to start some great project, anything to distract me from the dullness of where I was, although I couldn't conceive of what it should be or muster enough energy even to try to do so.

My strongest memory is of sitting in the bathtub, so thin that I couldn't get comfortable because there wasn't enough padding on my ass to cushion me against the hard porcelain. I was still losing weight at this point; I went from 85 pounds down to 80 in my seven days in the hospital. I lay in the warm water, often twitching with nervous discomfort, trying to remain in the soothing liquid for as long as I could. And when I stepped out and felt the air on my body, for one precious moment, I felt okay: warm and protected.

However, although detox certainly isn't fun, its physical horrors have been seriously exaggerated—perhaps because for so long, the importance of psychology and learning in addiction was ignored. First of all, detox is certainly far less painful than common symptoms or treatment side effects associated with cancer, AIDS, or hepatitis. It really isn't much worse than a bad flu. That doesn't mean it's not wretchedly awful, however. It's just that the physical symptoms aren't the main problem. What makes drug withdrawal hard to take is the anxiety, the insomnia, and the sense of losing the only thing you have that makes life bearable and worth living, not the puking and the shaking. It's the mental and emotional symptoms—the learned connection between drugs and relief and between lack of drugs and pain—that matter.

In fact, it's this type of existential terror and anxiety that actually gives the most excruciating flourish to any bad experience: the meaning of pain profoundly affects how it is felt, and the more worry and fear involved, the worse the suffering. Pain that is viewed as life threatening literally feels more intense and agonizing than pain with a known, nondangerous origin or time frame; studies find that people who believe a particular type of pain means that worse is to come actually rate it as more severe on pain scales than they do if they have the exact same physical experience but can be reassured that it's nothing to fear.

Gavril Pasternak, a pain specialist at New York's Memorial Sloan Kettering Hospital, once described a classic case of this phenomenon to me. A patient he'd treated for breast cancer pain returned years later with an aching back. She thought that her back pain meant that the cancer had returned and would soon kill her. She wanted opioids. But, in fact, there had been no recurrence. After testing ruled out cancer and Pasternak told her that she had garden variety disc pain, her relief was immediate. She no longer felt the pain to be unbearable and didn't want any kind of drugs. It was her analysis of what the pain meant that had tormented her. The psychology and the experience of pain are inseparable. Moreover, no matter whether the source of pain is obviously "physical" or "just psychological," the unpleasant aspect of the experience is processed by many of the same parts of the brain. Not coincidentally, these regions are rich in opioid receptors.

Without heroin, my receptors were screaming for relief, the addiction having reduced or otherwise disabled my natural supply of endorphins and enkephalins, which are the brain's heroin-like neurotransmitters. Mostly, I felt utterly stripped of safety and love. And so, what tormented me most as I shook through August of 1988 wasn't the nausea and chills but the recurring fear that I'd never have lasting comfort or joy again. I needed to understand what drugs had given me and what addiction had taken away.

ON THE SURFACE, addiction looks easy to define. Everyone thinks they know what an addict is. Should you attempt to characterize the condition in a precise fashion, however, you rapidly discover that most definitions feel more like a version of Potter Stewart's famous quote about pornography, "I know it when I see it," than like science. In other words, the key characteristics turn out to be difficult to delineate.

Some of this is due to the fact that criminalization of drugs like heroin, cocaine, marijuana, and even, briefly, alcohol resulted primarily from racist stereotypes and media exaggerations of the drugs' menace, not specific drug effects. Other aspects relate to the "defects of character" unfortunately claimed by some promoters of 12-step programs like Alcoholics Anonymous to typify all addictions, such as dishonesty and selfishness. A final and critically important factor is the sheer complexity of defining any mental illness or condition where the symptoms involve behavior and no unique neurological or genetic pathology has been identified.

Many people think this definitional debate is only seen around addiction. But the same is true in schizophrenia, bipolar, autism, depression, and all the other conditions in psychiatry's diagnostic manual. In none of these diagnoses has a single pathology, present in all affected people and absent in others, been found. In fact, once a specific genetic profile or a unique physiology can be detected reliably, it may be redefined as "neurological" rather than psychiatric and removed from the manual, which leaves psychiatric diagnosis open to ongoing controversy.

As I've discovered in my own journey, the developmental nature of neuropsychiatric disorders means that any particular diagnosis may be the ultimate outcome of many different causes. Variations in the timing of experience can also mean that the same gene produces risk for different disorders, which may emerge depending on when in life a specific trouble is encountered. This makes diagnosis and treatment—and getting people to understand that a condition is real and not just a "choice"—quite tricky. Addiction is the classic case of this problem, and failing to see it as a learning disorder has contributed to that.

Moreover, some characteristics that seem necessary to addiction actually aren't—and some that seem sufficient don't turn out to be necessary. Take those withdrawal symptoms, which conventional wisdom still puts at the center of the disorder. The shaking, vomiting, pallor, sweating, and diarrhea I suffered during detox are physically obvious. No one can say such symptoms are "all in your head." But, as noted earlier, cocaine and methamphetamine withdrawal produce few such objective signs. Although in the past this lack of a clear withdrawal syndrome led to claims that stimulants aren't addictive, no one can take that idea seriously post-crack.

Conversely, there are blood pressure medications that have potentially fatal withdrawal syndromes—if you are physically dependent on them and you miss too many doses, it can kill you. That sounds like it should be the ultimate addiction. But people don't crave these drugs and if they don't know how dangerous missed doses are, they fail to urgently seek them and even forget to take them. Also, because these hypertension medications improve health rather than do harm, it's difficult to see patients who rely on them as addicted.

Similarly, some antidepressants can produce a wicked withdrawal syndrome when stopped abruptly—but they, too, have no street value and no one has been known to rob drugstores to get them. People can and often do forget to take their antidepressants—but no addict ever forgets to take her heroin. That's why even though some depressed patients are not able to function psychologically if denied medication, it makes little sense to label this situation addiction, either. The drugs make their lives better, not worse. And if addiction is simply needing something to function, we are all food, water, and air addicts and the term is pretty well meaningless.

SO WHAT *IS* the essence of addiction, if not physical or psychological dependence? The basics aren't far from how the condition, now called "severe substance use disorder," is characterized in psychiatry's latest diagnostic manual, the *Diagnostic and Statistical Manual of Mental Disorders,* 5th edition (DSM-5). That is, addiction is best understood as compulsive use of a substance or compulsive engagement in a behavior despite ongoing negative consequences. The National Institute on Drug Abuse sums it up this way: "Addiction is defined as a chronic, relapsing brain disease that is characterized by compulsive drug seeking and use, despite harmful consequences." But although this describes some cases of addiction, it doesn't explain its nature and origins. Moreover, the term "brain disease" is both vague and stigmatizing. It doesn't capture the critical role of learning in addiction.

I will argue here instead that, fundamentally, addiction is a learning disorder. There are three critical elements to it: the behavior has a psychological purpose, the specific learning pathways involved make it become nearly automatic and compulsive, and it doesn't stop when it is no longer adaptive. I will describe these in greater detail later.

First, I want to make clear that I am far from the first to characterize addiction as learned behavior. The idea that learning matters in addiction is uncontroversial and has been accepted by every type of scientist who studies the condition and by many of those who treat it for decades. There is no theory—from the "brain disease" posited by the National Institute on Drug Abuse, to the sociology of Alfred Lindesmith, Lee Robins, and Norman Zinberg, to the psychological approach favored by Stanton Peele—that says that learning is *not* involved. In fact, addiction pioneer Lindesmith wrote in his 1947 book, *Addiction and Opiates,* that "addiction is established in a learning process extending over a period of time." Countless researchers and theorists—notably, Peele, Nora Volkow, Kent Berridge, Terry Robinson, Iain Brown, George Ainslie, Gene Heyman, Roy Wise, David Duncan, Larry Young, and Edward Khantzian—have made critical contributions to the conversation, some complementary and some conflicting with one another.

But the *implications* of framing addiction as a learning disorder have received far less attention. This book is my attempt at a popular synthesis of these ideas, which offer insights for treatment, prevention, and overall drug policy. To start, I want to stress that I do not mean that there is no biology involved, nor do I mean to imply that medical treatment—including medication—is not often useful and sometimes critical. I also don't mean to imply that addiction is driven by ignorance. The problem with our current understanding of addiction is that by ignoring the role of learning, we have tried to slot it into a category of medical illness or moral failing where it doesn't quite fit and then tried to ignore the round peg forced into the square hole.

Our society doesn't deal well with conditions that cross boundaries between mind and body, medicine and education, psychology and psychiatry, psychiatry and neurology. Instead, we tend to ignore the aspects of the disorder that don't fit our preferred view rather than recognize this complexity. Consider the endless controversies over medication versus therapy for depression or the ongoing fight over whether ADHD is real or simply a case of schools failing children with excess energy. Consider the difficulty parents have in determining whether to call a doctor, a tutor, or a therapist when a child has ongoing behavior problems that affect academic achievement. Consider the ongoing battle over whether any mental illnesses even exist or are simply cultural designations of deviance.

Too often discussions in this area become semantic battles that push stigma around: my condition is neurological while yours is psychiatric; my kid has a brain disease but yours has a developmental disability. In the addictions world, the word *disease* itself is an ideological front line, weighed down with far too much historical and moral baggage. I'd like to stop fighting over language (if you want to say it's both a learning disorder and a disease, I won't try to stop you, nor do I mind if you want to see these notions as completely distinct). Instead, I want to focus on how learning and development can allow us to stop trying to square circles.

From this perspective, like schizophrenia, depression, and autism, addiction has neurodevelopmental roots: some brains are more vulnerable to it than others as a result of genetic predispositions, which affect development in utero and beyond. Predispositions to addiction also tend to carry risk of other mental illnesses and developmental disorders: at least half of people with addictions also have another condition, like depression, anxiety disorders, bipolar, ADHD, and schizophrenia, with some studies finding rates of co-occurring disorders as high as 98%; and around 50% of people with one type of addictive behavior also engage in others. All of these propensities interact with early life experience, particularly trauma, over time to produce risk. Addiction doesn't just appear; it unfolds.

And as with other psychiatric and developmental disorders, addiction itself is not a choice. But more so than with conditions like schizophrenia and autism, it does involve choices made both consciously and unconsciously as children and adolescents face and manage life problems over time. Consequently, it is profoundly affected by cultural factors and by the way individuals perceive their own experience, particularly very early in life. This means that while addiction can certainly impair moral decision making, it does not eliminate free will, and the level of impairment varies widely from person to person and even from situation to situation in the same person. During my addiction I frequently made choices about using—for example, I never did so in court or where the police could see me—but my values were certainly skewed toward prioritizing it over school and relationships, which I ordinarily would have valued more. It's not that I had no control over my behavior, but I had less.

Further, as a developmental disorder, addiction is more likely to appear in some stages of life than others. In fact, a tight connection between a key period of brain development and the onset of symptoms is a defining characteristic of developmental disorders. Schizophrenia, for example, most often manifests itself during the late teens or early 20s, while the average age of onset for autism symptoms is in early childhood and the typical case of depression begins in one's early 30s.

In addiction, adolescence is the high-risk period because this is when the brain changes to prepare for adult sexuality and responsibilities and when people begin to develop ways of coping that will serve them for the rest of their lives. For example, the odds of alcoholism for those who start drinking at age 14 or younger are nearly 50%—but they drop to 9% for those who start at age 21 or later. And the risk of rapidly developing addiction to marijuana, cocaine, opioids, and pills like Valium is two to four times greater for those who start using at age 11–17, compared to those who start at 18 or later. If you manage to make it through adolescence and young adulthood without developing an addictive coping style, your odds of developing one later, while not nonexistent, are dramatically reduced.

Addiction, then, is a coping style that becomes maladaptive when the behavior persists despite ongoing negative consequences. This persistence occurs because "overlearning" or reduced brain plasticity makes the behavior extremely resistant to change. Plasticity is the brain's ability to learn or change with experience. Lowered plasticity means this ability is compromised, and when a pattern of activity is locked in, it is "overlearned."

The capacity for such overlearning is a feature of the brain's motivational systems, which evolved to promote survival and reproduction. The strong drives that these systems create can be useful when they spur persistence in love, work, and parenting. However, their intense resistance to change becomes a "bug" in our programming when drug taking or other unhealthy activities continue in the face of ongoing harm.

Moreover, unlike ordinary forms of learning, addiction involves interference with the brain processes that themselves guide decision making and motivation by determining the emotional weight of various options. It alters the way the brain decides what it values—for example, by making cocaine seem more important than college or by making all other pleasures pale. This can occur either because drugs themselves change this chemistry and circuitry or because these brain systems are inherently vulnerable to being altered by certain patterns of experience—or as a result of some combination of both processes. It's complicated, but it can be understood by following these processes as they unwind over the course of development.

THERE IS A Polaroid picture of me taken the day I entered rehab, on August 11, 1988. My pupils are so large from withdrawal that my blue eyes look black, and because I'm so thin, they are deeply sunken into my face. I have dark roots showing under my bleached blond hair and it is still scraggly and patchy from the hair pulling I'd done while using coke. My nose looks like I'd taken a punch—I'm not sure why, but it is swollen well beyond its normal size, even though I'd shot all my drugs rather than snorted them for years by that point. I'm smiling but I look utterly vacant and tentative. I'm so frail that I look old; except for the smoothness of my face, you'd never guess I was in my 20s. The muscles in my neck and my collarbone stick out, and on the back of the hand that's draped on the counter, you can make out the red lines of my tracks if you look closely. Indeed, 26 years later, I still have faint scars there.

Just looking at it now to write this is painful, especially as I keep it in a box full of photos that signify far happier memories. To find it, I had to flip through pictures of my high school and childhood self and wonder yet again how I got into—and ultimately out of—that distressing state. Where had the seeds of addiction been planted for me? To find out, I had to dig into my own developmental history.

FOUR

Intense World

That drive to return to the past isn't an innocent one. It's about stopping your passage to the future, it's a symptom of fear of death, and the love of predictable experience. And the love of predictable experience, not the drug itself, is the major damage done to users.

—ANN MARLOWE

THE FIRST DRUG I EVER TOOK was Ritalin—and it was prescribed to me. It was 1968 and I was three years old. I was not partaking of that era's freewheeling drug culture, although I would later wish that I could have. Instead, my mom had taken me to see a child psychiatrist on the recommendation of Mrs. Darling, my aptly named nursery school teacher. At the time, I attended a progressive—indeed somewhat hippieish—private school for gifted kids on the Upper East Side, which I had tested into and been given a need-based scholarship. My mother had wanted me educated in a creative, stimulating, and racially integrated environment. Earthy-crunchy New Lincoln School, which her younger brother had attended, aimed for such racial and economic diversity with its scholarship policies.

But now Mrs. Darling was worried. She had become so concerned about me that she asked my mother to come in for a day to observe my behavior. Far more than the other three-year-olds, I had difficulty sitting still. Sometimes, I'd be utterly engaged in an activity like finger painting or coloring—outside the lines, of course—by myself. Alternatively, I was hard to corral. I cried frequently. I couldn't deal with transitions. I did not play well with others; I preferred to be alone. Seeing how much I stood out, my mom agreed to have me evaluated. My Ritalin prescription was the result.

While I don't recall anything about my first psychiatric examination, according to my mother the doctor was a kindly looking and gentle older woman at Columbia with lots of toys in her office. She watched me play, talked with me, and listened to my mom's descriptions of my behavior in school and my difficulties with socializing. Ultimately, her evaluation found me "hyperesthetic"—not a compliment on my artistic skill or taste, but a clinical term meaning that I was abnormally oversensitive to various types of stimulation. I was also labeled "emotionally labile"—unusually reactive to feelings, in addition to my hypersensitivity to sights, sounds, smells, tastes, texture, and touch. Moreover (and unbeknownst to me until I started researching this book) the psychiatrist also diagnosed me with ADHD. If I had been a child evaluated today, however, I would probably have been placed on the autism spectrum.

At the time, though, Asperger's syndrome—the so-called high functioning form of autism—had yet to make its brief appearance in the DSM. First described by Austrian psychiatrist Hans Asperger in 1944, the condition that takes his name is marked by problems with social connection, sensory differences, difficulty with change, repetitive behaviors like flapping hands or twirling hair, obsessive interests, and, often, excellent memory, with average or above average intelligence. When I was a child, however, if autism was even diagnosed, it was a label only used in cases of profound disability—for kids who didn't speak at all, for example, or for those whose behavior was mainly limited to repetitive actions and tantrums when their routines were interrupted. The diagnosis was also rarely given to girls—in the 1960s, only the most severe cases were identified in females.

Indeed, my ADHD diagnosis was made in an entirely different climate around child mental health. Seeing a child psychiatrist in the 1960s carried a much worse stigma; it just wasn't done back then if it seemed possible to avoid it. Unlike today, when many parents actively seek labels to get services for their kids, at that time children's problems, even when extreme, tended to be dismissed as phases or simple misbehavior. Now, stimulant medication is being prescribed to more than six percent of schoolchildren—but in the '60s, it was not used by more than a fraction of a single percent. It would also be many years before Big Pharma overwhelmed psychiatric practice with direct-to-consumer marketing; the drug panic of the day was about dirty hippies dropping acid, not preschoolers taking Adderall.

So I was not a victim of what some see as today's rush to medicate and label. Instead, I was very obviously far more reactive to common experiences and emotions than other children were and it was not clear what to do to help me. My earliest memories are of terror—or ecstasy. And without a diagnosis to make sense of them, my oddities were alternately explained to me as either "giftedness" or "selfishness." My contradictions confounded me—not to mention those around me.

CLASSICALLY, DEVELOPMENTAL DISORDERS have been defined as conditions that first appear in early childhood, like ADHD, autism spectrum conditions, dyslexia, and Down syndrome. More recently, however, virtually all psychiatric disorders—including schizophrenia, bipolar disorder, personality disorders, and yes, addictions—have been found to be profoundly shaped by learning during development, and so these, too, are now under the neurodevelopmental umbrella. As a result, understanding how learning molds the brain over time is crucial. Brain development cannot progress properly without experience; therefore, almost any disorder that affects mental functioning in childhood and adolescence will have a learned component.

When the importance of learning and development in addiction are understood, many of its contradictions are resolved and it is easier to see why it is neither a moral failing nor a brain disease in the traditional sense. That's because by definition, learning is a change in the brain, one that represents experience. Such change is not harm or damage—it is what the brain is built to do. Consequently, not all brain alterations are pathological, even when they occur during addiction. Many, if not most, of the individual differences that are seen in neural tissue as we develop are actually alterations made to represent what we have learned—in other words, our memories. These brain changes are traces of our life stories and input for algorithms to guide future actions. They are not necessarily marks of pathology or scars (though extreme stress can actually cause physical harm to brain cells). If we really want to make sense of addiction, then, our memories, their social context, their patterns, and the way we have learned them are of profound importance. They make each brain—and each addiction—unique.

Fundamentally, learning or developmental disorders—the terms are used interchangeably—have four important features. First, they start early in life, caused by variations in brain wiring driven by innate genetic programming. Since most brain development depends on experience, environmental influences in childhood ranging from parents and peers to chemical exposures can determine whether wiring differences become disorders, disabilities, advantages, or some mix of all three.

Second, developmental disorders are not necessarily associated with global deficits. While Down syndrome *is* linked with an overall reduction in IQ, autism, dyslexia, ADHD, dyscalculia (a specific difficulty with mathematical learning), addictions, and many mental illnesses are not. In fact, all of these conditions except Down syndrome can be accompanied by high IQ, with learning impaired only in specific domains. (This point will be important in considering addiction: many treatment providers argued erroneously for years that during active addiction, learning in general is impaired if not impossible.)

Third, timing matters in learning disorders. Because healthy development unfolds in a precise pattern, the sequencing of environmental influences, particularly social ones, is of critical importance. Missing an experience may be trivial at one stage of development but can derail it at another. Indeed, at certain stages of development, known as sensitive periods, the brain expects particular experiences and if these are not delivered at the right time in the right order, development can go awry. The two main sensitive periods in human brain development are infancy and adolescence. When specific input that is important at one of these stages is missed, it is difficult for someone to catch up later.

Lastly, because development is sequential, well-timed intervention can change its course. Take language development, for which the sensitive period is infancy. Before infant screening for deafness became widespread, early hearing loss was commonly misdiagnosed as intellectual disability. Because deaf children didn't learn language when their brains were most responsive and receptive to it, they appeared to have innate difficulties with grammar and other skills, even in sign language. Many wound up in institutions because their hearing problems were never detected and their language exposure only came after the sensitive period had passed.

But early hearing screening, along with being raised by parents fluent in sign language, changed this dramatically. It eliminated many cognitive disabilities that had previously been believed to be linked to deafness itself. These disabilities, however, came not from lack of brain capacity but from lack of appropriately timed exposure to complex language. (Far less problematically, sensitive periods also account for why if you don't become fluent in a foreign language before adolescence, it's almost impossible to speak it without an accent.)

Autism research also illustrates how timing matters in developmental disorders. Here, early intervention can significantly reduce associated disability and can sometimes even eliminate handicaps entirely. As with deafness and intellectual disability, some of the social disabilities seen in autism may be linked not with lack of capacity but with lack of exposure to appropriately timed developmental experience. For example, it's hard to learn to make friends when you feel overwhelmed by sensory experience, which is a common problem in autism. This can lead to isolation due to tantrums or other behavior that frightens peers or to your own social withdrawal as you try to soothe yourself with repetitive motions like rocking. Once you have less problematic ways of addressing sensory discomfort, however, you can more readily connect and catch up on the social skills the other children learned earlier, although you may need extra help to do so.

Even with effective early help, of course, autistic adults and teens will still have autistic traits like obsessive interests, sensory differences, and specific talents related to their unique wiring. However, those are differences, not disorders. If you can intervene effectively during a sensitive period in a developmental disorder you may entirely prevent problems that previously were seen as intrinsic to it but that actually result from missing important developmental experiences. Although we don't know for sure yet, the same may be true for early interventions for addiction.

There's also another important set of implications of recognizing addiction as a developmental disorder. For one, many of these conditions can actually be outgrown, as is often seen with ADHD. Because they often involve delays in the timing of brain maturation—not permanent lack of ability—children with developmental disorders can sometimes catch up with and even surpass typical peers. Alternatively, some developmental disorders may be the result of slowed maturation of some brain regions, which lag for unknown reasons but eventually come online. This may account for why some people outgrow addictions on their own, while others require treatment.

And all in all, the importance of sequencing in all developmental disorders cannot be overstated. Later aspects of development often rely on the previous stage being successfully completed. If you don't have a foundation, proceeding with the rest of the building process will obviously not be successful.

As a result, many developmental processes are similar to those described in chaos theory in physics, where complex systems can sometimes become wildly different from each other, depending on what would otherwise be inconsequential changes. This is known as "sensitive dependence on initial conditions." The classic example of such a system is the weather, where, to paraphrase the cliché, a butterfly flapping its wings in Beijing may occasionally make the difference between a hurricane and a minor rainstorm in Brazil, while ordinarily having no effect. At certain times, in systems that have sensitive dependence on initial conditions, tiny things can have a massive and disproportionate domino effect.

A similar phenomenon is seen in technology design, in which an apparently innocuous early choice—such as the layout of a keyboard—can lock development onto a pathway that will later become highly resistant to change. For instance, once a device like the QWERTY keyboard becomes popular, overcoming the enormous inertia of large systems and making the huge number of alterations in training and equipment needed to try something different becomes virtually impossible or simply not worth the effort, even if it would ultimately produce a superior technology. Here, the phenomenon is called "path dependence." In any area where these kinds of sequencing effects are seen, it is almost impossible to understand the present state of a system without knowing its particular history. The exact order of events matters.

Likewise, in infancy, small, early tendencies receiving mild, initial encouragement can turn tiny differences in ability into huge variance in talent as they build on each other over time. For example, telling a small child she is good at artwork will encourage her to do more of it. Then, the more she does, the better she'll get and the more praise she'll receive. Seeking even more attention and now feeling confident about her skill, this young artist will be more likely to persist through boredom or challenges. That in itself will further accelerate her talent. Lather, rinse, repeat: now what once was an almost insignificant elevation in aptitude has become an enormous advantage and what once were neural pathways that could have developed in many different ways are tuned to one particular skill. Give the same kind of praise to an adult, however, and it is far less likely to have this big an effect.

Unfortunately, in the same fashion, an early identification with a negative quality or experience can have a much bigger impact than later reactions. Childhood ideas about oneself shape later self-concepts, in ways that can either increase or decrease resilience. Predispositions toward developmental disorders can either grow or wither as children, their peers, and their families interact with each other. Early interactions shape later ones, and over time, families aren't just reacting to the person's current behavior, but to their own interpretations of the person's past behavior and the results of this whole iterated process. (Just consider how you or your siblings behave when you go home for the holidays.) While the specific influences and their spiraling interactions can vary tremendously, the role of timing and development matters greatly in any given story.

ON MY FOURTH birthday, I just couldn't wait to go to the party room of the building where my family lived in Washington Heights. Before the party started, I vibrated with anticipation. I probably wore my favorite blue dress, with red-and-white appliqued flowers running in a stripe down each side. My mom had invited the other preschoolers in the building, as well as my New Lincoln classmates, for a total of about 15 kids. As the guests arrived, I circled the table. We had pointy party hats, but I wouldn't wear one because I didn't like the feeling of the rubber string cutting into my neck. But I was eager for dessert and presents, kicking my legs to disperse nervous energy as I sat.

Finally, my mom brought the cake in and the singing started. She placed the large conical confection on the table. What had been a noisy, chaotic room rapidly went surprisingly still and quiet: my cake was so bright and fiery that more than two decades later, when I was contacted by a girl I'd known at the time, it was the first thing she mentioned. While other girls might have chosen pink with petite rosettes or princess candles, my cake was shaped, per my request, like a chocolate volcano.

I don't remember exactly how, but my father, a creative chemist with an eager sense of fun, made the cake erupt. It was topped with a little dome of orange powder on a small plate or bowl. Once ignited, colored flames leapt up and red lava-like liquid burst out. It flowed down the sides of the little powder mountain creating a crater at the tip, just like on a real volcano. Everyone—parent and child—was enthralled, especially me.

I loved volcanoes, you see, and would tell anyone within hearing distance in great detail about their bright-red lava, which was called magma before it reached the surface of the earth. I'd recite names, locations, and defining characteristics: I liked the shield volcanoes of Hawaii and Iceland especially because their lava was so colorful and flowed rapidly. I also loved the litany of words and images associated with volcanoes: igneous, cinder cone, pumice. I found thinking about these manifestations of nature's power strangely soothing.

This tendency to lecture others and my obsession with volcanoes were classic Asperger's: when the condition was first defined, its namesake described affected children as "little professors." And the social isolation and distress such traits caused me helped land me on the path to addiction. Of course, none of us knew this at the time I was growing up—nor were we aware of the many other factors lurking in our family history that would put me at further risk.

TO START ON my father's side: my dad was a Hungarian Holocaust survivor. I inherited his gray-blue eyes, thick curly hair, and, probably, propensity for depression. Hungarians have an elevated rate of depression, in part for genetic reasons—a risk that may oddly enough be shared with the Finns. The two groups apparently were once one tribe that split up and headed in very different directions, creating two difficult but related languages and two cultures with high suicide rates and a love of alcohol. My father's family is Jewish, so it's hard to know what genes, if any, were shared with their countrymen. But my paternal grandfather, previously known for his skill as a rabbinical scholar, is reputed to have married my grandmother to get hold of her meager dowry to pay off a gambling debt. His gambling problem suggests a specific, potential genetic risk for addictive behavior in a close relative.

My father also had overwhelming environmental risks for depression. He was born in 1938, the year before World War II started. He grew up extremely poor, which alone at least doubles the odds of depression. His father was taken away and conscripted as a slave laborer for the Hungarian army when he was just a toddler; his family had no idea what had happened to him and when my grandfather returned years later, my father didn't recognize him. Early parental loss, not surprisingly, also increases risk.

And in 1944, when my dad was five or six, the Nazis began rounding up Hungarian Jews. Ultimately, he, his mother, and his two-year-old sister were taken to the Strasshof concentration camp in Austria. They were on the train to the death chambers of Auschwitz—starving, packed into an unbearably crowded car, without food, water, or bathrooms—as the Allies began moving in to liberate them. My father was so traumatized that when he started first grade, his teachers thought he was intellectually disabled. It's difficult to imagine how someone could emerge from a childhood like this *without* depression.

My mother's childhood was completely dissimilar, although it, too, shows potential risks for mental illness, both genetic and environmental. Both my maternal grandfather and his sister struggled with depression; my great-aunt's case was so severe that she was treated with electroshock therapy. And while my mom grew up in an upscale Westchester suburb of New York called Croton-on-Hudson, her early life was difficult in its own way.

Her father was a business school professor at Baruch College who once ran for vice president of the United States as a socialist; her mother was a homemaker. On the surface, it was a normal, if politically quite left wing, suburban Jewish family. But my mom had been born with a serious heart defect. Although the condition is now easily treated with routine surgery, it was then a possible death sentence. At age 15, she underwent one of the first ever open-heart surgeries, which itself carried a high mortality risk. Her recovery required a month-long stay in the hospital. Two years earlier, her own mother had died of stomach cancer. One month after my mother lost her own mother, in the same temple where the coffin had just lain, the family proceeded with my mom's bat mitzvah.

My mother was 20 when I was born; I was her first child. My parents had met in a Russian class at City College. My father was a determined young immigrant who enjoyed classical music; my mother was a slender, brown-haired, green-eyed beauty who had participated in Martin Luther King's 1963 march on Washington and preferred Pete Seeger and the Beatles. Their youth, their wildly different backgrounds, and perhaps their varied early trauma provided them no common frame of reference for raising kids.

As a result, it was not immediately clear to them how odd I was. Even as an infant, for example, people remarked on what seemed to be a preternatural alertness. Words came early to me; my first one was "book." Without being explicitly taught, I was reading by the time I reached my third birthday. I chanted the Mah Nishtanah—the four questions the youngest child at a Passover seder is supposed to ask—in Hebrew, flawlessly from memory, around the same time, impressing my relatives forever. I had what people then called a photographic memory and could do things like recite much of the periodic table, which also wowed the adults.

From the start, I hungered for information, in much the way I'd later crave drugs. Among those who study gifted children, this is called a "rage to learn," but it can also be a symptom of asynchronous or uneven development and is often found in autism. Early reading, now labeled "hyperlexia," can also be symptomatic of both giftedness and autism. Around 5–10% of autistic children are hyperlexic, which is defined as learning to read without being formally taught, typically before age 5. This rate is far higher than for typically developing children.

I also had sensory and social differences. Groups of children made me anxious; I cried frequently and often couldn't refrain from wailing in unison if another child burst into tears. I wanted to wear the same outfit over and over and did not like any other type of change. Typical childhood behaviors like a desire for sameness and repetition were exaggerated. Before I'd started school, there had been hints that sensory experiences overwhelmed me: bright lights, loud noises, itchy materials, new tastes, slimy food textures—all of these upset me while leaving most other children only mildly bothered, if at all. I remember feeling that the world was always on the verge of erupting into chaos and engulfing me.

These kinds of sensory issues are now recognized as a key diagnostic symptom of autism. In fact, neuroscientists Henry and Kamila Markram (who have an autistic son from his first marriage) and their colleague Tania Rinaldi Barkat have theorized that the social withdrawal, repetitive behaviors, desire for sameness, and difficulties with communication and relational signaling that are typically seen as characterizing the condition are actually a result of attempts to deal with this overload, rather than the root problem in autism. They call this hypothesis the "intense world" theory and it is receiving growing attention in autism research.

Imagine being born into a world of bewildering, inescapable sensory overload, like a visitor from a much darker, calmer, quieter planet. Your mother's eyes: a strobe light. Your father's voice: a growling jackhammer. That cute little onesie everyone thinks is so soft? Sandpaper with diamond grit. And what about all that cooing and affection? A barrage of chaotic, indecipherable input, a cacophony of raw, unfilterable data. Just to survive, you'd need to be excellent at detecting any pattern you could find in the frightful and oppressive noise. To stay sane, you'd have to control as much as possible, developing a rigid focus on detail, routine, and repetition. Systems in which specific inputs produce predictable outputs would be far more attractive than human beings, with their mystifying and inconsistent demands and their haphazard behavior.

This, the Markrams argue, is what it's like to be autistic. The behavior that results is due not to cognitive deficits and deficient empathy circuitry—the prevailing view in autism research circles today—but to the opposite, they say. Rather than being oblivious, autistic people take in too much and learn too fast. While they may appear bereft of emotion, the Markrams insist that people with autism are actually overwhelmed not only by their own emotions but by the emotions of others. The developmental disorder now believed to affect around one percent of the population is not caused by problems with empathy, the Markrams say. Instead, social difficulties and odd behavior result from trying to cope with a world that's just too much.

While this may not be the case for all autistic people, it is pretty much what it was like for me as a child. For instance, I was terrified of the loud bell that rings to signal the end of the carousel ride in Central Park. I loved that merry-go-round, with its cheery music, jaunty wooden horses, and rocking, repetitive motion. But I would spend the last part of the ride trying to anticipate the jarring clanging so that it would not startle me and ruin the experience. Because everything seemed so overpowering—as the Markrams' theory predicts—I spent a lot of time trying to foresee and control what would happen next, attempting to manage my responses in advance. Unfortunately, that meant that I had few attentional resources left over to consider the thoughts and feelings of other people. This was one of many initial tendencies I had that began small but grew over time with repetition.

Some of my sensitivities were extreme: I found songs in minor keys so upsetting, I would cry just because a melody seemed sad; I generally refused to wear anything confining, like a snowsuit. I lived on bland foods like peanut butter and pizza crusts, refusing to try new tastes. I couldn't watch cartoons or violent films or TV, because I would always imagine what it would be like if the violence was perpetrated against me. It was as though the volume knob on all of my senses was turned up to a real *This Is Spinal Tap* 11. When you added in social contact—even a few other children felt overwhelming—it was just too much to take.

And after my sister, Kira, was born in May 1968, another sensory symptom became obvious. She was an especially affectionate baby, always wanting hugs and cuddles. By contrast, my parents now saw that I was not. As with my other senses, I had a difficult relationship with touch: I often longed to be held and comforted, but I hated feeling out of control and some types of pressure made me feel like I was going to suffocate. So I resisted and pulled away from many hugs. My family members learned not to touch me unless I initiated it or otherwise indicated that it was okay. Paradoxically, this sometimes made me feel rejected because I couldn't understand why I was so different.

All in all, the word "intense" was not only an extremely accurate way to describe how I experienced the world—it was also one of the adjectives most commonly used by others to describe me or my reactions. My teen years would become a constant battle, in '70s terms, to try to go from "intense" to "mellow."

WHILE ASPERGER'S OR some variant of it may be an unusual route to addiction, oversensitivity, social anxiety, genetic tendencies toward depression, ADHD, and an inability to regulate emotion and behavior are not. Indeed, each of the latter factors is linked with increased risk by itself, let alone in combination. In fact, it's hard to find a story of addiction that doesn't include at least one of these problems. For example, Lisa Mojer-Torres, a former heroin addict, lawyer, and pioneering advocate for addicted people (who, unfortunately, died a few years ago from ovarian cancer), described her childhood to me this way: "My mother used to say that I was more finely tuned than anyone. I heard noises louder or saw colors more vividly. I walked through the world differently." In a story of his alcoholism, The Who's guitarist Pete Townshend noted, "Most people for whom alcohol becomes a problem are running away from something. . . . Usually, what they are running away from is feelings and the inability to deal with the intensity of their feelings."

A recent online discussion among people grieving for victims of overdose focused on the childhood temperaments of their lost loved ones—long before drugs. With the number of people reporting ritualized behavior and oversensitivities like finding clothing tags unbearable and being overwhelmed by reactions to other people's pain, it could have been mistaken for a group discussing autistic sensitivities. One mom wrote that she had to buy seamless socks in order to get her child to wear them; several others noted the same issue.

In reading and listening to hundreds of these stories over the years, I've found that many include a sense of being fundamentally different and uncomfortable in their own skin. Here are just a few examples, from classic accounts of addiction:

"I was always at odds with the entire world, not to say the universe. I was out of step with life, with my family, with people in general. . . . I was a mess." (Alcoholics Anonymous, *Big Book*)

"From the time I was a little girl, I can remember feeling like I didn't quite belong. I thought I must be an alien from another planet." (Narcotics Anonymous, *Big Book*)

"[Codeine, an opiate] made me feel right for the first time in my life. I said to myself, 'Aha, this is what I've been missing.' I never felt right from as far back as I can remember and I was always trying different ways to change how I felt." (Anonymous user, quoted in *From Chocolate to Morphine*)

The particular symptoms that people use drugs to cope with may be infinitely varied, but the desire to feel accepted and secure when you typically feel alienated, unloved, anxious, and in danger is common. Alternatively, some people's temperaments make them feel understimulated; here a sense of exploration, a desire to push limits, and a love of risk and adventure may underlie addiction. Either way, the temperaments that make people stand out begin affecting them from birth.

As a learning disorder, addiction results from a dysfunctional coping style in a way that is profoundly affected by development. While conditions like autism, ADHD, and dyslexia become apparent during early or midchildhood, addiction itself typically doesn't appear until adolescence and early adulthood. The sensitive period for developing addiction—as with schizophrenia—is adolescence, not early childhood. Like schizophrenia, addiction can, sometimes, have an earlier or later onset—but, in most cases, adolescence is the most critical time.

That doesn't mean that early childhood doesn't matter, however. During this period, while few kids take nonprescribed drugs in elementary school, let alone preschool, our sensory reactions and our predispositions are shaping the way we respond to the world and the way it responds to us. This learning creates patterns that will echo down the years. A timid child, for instance, might learn that she can conquer her fears and that it will be all right—or he might discover that it's even worse than he expected. A bold child may continue to seek greater risks and bigger challenges—or be made more cautious via injury.

Our social environment and our reactions to it teach us how to be, and whether the world around us suits our personalities and inclinations or clashes with them helps determine our direction. In the context of addiction, the learning that occurs during preadolescence either enhances or diminishes the emotional pain we feel and our sense of either connection or isolation. The more discomfort, distress, trauma, and pain we feel—whether due to overreaction to ordinary experience, underreaction to it, or normal reaction to traumatic experience—the greater the risk.

MY MOTHER THOUGHT the Ritalin would give me a break from my intense world, much of which she didn't even realize I was experiencing. I have no conscious memory of the drug or its effects, however. My father disapproved of medicating children and my mother began to fear that the drug might dull my creativity, so they stopped giving it to me in the summer when preschool ended. Since I seemed far better able to cope when I went back to school, they never considered the drug or the diagnosis again during my school years.

Some would argue that it was this very early drug exposure alone that set me up for addiction, subjecting my brain at a critical period of early development to a potentially addictive stimulant. I think that the temperament that prompted the prescription at a time when young children were rarely medicated almost certainly played a much more important role. Having ADHD alone, in fact, nearly triples the odds of illegal drug use disorders and increases the risk of alcohol disorders by 50%, according to one review of the research. Individual studies have found even greater odds for the most severe addictions.

Somewhat surprisingly, however, research finds few effects of medication, positive or negative, on the later risk for addiction in children with ADHD. The idea that simple childhood exposure to a potentially addictive drug will automatically produce addiction is not supported by the Ritalin data—or, for that matter, by research on therapeutic use of other potentially addictive drugs like opioid pain relievers in children.

But sadly, the idea that early therapy for ADHD will help *prevent* negative consequences like addictions isn't supported either, although some studies do suggest benefit. The theory here is that stimulants will help kids with ADHD by allowing them to do better academically and socially. This, in turn, will reduce the risk of self-medication with illegal drugs or alcohol to cope with such difficulties during the teen years. Unfortunately, it doesn't seem to work that way: most studies find the risk is the same. And whether there are specific sub-types of ADHD—some likely to be harmed by medication, others helped—remains unknown.

Studies on temperament, on the other hand, do suggest profound effects, as we'll see in the next chapter. When children are followed from infancy into adulthood, those who go on to have some type of behavioral or psychiatric problem are often discovered to have been measurably different, sometimes from birth. Notably, these differences aren't all the same; in fact, some are opposite to each other, like extreme boldness or extreme fearfulness; reckless impulsiveness or steely rigidity. Opposing extremes can even alternate in the same child—as in my case where, for example, I could be either utterly absorbed or totally distractible.

Indeed, although people with ADHD are much better known for their spaciness and distractibility, an ability to be completely entranced when your interest is engaged by something is also a classic symptom of the disorder. Interestingly, there's also a large overlap between autism and ADHD: between 30% and 50% of autistic people have ADHD symptoms and two thirds of people with ADHD have elevated levels of autistic traits. And around 20% of people with drug use disorders have ADHD, a rate at least four times higher than that seen in the general population.

Of course, as a child, I was completely unaware of these diagnoses and characteristics or of their connections to each other. Nonetheless, I was trying to learn to cope with them—and changing the path of my own development in the process.

FIVE

The Myth of the Addictive Personality

Addicts are just like everyone else—only more so.

—ANONYMOUS

METHANE. ETHANE. PROPANE. BUTANE. I CHANTED the words to myself, thinking not really of the chemistry my father was teaching me (the alkane series, for those interested) but more of the soothing sounds of the words. I was alone, on the swings in my elementary school playground, trying to calm myself with the familiar syllables and the patterned beat of my movement. I didn't know how to cope with the other children and their shrieking voices, sudden actions, and incomprehensible social world. I must have been around six or seven.

I closed my eyes, only occasionally peeking to be sure someone wasn't sneaking up on me. With the rhythmic motion and silent mental chant, I aimed for a kind of a trance, though I didn't know that's what I was doing. I imagined myself flying off the swing, in a perfect arc, soaring over the fence at the edge of the enclosed playground, landing softly in a meadow where there were no other children to torment me, just brilliant sunlight, lush green grass, flowers and hills. But I couldn't totally let myself go, as I would do when I played this game on the much smaller swing set in our yard. Here, there was always a chance that someone would come over and bring me back to earth, rudely and abruptly.

I hated recess. At least in class the teachers would keep people from making fun of my strange interests and odd actions; at least in school there was structure to hold on to and books to read. And at least the teachers seemed to like me: unlike other kids, they told you exactly what you were supposed to do if you wanted their approval. As long as I could restrain myself from shouting out answers or attempting to be called on every single time, I was able to please them. However, as a result of my inability to always succeed at this, for parts of the day, the school had devised a makeshift gifted program for me. Because I was so advanced in reading, my boredom was obvious and probably distracted the teachers from helping those who really needed to learn.

At those times, I was either allowed to read on my own in the teachers' lounge, supervised by Mrs. Quackenbush, or, alternatively, I'd do craft projects. I distinctly recall sitting with the coffee-guzzling teachers and cutting colored construction paper with pinking shears, which fascinated me with their zigzag blades. When I wasn't with the teachers, I'd be placed with the learning disabled children—again, so that I could work at my own pace without being bored or distracting others. I did not like this as much: some of the boys scared me and I was also acutely aware of the stigma attached to them and, by extension, to me. Since I believed that the best thing about me was that I was smart, I especially didn't want people to connect me with the kids that were at the time labeled "retarded."

When I was six and about to start first grade, my family had left the city for a small village about an hour upstate, on the shore of a beautiful nine-mile-long lake. Greenwood Lake was close enough to the city that it should have been a suburb. However, because the public transportation linking it to Manhattan was so poor, it was actually quite rural. My parents had seen the house they decided to buy in the summer, when the town was filled with well-off visitors. The "townies" were quite different: overwhelmingly Irish, German, and Italian working-class families. Many of the men were New York City police officers or firefighters; most of the moms stayed home. We were the only Jewish family in our neighborhood. Not only did I stand out because of my temperament, I also stood out for class, religious, and cultural reasons that were not comprehensible to a child.

And I couldn't do anything about it other than start looking for ways to escape.

A WEIRD LITTLE girl on the swings engaging in compulsive behavior to soothe herself is probably not what you picture when you think of an addicted person or her background. Our cultural images of addiction tend to be much less likely to engender sympathy. For one, they are racialized—so even though black and Hispanic people are not more likely than whites to become addicted, those with dark skin tend to be pictured in American media stories about addiction. And when whites are shown, we are typically described as not being "typical."

Second, in part as a result of the racism that has driven our drug policies, these images tend to depict people with addictions as "fiends" or "demons" whose debauchery is driven by a ravenous hedonism, not a human and understandable search for safety and comfort. The "addictive personality" is seen as a bad one: weak, unreliable, selfish, and out of control. The temperament from which it springs is seen as defective, unable to resist temptation. Even when we joke about having an addictive personality it's usually to justify an indulgence or to signal our guilt about pleasure, even if only ironically. To understand the role of learning in addiction and in the temperaments that predispose people to it, we have to examine the relationship between addiction and personality more closely.

Although addiction was originally framed by both Alcoholics Anonymous and psychiatry as a form of antisocial personality or "character" disorder, research did not confirm this idea. Despite decades of attempts, no single addictive personality common to everyone with addictions has ever been found. If you have come to believe that you yourself or an addicted loved one, by nature of having addiction, has a defective or selfish personality, you have been misled. As George Koob, the director of the National Institute on Alcohol Abuse and Alcoholism, told me, "What we're finding is that the addictive personality, if you will, is multifaceted," says Koob. "It doesn't really exist as an entity of its own."

Fundamentally, the idea of a general addictive personality is a myth. Research finds no universal character traits that are common to all addicted people. Only half have more than one addiction (not including cigarettes)—and many can control their engagement with some addictive substances or activities, but not others. Some are shy; some are bold. Some are fundamentally kind and caring; some are cruel. Some tend toward honesty; others not so much. The whole range of human character can be found among people with addictions, despite the cruel stereotypes that are typically presented. Only 18% of addicts, for example, have a personality disorder characterized by lying, stealing, lack of conscience, and manipulative antisocial behavior. This is more than four times the rate seen in typical people, but it still means that 82% of us don't fit that particular caricature of addiction.

Although people with addictions or potential addicts cannot be identified by a specific collection of personality traits, however, it is often possible to tell quite early on which children are at high risk. Children who ultimately develop addictions tend to be outliers in a number of measurable ways. Yes, some stand out because they are antisocial and callous—but others stand out because they are overly moralistic and sensitive. While those who are the most impulsive and eager to try new things are at highest risk, the odds of addiction are also elevated in those who are compulsive and fear novelty. It is extremes of personality and temperament—some of which are associated with talents, not deficits—that elevates risk. Giftedness and high IQ, for instance, are linked with higher rates of illegal drug use than having average intelligence.

Whether these extreme traits lead to addictions, other compulsive behaviors, developmental differences, mental illnesses, or some mixture depends not just on genetics but also on the environment, people's own reactions to it, and those of others to them. Addictions and other neurodevelopmental disorders rely not just on our actual experience but on how we interpret it and how our parents and friends respond to and label the way we behave. They develop in brains designed to change with experience—and that leaves us vulnerable to learning things that create damaging patterns, not just useful habits.

The impact of all these factors together can be seen most clearly in studies that follow participants from infancy into adulthood (which are rare because they take so long to conduct and are thus very expensive). In these types of data, some strong patterns emerge. One of the earliest and best known longitudinal studies related to drug use followed 101 children—mainly middle class, two-thirds white—raised in Berkeley in the 1970s.

Conducted by psychologists Jonathan Shedler and Jack Block, then at the University of California, the research was published in 1990 and its main finding generated much controversy. The authors discovered that the most mentally and psychologically healthy teens were *not* those who abstained entirely from alcohol and other drugs, but rather the kids who experimented with weed and drinking, but didn't overdo it. In this study, occasional teen drinking and marijuana use was normal adolescent behavior. However, while it was common, it was typically not problematic.

Unsurprisingly the teens who became frequent users and drinkers had the problems you might expect, like depression, anxiety, and delinquent behavior. Then again, many of the same psychiatric problems were also seen in the adolescents who *rejected* the idea of drinking and drugs entirely. That's probably because, in order to avoid *any* experimentation as a kid growing up around the Berkeley campus in the '70s (when nearly two thirds of high school seniors nationally reported at least trying marijuana), you'd have to be either a loner with few friends or a person who was unusually fearful and/or resistant to peer pressure. Not using drugs may well have been a wise choice for these youth— but good decisions aren't always made for healthy reasons.

And indeed, that's exactly what the study found. The youth who abstained did not tend to do so because they rationally recognized the risks. Instead, they were overly anxious, uptight, and lacking in social skills; some may not have had to say no because they didn't even get the chance to say yes. Similar data have been published on teen drinking as well. Moderate drinkers—not nondrinkers—are the most well adjusted, at least in countries where drinking is a social norm. The healthiest patterns are found in the middle of the curve, not at the extremes.

To understand how having these outlying traits increases risk for addiction, we have to look at how they affect development. Critically, in Shedler and Block's data, the traits that marked both abstainers and heavy users could be seen long before drug use began. After all, the authors had started following these children in preschool. Once they knew how the participants behaved in adolescence, they could look back and see what early traits were linked to particular problems.

LONGITUDINAL STUDIES LOOKING at addiction risk have found three major pathways to it that involve temperamental traits, all of which can be seen in nascent form in young children. The first, which is more common in males, involves impulsivity, boldness, and a desire for new experience; it can lead to addiction because it makes it hard for people to control their own behavior. The second, which tends to be seen more in women, involves being sad, inhibited, and/or anxious. While these negative emotions can also deter experimentation, when they do not do so, people may find themselves on a "self-medicating" path to addiction, where drugs are used to cope with painful feelings.

Being bold and adventurous and being sad and cautious seem like opposite personality types. However, these two paths to addiction are actually not mutually exclusive. The third way involves having both kinds of traits, where people alternatively fear and desire novelty. Behavior swings from being impulsive and rash to being compulsive, fear driven, and stuck in rigid patterns. This is where some of the contradictions that have long confounded the study of addiction come into play—namely, some aspects seem precisely planned out, while others are obviously related to lack of restraint. My own story spirals around this paradoxical situation: I was driven enough to excel academically and fundamentally scared of change and of other people—yet I was also reckless enough to sell cocaine and shoot heroin.

If we look more closely, however, the paradoxes disappear. All three pathways really involve the same fundamental problem: a difficulty with self-regulation. This may appear predominantly as an inability to inhibit strong impulses, it may be largely an impairment in modulating negative emotions like anxiety, or it may have elements of both. In any case, difficulties with self-regulation lay the groundwork for learning addiction and for creating a condition that is hard to understand. The brain regions that allow self-regulation need experience and practice in order to develop. If that experience is aberrant or if those brain regions are wired unusually, they may not learn to work properly.

The importance of self-regulation is evident in the Shedler and Block data. From the very start, the children in the study who grew up to be heavy drug users were, as they put it, "visibly deviant from their peers, emotionally labile, inattentive and unable to concentrate, not involved in what they do," and "stubborn." This is a picture of emotional dysregulation—and it could have described me as a child, except for "not involved in what they do."

But while such children can be summed up as having "low self-control" or "impulse control problems"—and in the study, these kids tended to have lower grades—this doesn't account for the compulsive side of addiction. In my case, when it came to schoolwork, I didn't shirk. Indeed, I was desperate to be a good student and terrified of getting in trouble. Here, I had trouble *stopping* intellectual engagement, not *starting* it.

Obsessiveness like this, however, also involves impaired self-regulation—in this case, at the other end of the spectrum. It's a problem with stopping what has already been started, rather than starting an action that should have been stopped. In other words, while impulsiveness involves too little behavioral inhibition and a failure to prevent reckless behavior, obsession and compulsiveness is a problem with too much inhibition, a difficulty with getting out of a rut, rather than with preventing actions from being initiated. Further, inability to modulate fear and other emotions also involves a reduced capacity to self-regulate.

In their studies, Shedler and Block found that the *abstaining* youth were "fastidious, conservative, proud of being 'objective' and rational, overly controlled and prone to delay gratification unnecessarily, not liked or accepted by people," as well as "moralistic," "not gregarious," and "basically anxious." Most of that could also have described me as a three-year-old. Indeed, it reads now as a somewhat judgmental description of the key traits of children with Asperger's.

My own behavior as a young child and elementary school student swung between the poles of being overly controlled to being out of control. Both behavioral extremes, however, result from a failure in self-regulation. And neuroscience now strongly suggests that such dysregulation plays a key role in addiction. In fact, similar brain circuits are involved in both addiction and obsessive-compulsive disorder (OCD): whether the problem is failing to stop an impulsive action or failing to end a habitual routine, many of the same regions are engaged. It is here that addiction is learned.

The relevant areas of the brain include the prefrontal cortex (PFC), which imagines possible futures and plans and makes decisions accordingly. Of particular importance within the PFC is the orbitofrontal cortex, which helps determine the relative emotional and psychological value of your options and, therefore, your level of motivation and your tendency to make particular choices. The PFC works in concert with the nucleus accumbens (NAC), the region famed as the brain's "pleasure or reward" center. This area is involved in determining the desirability of particular options and how much you want to seek or avoid them. Another region related to reward and motivation, the ventral pallidum, is also part of this brain system, as is the habenula, which seems to be involved primarily in aversion and disliking.

The insula, which processes emotions like lust and disgust and also monitors internal states like hunger and thirst, is another node in this circuitry. So is the anterior cingulate, which looks for conflicts and errors and changes emotion accordingly. The anterior cingulate seems to be especially important for obsessive behaviors, perhaps because it creates a sense that things are "not right" until they are perfect or complete. In OCD, it may wrongly detect errors, which could cause constant anxiety. Finally, the amygdala is also in the loop. While best known for its role in processing fear, the almond-shaped amygdala is also involved in a variety of other emotions, including positive ones.

Together, this whole neural network sets values, priorities, and goals. Crucially, parts of it can also simplify repeated behavior into programs for habits that can be engaged or disengaged with little conscious thought. Indeed, research shows that as a behavior is learned and becomes more automatic, it engages different parts of the striatum, which is the broader area that contains the nucleus accumbens. As a behavior moves from being a conscious choice to a habit, brain activity changes, moving up toward the top or "dorsal" portion of the striatum and away from the bottom or "ventral" area. In addiction and other compulsive behaviors, brain activity that is increasingly dorsal in the striatum seems to be linked with reduced ability of the prefrontal cortex to stop or control the behavior.

One critical aspect of addiction, in fact, is an alteration in the balance between brain networks that drive habitual behavior and those that determine whether or not to execute those routines. Again, all of these regions are made to change with experience and are, as a result, developmentally vulnerable both in early childhood and adolescence. With any activity, as it is learned, it becomes easier, more automatic, and less conscious. This is essential when you are learning to play the piano or throw a ball—and it allows "muscle memory" to develop and hone your skills. However, it's not such a great capacity to have when you are learning addiction because, by definition, more reflexive behavior is less under conscious control.

It seems that the same regions that gave me my intense curiosity, obsessive focus, and ability to learn and memorize quickly also made me vulnerable to discovering potential bad habits and then rapidly getting locked into them.

TRAITS THAT MAKE children stand out from their peers at the earliest stages of life are clearly linked to biology: either something in the genes or womb or both must be responsible for them. But while biology shapes these predispositions, they are far from destiny. Estimates of the genetic heritability of addiction center around 40%–60%, which leaves a large opening for the influence of other factors. The environment—and whether it's nurturing and safe or marred by trauma like loss, violence, or other uncontrollable stresses—plays perhaps an equally large role in determining how a trait will express itself. To make it even more complicated, the effects of the environment are subjective, not objective: one person may perceive trauma and rejection where someone else sees minor stress and support, so the interpretation of experience also matters, as does how other people react. In some cases, these environmental factors alone may be enough to push someone with little predisposition to addictive behavior over the edge.

In my case, there was little obvious trauma in my early life, though I was certainly exposed to secondary trauma due to my father's experience as a Holocaust survivor and my mother's maternal loss. But for many, if not most, people with addiction, trauma is perhaps *the* critical factor that causes the problem. Over the years I've interviewed hundreds of addicted people and heard thousands of stories of addiction in support group meetings. It is impossible to do this and not be struck by the amount of pain and heartbreak that precedes most cases of addiction.

To pick some fairly typical examples, I once spoke with a heroin and cocaine addict who not only was regularly beaten with an extension cord but also saw his father punch his mother's teeth out. At age seven, he watched his mother slit her wrists. I talked with a woman with alcoholism whose mother often told her that she was so ugly as a baby, she wanted to put her bonnet on backward— and that was far from her only experience of abuse, emotional and physical. I spoke with a crack-addicted woman who called child welfare authorities on her own parents. I've also spoken with dozens, men and women, who walked on eggshells during screaming family fights complete with crashing furniture and blood, who've witnessed stabbings and shootings at incredibly tender ages, and who've spent their childhoods considering themselves worthless and hopeless.

And I've interviewed far too many people with addictions who were repeatedly raped starting before puberty, including some who were violated as toddlers. One woman, whose preferred drugs included alcohol and crack, discovered only after going through treatment that the uncle who repeatedly raped her when she was seven or eight had done the same to her mother—but her mother did nothing to protect her. I've spoken with addicts who've lost multiple family members to violence before they finished elementary school—basically, people who have had the unspeakable and unthinkable happen very early in life, and often, more than once.

It's hard for those who have not had traumatic experience or studied trauma to imagine the pain of these lives. Indeed, when I was interviewing a group of women who fall into the most despised category of addicted people—crack-addicted mothers who'd smoked during their pregnancies—I remember thinking that the horror of just hearing these stories made me want chemical escape and how much worse it must be to have lived them. Their childhoods were a litany of sexual abuse, physical abuse, neglect, death, violence, disease, poverty, bullying, and just loss after loss after loss. These stories are not the exception, they are the rule. At least two thirds of addicted people have suffered at least one extremely traumatic experience during childhood—and the higher the exposure to trauma, the greater the addiction risk.

Further, the more extreme the addiction, generally, the more extreme the childhood history of trauma. In fact, one third to one half of heroin injectors have experienced sexual abuse, with the sexual abuse rates for women who inject roughly double those for men. And in 50% of these sexual abuse cases, the offense was not just a single incident but an ongoing series of attacks, typically conducted by a relative or family friend who should have been a source of support, not stress. The same proportion—50%—of heroin addicts have suffered emotional abuse and physical neglect, leading one research group to characterize the typical preaddiction experience as a "shattered childhood."

This doesn't mean that all addiction comes from trauma: in my own case and perhaps one third of others, there is no significant trauma history. Moreover, most people who do experience childhood trauma do not develop substance use disorders—humans have a remarkable tendency toward resilience. Substance use is just one of many ways that people learn to cope. And since coping behavior is essential to psychological survival, the coping methods learned during childhood and adolescence become deeply engraved in the brain.

Nonetheless, the fact that trauma and addiction are linked is incontrovertible. Studies of what are known as "adverse childhood experiences" (ACEs) show a linear relationship between the number of such traumas and addiction risk. Even just one extreme adversity—like losing a parent or witnessing domestic violence—before age 15 doubles the odds of substance use disorders, according to a study of the entire Swedish population. Other ACEs include divorce; verbal, physical, and sexual abuse; neglect; active addiction; or symptomatic mental illness in the immediate household and having incarcerated family members. If you learn that the world is not a safe and stable place—and that others are unreliable—when you are young, it can shape the trajectory of your emotional learning and the way you cope for the rest of your life.

Research involving tens of thousands of patients in the Kaiser Permanente health system in San Diego demonstrates the terrible impact such experiences can have. For instance, a child with five or more ACEs has a risk of illegal drug addiction 7 to 10 times greater than one who has none; one study suggested that 64% of risk for these addictions could be attributed to child trauma. The risk of heavy smoking is nearly tripled for people with five or more ACEs—and alcoholism risk is increased by a factor of 7 for those with four or more. Although some ACEs may also represent genetic risk factors (e.g., having an addicted or mentally ill parent can both signal genetic risk and create an unstable home environment), the strong dose/response relationship between the amount of traumatic experiences a person has and his or her risk for addiction is undeniable.

AND NOW RESEARCH is also demonstrating what novelists, poets, and playwrights have known forever: trauma doesn't only harm the person who is directly affected, it can also be passed down to the next generation. This may happen not just because trauma can affect the quality of parenting, but because it can also lead to certain changes in chemicals that regulate genes. These alterations can biologically influence parenting behavior in traumatized parents—and they may be able to affect sperm and egg cells, directly influencing the resulting child's brain development. Such alterations are known as "epigenetic" changes—and research here offers new insight into how our lives both alter and are affected by our genes. Epigenetics starkly illustrates that nature and nurture are not separate: they are intimately entangled and interact repeatedly over the course of development.

Basically, epigenetics involves changes to molecules that determine which genes will be turned on and which will remain silent. These changes don't alter the DNA that transmits genetic information itself. Instead, they affect the structures around the DNA that determine how it will be read. That, in turn, alters how active or inactive particular genes will be. Some of these "reading instructions" can be passed down along with the genes—although these changes only seem to affect two generations and then disappear. What this means, however, is that trauma your parents or even grandparents experienced—and even their diets and their chemical exposures—may potentially affect your brain development and, therefore, your risk for addiction.

Studies on the offspring of Holocaust survivors suggest that our parents' experiences may be written in our genes. The results often vary depending on which parent was affected—and the data here is intriguing, but still considered preliminary. One study showed, for example, that having a mother with post-Holocaust posttraumatic stress disorder (PTSD) increased the effects of a gene involved in amplifying stress signaling in the brain. However, having a father with the same disorder had the opposite effect. Oddly, this means that in this case, the father's negative childhood experience might actually be protective—at least with regard to PTSD. The actual pattern of inheritance of PTSD following the Holocaust bears this out: children born to mothers traumatized by the Nazi death camps have overresponsive stress systems and a greater risk for PTSD. But the same is not true if the father was the only affected parent, as in my case.

However, depression—which both my father and I have suffered from—seems to be a different story. Although there are no human studies yet, mouse studies find that babies born to fathers who suffer extreme, uncontrollable stress in infancy have female offspring with more depression-like behavior. This occurs even though the mouse mothers are normal and even though male mice aren't involved in parenting. Bizarrely, while male pups fathered by these traumatized dads are themselves unaffected, their own sons (the third generation) show the same increase in depression-like stress responses seen in their aunts. This research also shows that changes in methylation, which is one epigenetic mechanism, are found in the sperm of these mice in genes that affect mental health and stress responses. And these changes persist for two generations after the initially stressed one, before reverting to their prior state.

Changes in sperm and egg cells, however, are not the only way epigenetics can influence development. Overwhelming stress and trauma can deeply influence how parents interact with infants. That, in turn, can change the babies' gene expression, during a phase of life when infants' developing brains are looking for signals about what type of environment they will face. This influences their growth patterns, both physical and psychological, which is why abused and neglected children are often small for their ages. Epigenetic signals particularly affect the stress response systems, which are critical to addiction risk, since drug use is often an attempt to manage stress.

Rat research, for example, shows that pups that receive the highest levels of nurturing and maternal grooming have calmer stress responses and are better functioning and smarter in most environments. When females raised by these affectionate mothers become moms themselves, they, too, are more nurturing to their babies, at least compared to other rats. But this nurturing style is not passed down by changing any genes.

If an infant pup born to a highly nurturing rat is "fostered" by a mother that is neglectful, that pup will tend to have a dysfunctional stress response and be generally less intelligent, except under conditions of high stress. Moreover, it will raise its own pups in the neglectful style in which it was raised. Nurture or lack of nurture itself is what determines the epigenetic outcome here: good parents literally activate a more optimal set of genes for their children, which allows them to perform well in the situations they are most likely to encounter.

Importantly, however, the changes that result from being raised in a stressful environment aren't all negative. The reason that early environment affects later development is to allow organisms to prepare for the life they expect to experience by turning on or off certain genes. This is nature's way of adapting children to the environment they are most likely to face, given the genes they have. So, if a child is brought into a wildly stressful, harsh world, genes that will help him thrive in such situations are turned on, while those that would be best in a calmer, safer place are muted. This affects the developmental trajectory of the brain's stress system, which has a profound effect on mental and physical health. It also influences cognition because the brain's ability to reach peak performance in areas like abstract thought is diminished under severe stress. Adaptations for stress, sadly, can come at the expense of intellectual talent, though, fortunately, these kinds of changes aren't necessarily permanent.

Still, reactions that are adaptive in a stressful world—like rapidly perceiving and responding to even small threats—can also be handicaps in a calmer environment. Being wired for a future in a threatening world can create a "live fast, die young" mind and body—and this increases addiction risk. In an uncertain world, it's rational not to count on anything too far in the future. However, this type of short-term thinking can lead to impulsive choices like eating one marshmallow now rather than waiting for two later or taking drugs rather than going to school. Alternatively, it can create a rigid desire for control, to try to minimize the chaos. Either way, it makes evolutionary sense that the environments faced by parents and even grandparents would affect the genes of their offspring: being appropriately attuned to the expected level of resources and stress certainly can affect survival. The difficulty comes when those stress settings and the world are mismatched—or when ordinary stress is magnified by an oversensitive brain into a sense of uncontrollable stress.

*　*　*

LONG BEFORE MY addiction began, I had already learned obsessive-compulsive behavior. For me, it seemed to be linked to fear of death, which has obsessed me ever since I first heard about mortality when I was three or four. I'm not sure what prompted the conversation: it may have been the demise of our pet hamster at nursery school (it had a large tumor that I can picture, even now), or it could have been a result of my curiosity, which led to constant questioning of adults. My immediate reply when told that humans die was, "Oh, and then you're born again, right?"—demonstrating either a very early propensity for denial or some spiritual knowledge that has sadly since left me. As far as I can tell, however, it was the staggering anxiety that this information provoked that initially sparked my compulsive behavior, which would eventually be seen in actions like my ritual on the swings in elementary school.

When I gained that painful knowledge, fear of nonexistence became a daily challenge. As a child, I would lie awake at night, overcome with fear about the end. I would try to pray, hoping that there was some sort of afterlife, but the Judaism I was being raised with was conspicuously silent on this issue. My parents had mentioned "living on in other people's memories"—but that was not at all comforting. I'd try to soothe myself with rationalizations about why some sort of eternal soul was probable—or with ideas about becoming a scientist who would cure death. Still, I would often wind up shaking with dread, kicking and thrashing as if to physically distance myself from the awful thoughts of the void.

The idea that I was powerless over the whole thing constantly upset me. Worrying that I would soon be gone forever seemed perfectly natural. I couldn't understand why other people were so nonchalant about it. I didn't realize that everyone else didn't have these types of obsessive fears—or that a focus on death itself is often a symptom of the obsessive behavior that is seen in both Asperger's and OCD. While Ernest Becker would later write in his Pulitzer Prize–winning book *The Denial of Death* that fear of mortality "haunts the human animal like nothing else" and is "the mainspring of human activity," for me, literal thoughts about dying were way too close to the surface.

Perhaps this isn't surprising in the child of a Holocaust survivor—and I suspect now that it might also have been influenced by the fact that my mother was undergoing treatment for cancer during my preschool years. She had always been haunted by her own mother's death from that disease—and while I wasn't told what she had, I did know that something was wrong. The cancer was on her neck and the first surgery she had to treat it didn't heal properly. I could see quite clearly that she was wounded. Fortunately, a second surgery cured it. However, I probably picked up on the current of fear and anxiety in our little family, painting pictures of "the bump."

And over time, my obsessive fears translated themselves into hidden compulsive rituals, which I would repeat to attempt to pacify my fear. Counting, memorization, certain acts like having to be the last person to go to the bathroom before I left somewhere all became secret defenses against my anxieties, which included both agoraphobia and claustrophobia. Even now, I am terrified of being trapped in tight places or in crowds of people or, especially, the two combined. To cope, I need to be near a door or exit, which led to frequent fights with my siblings when I refused to sit in middle seats. I am also afraid of being too far away from a bathroom—so car rides make me especially panicky.

I did not figure it out until much later, but these specific fears eerily replicated the worst experiences my father had as a child during the Holocaust. My father never talked explicitly about any details when I was a child, but he seems to have somehow transmitted his terror to me. Whether this is an example of epigenetics—one intriguing mouse study suggested that specific fears can be passed down via changes in the reading instructions of DNA—I don't know. But it feels uncanny to me, even just writing about it.

Further, because I was ashamed of my compulsions, I hid them as much as possible. This shame itself, it turns out, is also symptomatic of compulsive behavior, whether it occurs in autism, OCD, or addictions. Perhaps because of the visible loss of control they induce—or perhaps because the affected brain regions are also those that process emotions like disgust—these conditions are all marked by powerful feelings of self-loathing. Either way, the mental and physical ritual I engaged in on the swings at my elementary school was far from my only unusual routine.

And although no drugs were involved, the repetitive pattern of my behavior itself was enough to etch it deeply into my brain. As with any habit, repetition both strengthens the memory of what is done and automates the processing involved, making further repetition both easier and more pleasant. Repetition alone is especially rewarding to the developing brain; just ask any parent who has been driven mad by having to read the same book or listen to the same children's music over and over and over again. If you repeat something that seems to relieve anxiety, it is all the more attractive.

But in OCD and Asperger's, as in addiction, over time the behavior stops serving its intended purpose. Rather than making things better, it starts to make them worse. Unfortunately, by this time, habitual responses are already deeply engrained and well learned; and even if you know the ritual you perform to try to make you less anxious will actually make you more so, you don't believe it. You feel compelled to repeat it, even when you are utterly sure that it will not help. This is the heart of why addictions—and, not incidentally, OCD—are learning disorders.

It also has critical implications for drug policy. OCD is not driven by the wonderful anxiety-reducing properties of, say, hand washing. There is not something hidden in soap or water that makes people want to wash and wash again. People don't "catch" OCD by simply washing their hands—and, by the same token, they don't develop drug addictions by just taking drugs. Preexisting differences in temperament and in negative experiences are what drives the learning of addiction.

Though some drugs obviously have risks other than addiction (such as overdose), if we look for the cause of addiction in the chemicals themselves, we miss these connections. Seeing addiction as "caused" by the availability of specific substances blinds us to what drives it more generally. It creates instead a serial focus on an ever-expanding list of potentially dangerous compulsions, from new drugs sold online to the Internet itself, from mobile phone use to gaming and porn. The problem isn't the existence of activities and substances that offer escape; it's the need for relief and the learned pattern of seeking it that matters.

Consequently, trying to fight addiction by criminalizing particular drugs is like trying to suppress repetitive behaviors in autism by punishing them. A child with autism may seem more typical if he stops flapping his hands—but this by itself doesn't alter the underlying reason for the activity or change his autism. As a result, the behavior will either be hidden or replaced by a substitute action if the real driver of the behavior isn't addressed. Indeed, many autistic adults describe being traumatized by therapies aimed at suppressing their self-soothing and self-stimulating behavior because this left them exhausted, upset, and without alternative means of coping.

The same is true for drugs: banning new ones or cracking down on old ones may make policy makers feel like they are "doing something," but it ignores the real problem, which is the distress that makes people seek escape. Worse, this penalizes people whose condition itself is marked by failure to change when punished. Unfortunately, however, treating the symptoms as the disease has a long and undistinguished history in addiction policy. And it comes, at least in part, from a failure to understand how learning affects addiction.

SIX

Labels

The reason diagnostic labels are so important . . . is that without those labels, we only have the labels we got in the streets, which are hateful.

—JOHN ELDER ROBISON

MY SISTER KIRA AND I, AGES approximately four and seven, were seated on the floor of the basement of our house in Greenwood Lake, which had blue and gray speckles and a strange, bumpy texture. (My dad had been experimenting with formulas for epoxy flooring and had coated the whole basement in it.) We were playing with a large selection of toys spread out before us. We loved the small wooden and plastic figures that Fisher Price sold as "little people." For some reason known only to my childhood brain, we called them "charms." We had many of their accessories—a garage, airport, schoolhouse, barn, house, and Ferris wheel. We'd spread them out like a small city, using lines of masking tape we placed on the floor to define where the roads were. (When I was in second or third grade, I even wrote to the company asking them to make a hospital so that the women would have somewhere to have babies—and received a cordial and encouraging reply from some executive. But as far as I know, they never made one.)

Mostly, Kira and I played happily, creating stories of events in our little civilization and pretending to be various characters. Of course, we'd also have the typical squabbles over who got to go first and the like. Otherwise, there was only one common bone of contention: the question of evil. I often wanted bad things to happen in order to make our narrative more interesting. Without troubles or disasters, there was no plot and I was bored.

For instance, I created a disease called "crackeopia" (it had nothing to do with the drug, which wasn't yet invented). Crackeopia affected the wooden figures when they got wet, causing them to develop cracks. It soon became a plague, killing many. Thankfully, however, the plastic figures were immune to the disorder and survived. I also engineered plane crashes and multicar pileups and construction accidents. In Kira's version of our universe, however, there was no disease or difficulty.

I CAN'T RECALL exactly when I decided—or believed that I had discovered—that I was a bad person, but by the time I started first grade I saw myself as somehow innately morally defective. It's hard to imagine a little child making such a harsh and lasting judgment, but research suggests that kids of preschool age often fixate on ideas of "goodness" and "badness" as they relate to themselves and tend to see the world in stark extremes of black and white. And if you think about it, it's not really that surprising, given the context of their lives. Young children constantly hear remarks focused on the moral nature of their behavior—like "good girl" or "bad boy." Nearly every action is subject to adult scrutiny and then categorized as right or wrong. Children's stories, films, and TV shows also tend to have blunt moral lessons—characters are permanently "wicked witches" or "evil stepsisters" or, alternatively, always kind and beautiful and good. Rarely do you ever see a "bad guy" change.

And so, inside the mind of what photos show as an apparently happy, carefree little girl with strawberry-blond braids, was a negative self-concept that, at first imperceptibly and later visibly, began to determine the course of my life. I didn't verbalize my negative beliefs and of course, my parents didn't deliberately teach me to see that way. Below the surface, however, my inchoate ideas about the self and *my* self in particular began to coalesce around the idea that I was irremediably flawed.

I had never been told about my earlier ADHD diagnosis or the reason for the medication I briefly took. But I didn't need any psychiatric jargon or official label to know that I was different. That was obvious to me in how much I stood out from other children—whether I was chanting chemical names in my head on the swings or rejecting hugs or simply being overwhelmed by noise and motion. The idea that I was somehow born bad and could never change played a big role in how my mental health problems grew over the course of development.

While much has been said about the dangers of psychiatric labeling and medicalization—how this can distress children and make them feel like "sick people"—far fewer scholars and journalists have investigated the inverse possibility. That is, without a label, children may come up with their own ideas about "what's wrong" with them—because it is plain from how other people react that something is. Certainly, a label like "autistic" can make some kids feel as though they are defective and can never have friends or a career. However, it can also make them feel relief that there are other people like them, that they aren't uniquely bad, and that there are ways that social skills can be taught.

It's hard to study the effects of the labels children give themselves, but fortunately, there is a growing and elegant literature in this area. From this research, it is clear that many kids do decide whether they are "good" or "bad" or "smart" or "stupid" at a very young age—and they also form lasting views about whether such traits are permanent or changeable. Children take in and react to what they are told about themselves, in unpredictable and often unintended ways. These initial self-concepts shape their choices and, in turn, their brains. If you think you are good, you may act one way, but if you think you are bad, you'll act another, for instance.

Moreover, deciding to see yourself as "bad" as a preschooler can hardly be said to be a free choice, since it occurs at such a young age. Few parents, for example, would knowingly burden their two-year-olds with decisions that could have lifelong implications for their mental health. However, kids do make life-altering choices about how they see themselves every day, which can have a huge impact because of the way these self-concepts filter their perceptions of themselves and shape the way they interpret other people's behavior. Parents and others who influence the way children see themselves are typically quite unaware of the power they have over this—and even the best intentioned cannot do it perfectly, nor can they predict how a particular child will react.

I remember only one experience of my own that I think helped form my self-concept, though I'm sure there were hundreds of others that I can't consciously recall. This one involved my maternal step-grandmother, whom my grandfather married when I was around a year old. Grandma Marge had short, straight red hair and lots of freckles. She loved to swim but was always wearing hats or making other efforts to ward off the sun. In her late 40s, she didn't know much about how to talk to or play with young children; she had no children of her own and she hadn't spent much time with kids. While she tried hard to be involved in her grandchildren's lives, her rigid ideas about appropriate behavior often weren't in line with our actual developmental capacities or my peculiar temperament. But she said things that somehow crystalized for me how my differences were perceived by others. This was one part of how I began to see myself in a negative light.

The fact that I was far more interested in ideas than in people disturbed Grandma Marge. She had what you might call the Barbra Streisand worldview: there were people who needed people, and people who didn't. Those who did, like my cuddly and more cordial sister Kira, were not just "the luckiest people in the world" but also good and kind; the others, well, weren't. This latter category included me, since I was always fixated on some intellectual obsession, ranging from the volcanoes that had prompted my erupting cake to dinosaurs, the periodic table, opera, and science fiction. I was constantly going on about whatever special interest I had at the time, heedless of anything or anyone else.

Apparently in response to my frequent rejection of hugs and intense focus on ideas and patterns, my grandmother said to me (or perhaps I overheard her describing me to someone else) that I was "not a people person." This, I could tell, was not a good way to be. While I'm sure she didn't intend for it to be perceived this way, I decided that being more focused on ideas than people was an irremediable character flaw. A "people person," I thought, was either something you were or you weren't—and I wasn't. I had what Stanford psychologist Carol Dweck later described as a "fixed" or "entity" mindset. Someone with this perspective views capacities and abilities as inborn and unchanging; if you are "born that way," you stay that way—like a cartoon villain. In contrast, the mindset she calls a "growth" or "incremental" perspective views capacities or tendencies as malleable and amenable to improvement via effort. The mindset that a child constructs by elementary school can have an enormous impact on the rest of her life.

No one really knows how children "decide" on a mindset. However, Dweck found one factor that has a large effect: the way that parents and teachers praise them. Kids who are praised for *being*, say, smart or athletic or artistic or musical tend to develop a "fixed" view, while those who are rewarded and encouraged for their *effort* in a particular area learn to see ability and character as something that can grow with experience. In my case, I was frequently told how brilliant I was—my early reading often prompted such remarks, even from strangers on the subway, for instance. I got the "gifted" label early as well. While such praise and positive labeling is obviously well intentioned, it can sometimes have unintended consequences.

Dweck's research offers important insights into how children's views of themselves develop and can shape their futures in profound ways, including those that influence their mental health. Dweck is best known for her work on the influence of fixed or growth mindsets on academic performance. This research finds that kids who perceive intellectual ability as fixed tend to underachieve in school because they fear that major challenges will reveal weakness. Such children often avoid the hardest problems because, in their minds, "you've either got it or you don't." Finding out that they can't do something, that they "don't have it," threatens to be psychologically devastating, and so instead, they just don't try or stick only to safe situations in which they are sure they will excel. When faced with difficult tasks, those with the fixed view tend to either give up—since they think the difficulty implies that they really aren't naturally talented—or sometimes, they cheat. In contrast, children who believe that it's the effort that matters are less threatened by setbacks or defeats because they know that it is almost always possible to try harder.

One 2007 study of nearly 400 seventh graders in New York City found that those who saw intelligence as malleable tended to have improvement in their math grades in junior high—whereas those who saw it as unchangeable plateaued. Another found that a whopping 40% of those who believed that intelligence was fixed actually lied to hide evidence of poor performance and said their test scores during the research were higher than they actually were.

An even more fascinating paper, published in the *Journal of Personality and Social Psychology* in 2014, included several further experiments. The first involved 158 ninth graders, all of whom separately played a video game in which they were apparently rejected and excluded by their classmates. (It was actually a computer representing the other players—and this was revealed to the kids when the experiment completed.)

Those who believed that character traits that determine social acceptance—like being a bully or victim—were fixed found the rejection much more distressing than did those who saw character as changeable. The fixed-belief group also reported greater stress and worse physical health during their first year in high school—and got lower grades, averaging a 2.62 GPA, compared to a 3.08 for the others, after controlling for prior academic achievement.

A second experiment involved 82 ninth graders who were starting algebra I in a high school in a middle-class neighborhood in California. Teachers had found that kids who haven't passed this class by the end of their freshman year tend to drop out of school, in part because most academically oriented students are already in algebra II by this point. The researchers wanted to find out if teaching the kids that effort matters and intelligence is not fixed at birth would make a difference. To make sure they weren't simply teaching optimism, they taught one group that *intellectual* ability can grow with hard work, while teaching the rest that *athletic* skill was primarily due to effort rather than natural talent.

The intervention worked. And, it had the biggest impact on the kids who needed it most. Of those who were taught that they could improve their academic ability, only 2% got a D or lower in the class—whereas 14% of those who were only told this about athletics got Ds or Fs. (They didn't measure athletic achievements, but it would be intriguing to know whether this improved similarly as the other studies predict.)

Even more impressively, a third experiment replicated these results in one of California's worst-performing high schools, which is attended primarily by poor and minority children, many of whom are not native English speakers. Here, the intervention primarily affected the children who initially believed that mental ability was fixed, rather than improving grades and health and lowering stress in everyone in the experimental group. A full 42% of the control group got Ds or below in algebra I—but this was lowered to 19% among kids who started with the belief that ability is fixed but were taught instead that effort makes a difference.

In another series of elegant experiments, Dweck and her colleagues devised what seems to me to be a highly plausible explanation of how an obsessive orientation toward achievement—like I had—can also be associated with depression. The researchers were investigating how the youngest children can come to view their own value as people as being dependent on having certain qualities—like being smart or being good at sports—even before they are aware of whether these qualities are innate or acquired. Dweck's team knew that before children are about nine or ten, whether they have a fixed or malleable view of intelligence doesn't predict their goals or achievements in the way that it does for older kids. However, they had also found that some younger kids gave up and acted helpless when challenged in exactly the same way that the older ones with a fixed perspective did. But why were these younger kids giving up if they didn't think trying harder was useless?

To find out what was causing this defeat response, the researchers looked at self-development in younger children. Around age two, kids normally begin to have a sense of self that they can compare to that of others. This can be seen in the fact that they begin to show a sense of pride when they achieve goals that are applauded by adults and a sense of shame when they don't live up to grown-ups' standards. While preschoolers and kindergartners don't tend to focus much on how their abilities compare to others at this age, they are, as one study puts it, "very concerned about whether their performance means that they are 'good' or 'bad.' ".

Unfortunately, little children can't clearly distinguish between moral goodness and other valued qualities like beauty, intelligence, diligence, tidiness, and physical ability; it's all wrapped up together for them. Indeed, fairy tales, cartoons, and children's movies often deliberately conflate characteristics like moral and physical ugliness or special talent and moral worth. And Dweck's research shows that this confusion can cause problems for children who get the idea that their value as a person depends on their ability to achieve certain goals. It's these children, her experiments show, who respond with helplessness or despair when a task is too challenging—but only if it is a task in an area they've decided matters. For example, a child who sees herself as an athlete might respond to a setback in sports by feeling ashamed and unworthy, whereas one who has decided that it is his English skills that make him valuable and not his soccer performance would simply shrug off the same experience.

And so, at the age when I was developing a sense of self, I determined that mine was bad. My fixed mindset helped me internalize the idea that "not being a people person" meant I would never be able to connect with others and that my fascination with ideas meant I'd always miss out on friendship and, ultimately, love. This self-diagnosis was exacerbated by other labels I got from my grandmother and other adults of "selfish" and "bossy."

I think that my reaction to these labels set off another one of the negative spirals of development that ultimately led to my addiction, by first causing depression. Because I was overwhelmed by sensory experiences, I tried to control my environment as much as possible. A weird facet of autism, in fact, is that sensory oversensitivity can both fluctuate and be influenced by how much control a person feels they have over their experience. The very same stimulus that is sometimes painfully aversive—loud music, for example—can actually be enjoyed when it is chosen and the volume can be controlled or when sensory experience is not being felt as overwhelming.

What that looks like from the outside, however, is bizarre: it seems as though the child is either lying about being bothered by the stimulation to "have their way" (otherwise, how could they sometimes like it?) or is simply obsessed with power and claiming to be distressed in order to exercise it arbitrarily. Similarly, the capacity for intense attention when intellectually engaged seen in autism and ADHD is often taken for willfulness: such children clearly have unlimited attention when they want to focus, but not when they don't, and this looks like it could be solved by "trying harder." However, since I didn't really know how different my experience was from that of others, I had no way of explaining what was going on.

The result was that I was seen as a bossy child as I tried to manage my sensitivities and compulsions. The fact that I also didn't like to be hugged or touched reinforced my self-conception as "not a people person," and my desire for control despite other people's preferences made me "selfish." My inability to see that other people might not share my intellectual obsessions also made me seem especially self-centered; this isolated me from my peers as well. And then my lack of shared interests only reinforced my preference for ideas and books over other children and that, of course, further increased my isolation, creating an escalating feedback cycle.

Dweck's idea that having a contingent sense of self-worth—in my case, the sense that the only thing that was valuable about me was my intelligence—can produce a sense of helplessness and despair is certainly borne out in my story. On the plus side, it gave me a laserlike focus on academic achievement, which powered me through grade school and high school and into Columbia. But simultaneously, it drove my anxiety, self-hatred, and depression, creating ways of seeing myself and the world that ultimately led to a desperate need for escape.

Of course, there are many other routes to addiction as well. However, while mine may appear especially idiosyncratic, it actually isn't as weird as it may seem. Addictions grow in the interaction between childhood temperament, childhood experience, and children's interpretation of their experience. While the specifics of my case are unusual, this interplay and the way it iterates over time are not. In my case, oversensitivity led to bossiness that led to self-hate; this combined with a lack of other coping skills led me to find drugs, and heroin in particular, overwhelmingly attractive.

But the same type of iteration is seen in the stories of other people with addictions. Ron, who runs a treatment center, for instance, told me how his heroin problem began. He grew up black and middle class, but stuttering made him feel insecure. He says he felt "tremendous pressure" to live up to his parents' dream that he become a doctor or lawyer—and that gave him a highly contingent sense of self-worth. When he tried heroin at 14, he said, "I didn't feel anxious, I didn't feel wound up, I didn't feel tense, afraid." The sense of well-being that he felt led him to use more and more and more.

In contrast, Violet (not her real name) had an experience largely driven by trauma. Before she was 13, she'd witnessed her father's death, lost her older brother in a motorcycle accident, been bullied repeatedly at school, and suffered at least four years of molestation by an uncle. She described her first experience of crack as feeling "like this weight was lifted off my shoulders." Her early life was filled with loss and a sense that she was worthless and powerless. This made a drug that creates a sense of power and ability especially attractive.

It was typically years before our experiences and our damaged sense of self led us to start drugs—and then at least months of repeated, daily use before we thoroughly learned addiction. But the condition originated in the way we learned to frame ourselves and the world and the way we learned that drugs could alleviate the resulting distress.

IT'S NOW CLEAR that most cases of addiction start long before affected people are ever exposed to drugs. Because only a small minority of those who try any recreational drug become addicted, mere exposure isn't what differentiates between those who maintain control over their use and those who do not. Drugs alone do not "hijack the brain." Instead, what matters is what people learn— both before and after trying them.

While this may seem hard to believe in light of claims that certain drugs are "instantly addictive," in fact, one of the reasons that addiction is best seen as a learning disorder is that it simply cannot take place if someone doesn't learn over time to associate a drug with pleasure and/or relief. Addiction is, first and foremost, a relationship between a person and a substance, not an inevitable pharmacological reaction.

A striking example of this can be seen in people who are treated for pain with opioid drugs like morphine or Vicodin for several weeks in the hospital following surgery or accidents. That is enough time for some people to develop physical dependence, although even something as seemingly biologically standard as the amount of time it takes to become tolerant and dependent is actually quite variable.

People who do become dependent will experience symptoms like nausea, vomiting, cramps, sleeplessness, and diarrhea when released to go home without medication. However, many of them never realize that this "hospital flu" is actually opioid withdrawal. They don't suddenly, out of nowhere, get an urge to "cure" the problem by buying painkillers or heroin on the street, because they haven't learned that lack of drugs is the source of their symptoms. Because they haven't identified drugs as the source of their comfort or as the best way to cope, they don't have the severe anxiety levels seen in people with addictions during withdrawal.

If you haven't learned that a drug "fixes" you, you cannot be addicted to it, even if your body is dependent on it. Past ideas about "physical" and "psychological" dependence, which still shape public opinion on drugs today, led us down the wrong road. The idea that physical dependence was "real" addiction but psychological dependence was a mere trifle led people to ignore the role of learning. Obviously, "physical addiction" is encoded in the brain—but so, too, is psychological addiction; otherwise, it couldn't affect the way you behave (and you'd have to see the mind as a force that acts independently of physics to change your behavior without any brain activity). The mind/body distinction is a false one: psychological needs and desires drive addiction and these change with learning and development. Physical dependence does make quitting drugs harder—but if it were the real problem, addiction could be cured simply by being forced to wait out withdrawal.

The failure to distinguish between physical dependence and the learning that creates addiction is also why, contrary to claims you may see made in the media, babies can't actually be "born addicted." Infants can be born with physical dependence on a drug like heroin or Vicodin, but since they do not learn the vital relationship between choosing to take the drug and feeling better, they can't crave it. Of course, they may feel uncomfortable and distressed if they undergo abrupt withdrawal—but this cannot translate into drug craving since they don't know what to crave.

Babies' inability to learn the association between getting the drug and feeling better prevents them from developing the habitual behavior pattern that's necessary for addiction to occur. Their early physical dependence may shape their reactions if they take drugs later in life—perhaps through unconscious associations with life in the womb—but this is very different from being "born addicted." Such children—like children of alcoholics and addicts more generally—are at increased risk for addiction, but nonetheless, the majority does not develop drug problems. For example, children of alcoholics have a risk of developing alcoholism that is two to four times greater than that found in the general population, even if they are adopted by nonalcoholics or raised by parents in recovery. But even on the high end of that risk scale, 50% still don't develop severe drinking problems.

And genetics interacts with learning in many ways to create addiction. Consider the fact that about one in seven people seriously dislikes opioids—the drugs themselves make them feel nauseous, dizzy, and uncomfortably, rather than comfortably, numb. While many people think that heroin or Oxycontin are heavenly for everyone, that's only true for a minority of users. Longtime CBS News anchor Dan Rather, for example, once took a shot of heroin for a news program. He said that he'd never want to repeat the experience and that it gave him a "hell of a headache." Two normal volunteers who injected heroin for several days (in a 1969 experiment that would now be considered unethical) also found it primarily unpleasant. "My personal view at present is just one made grey and utterly grim by heroin. The extraordinary thing is that it brings no joy, no pleasure. . . . At most, some hours of disinterest . . . how can people want to take this stuff? To escape to all this—life must be hell if they want to escape to all this."

A more recent study, which gave 228 healthy adult twins an intravenous infusion of alfentanil—an opioid around five times stronger than heroin— found that only 29% emphatically liked it. Most had mixed (58%) or neutral (6%) experiences—and 14% found it outright unpleasant. Because this study included both identical and fraternal twins, the researchers were able to determine whether genetics influenced the response to the drug, which it did. And obviously, if you inherit a negative response to a drug, this substantially reduces the odds that you will become addicted to it. Indeed, a gene that causes a hot-flash-like response to alcohol reduces risk for alcoholism by a whopping factor of 9. Although culture can override even strong protective factors like this—in Japan, when heavy drinking became de rigueur among businessmen, the proportion of alcoholics with this gene went from 3% to 13%—a person's experience of the drug obviously matters. This interaction shapes how addiction is learned—or avoided.

Long before children hit puberty, then, temperament has already started to become established, creating tendencies that can set a life course that becomes increasingly difficult to alter. As children learn about their own personalities and how others see them, they create their own sense of themselves. Their self-concepts then further influence the path of their development. If kids see traits and character as fixed, they may develop a self-defeating perspective; if they see change as possible and achievable, this is less likely. Often, what actually happens in children's lives is less important than how those children interpret their experience. These interpretations, made at such a young age that they cannot truly be said to be chosen, can promote resilience or increase vulnerability. They are, like addiction, learned and shaped by the process of development.

And if my early childhood left me with a multitude of risk factors for addiction, my elementary and junior high school years were probably what made these predilections into an accurate prophecy.

SEVEN

Hell Is Junior High School

The seventh and eighth grades were for me, and for every single
good and interesting person I've ever known, what the writers of the
Bible meant when they used the words hell *and* the pit. *. . . . One*
was no longer just some kid. One was suddenly a Diane Arbus
character. It was springtime, for Hitler, and Germany.

—ANNE LAMOTT

SHIVERING IN THE EARLY AUTUMN CHILL at a school bus stop in Greenwood Lake, I hid myself under the poncho that my mother had crocheted for me. Of course, I knew that the other kids could still see me, but at least I felt warmer and didn't have to look at them. And I could pretend that the worst part of my day wasn't just about to arrive. "Maia, your hiney wiggles." That's what the older boys would yell at me as I clambered up the steps onto the school bus to face another day at what I'd privately labeled "the torture chamber" but was officially known as Greenwood Lake Middle School. I sat in the front so I didn't have to walk past too many of them, but I never felt safe on that bus.

At 12, my body was beginning to show the swells of puberty and my mind was immersed in a fug of hormones. I felt constantly humiliated and embarrassed; I couldn't understand why boys seemed to both despise me and be expressing sexual interest. I didn't believe adults when they told me that boys do mean things when they like you. I knew I was a social outcast, and I felt there must be something deeply wrong with me to provoke such a reaction. Being scapegoated by my peers only confirmed my vicious view of myself.

It was in these early years of adolescence that what I now understand as my depressive tendencies combined with my social environment to elevate my odds of addiction. Before I hit my teens, I'd already begun a habit of rumination and self-loathing; research now shows that youth who learn these thinking patterns are more prone to adult depression. When you wear a groove of self-denigration into your brain, your thoughts simply tend to slide into it more often: like a muscle that gets repeated use, the pathway of negative thinking gets stronger. As with any other habit, it then becomes less conscious and more part of the background, like your mental furniture. Rather than something you have constructed, it comes to seem like "the way it is" rather than a potentially warped self-perception. And research suggests that for women especially, this is a classic route to addiction: from depression to drugs.

My sensory issues and inability to control my emotions also made me a target. The data are clear that kids who can't self-regulate—whether because of ADHD, trauma, autism, mood, or personality disorders—are prime candidates for both bullying and social rejection. Being dysregulated makes people stand out, whether in terms of impulsivity, sensitivity to experience, or inability to manage mood. Whether you can't stop yourself from shouting out answers, can't keep from crying when you are hurt, or can't hide your enthusiasm or rage, lags in the development of self-control are both obvious and disturbing to typically developing peers. And this is where the spiraling interaction among learning, self-labeling, and peer pressure really plays out. In many cases, addiction is learned as bullies confirm children's worst social fears and anxieties about themselves.

I was a bully's delight. My classic Asperger trait of taking everything literally, rather than realizing when people were teasing, made me the perfect target: I didn't "get the joke." I always reacted intensely and I couldn't stop myself. On top of that, I resisted fashion and social convention. And all of this was further worsened by the fact that my family had moved into a community where I was far more likely to stand out for class and cultural reasons than I had in Manhattan.

* * *

ALTHOUGH CONVENTIONAL WISDOM has it that peer pressure is an important cause of addiction, the reality is more complicated than the stereotype. Peers certainly can spur each other to try drugs. However, in most cases, this is done subtly—not by overt urging, but by doing it and making it look cool. Besides, use itself can only produce exposure, not addiction. The peer pressure that really drives people to heavy drug use is the type that causes social isolation and distress.

For me, it had been relatively easy to distract myself from my social reject status with intellectual pursuits before puberty. But as I got older, my inability to fit in made me increasingly unhappy. I thought social rules and signals were arbitrary and illogical and when I was very young, I'd paid little attention to them. I thought conformity was a curse. Why would you want to be like every-one else? I was proud of being different and, particularly, rational and intellec-tual. Now, I became an obsessive student of fashion and status signals.

On the bus and around the neighborhood, the boys continued to tease me, shoving me, looking at me threateningly. They would mockingly ask me out on dates. From all of this, I got the idea that I was so revolting, so unlovable that no one would ever desire me. I developed a deep sense of shame about myself as a young woman. My own sexual and social awakening made this ever more painful. I dropped my resistance to conformity and went overboard in the other direction. Even though I often sympathized with Spock on *Star Trek* (I, of course, was a rabid fan), I tried to reconcile myself with the apparently irrational social emotions.

At 13 or 14, I even considered trying drugs and alcohol, though the idea ter-rified me. I was desperately afraid of anything that could make me feel out of control. We had drug education in seventh grade, and I remember it vividly. The teacher put great emphasis on the force of peer pressure and on how danger-ous particular drugs were. What I got from the lesson, however, was probably not what the teacher intended. I did learn that some drugs were more risky than others—that seemed obvious. But I also intuited that since peers were likely to pressure you into using, this means that you will be socially unaccept-able if you don't at least try something.

At the time, I remember conducting a careful analysis of the risks and ben-efits of the most common substances. Alcohol? Too poisonous, and the high seemed to be too scary, too uncontrollable. I didn't want to lose motor control. Tobacco? There didn't seem to be any high, and it seemed terribly bad for your health. Why bother? But marijuana—now, this was something I could live with and it didn't seem to take you too far out of reality. I decided that if having friends required that I take a drug, pot would be the one I'd say yes to. However, at this age, I was so socially isolated that I didn't get the chance.

I would sit in the cafeteria, alone and aching with awkwardness. Sometimes I attempted to fit in to the lowest social groups by joining them at their tables, but even there I was a dismal failure. I'll never forget the last time I tried this. There were two groups of outcasts among my classmates: the nerdy, well-behaved ones who were the absolute bottom of the social ladder, and the bad girls who came from abusive homes and who drank and smoked and fooled around with boys. They were at least one or two ranks above the dweebs. I'd started with the nerds, but they were so prim and proper and so judgmental that I felt uncomfortable. Besides, I wanted to be cool, and they didn't even try.

So, I started hanging out with the bad girls. After a week, they walked me into the bathroom and told me point-blank, "you don't belong, you never will, so stay away from us." I just sat on the toilet and wept.

WHEN I WAS growing up, such experiences were seen as a normal and ultimately harmless, if unpleasant, part of childhood. "Boys will be boys" and "sticks and stones" were the watchwords of the day. But research now shows that being consistently and extensively rejected and bullied by peers can have large and lasting effects on health. In fact, one recent study found being a victim of chronic bullying can sometimes be even more damaging than being abused by your parents or caregivers.

Of course, nearly all children will experience friendship breakups, tortured romances, insults, and other social frictions: that's life. In contrast, what's truly dangerous is being on the absolute bottom of a crushing social hierarchy for years. Occupying a low and debased social position produces a level of stress that can take a serious toll on both physical and mental health. Though it can seem like a trivial problem, because of its ripple effects on children's social lives and self-images, protracted bullying can sometimes cause a lifetime of pain.

The most obvious way that bullying raises addiction risk is that being or feeling left out can lead to drug taking to demonstrate "coolness." Indeed, research later confirmed that the prevention messages of my era backfired, in part by essentially telling kids that drugs are a symbol of peer acceptance and rebellion. Unsurprisingly, this made them more attractive, not less so. For me, when I wanted to take drugs simply to be cool, I was too uncool to get them—so this was not part of my experience. In fact, when I actually did start getting high in high school, I deliberately went out looking for peers who used drugs, rather than being pressured to take them by people I already knew. (I always find the idea that "peer pressure" is to blame for addiction rather funny in light of this: no one ever claims to be the peer who does the pressuring. Every addict is, instead, always the "victim" of this pressure that, apparently, springs from nowhere.)

But those kids who do manage to be "cool" enough to get drugs in early adolescence are also at high addiction risk. A recent study published in *Child Development* explored what happens to the youth who rise to the social peaks by engaging in risky behavior ahead of their peers. These are often the children labeled definitively by their middle-school classmates as "popular"—even though they sometimes turn out not to be the kids with the most friends.

This distinction is important: what kids usually mean by "popular" is what evolutionary psychologists studying primate behavior would label "dominant." In other words, if you ask seventh graders to anonymously pick out the most popular kids in their class, they'll generally name the kids perceived to be at the top of the social pecking order. However, if you ask the same kids to name their best friends or the children they'd most like to spend time with, the ones who have the most friends and are the most desired companions are typically not those in the in crowd. In other words, while the nicest kids are actually the most popular by the numbers, they are typically not the most dominant.

A few rare middle schoolers can balance social dominance and true friendship—Kim Reinle was one such popular girl in my school who sometimes reached out to me and helped keep me from total despair. But, generally, those who rule the roost in junior high generally tend to be more feared than loved.

The *Child Development* study showed that this dominant group often exhibits what the authors label as "pseudomature" behavior. Basically they drink, drug, have sex, and do other things that piss off adults at rates that are pretty normal for 16- and 17-year-olds, but are rare at 13 and 14. Following 184 teens from ages 13 to 23, the authors found that the youth who took this path peaked young, socially: they were cool in junior high, but not for long. They impressed their classmates with their precocious sexuality and defiance—but they didn't go on to develop the social skills needed to maintain the status they sought.

I narrowly missed joining one version of such a group when I was rejected by those girls in the bathroom—but if I had joined, I likely would have simply followed a different route to addiction. The study found that kids who exhibit pseudomature behavior at 13 are 45% more likely to have drug problems at 23.

Whether it's early exposure to drugs alone or, as seems more likely, drug exposure in combination with a predisposition or early-life trauma that makes kids desperate enough to try them that young, taking drugs at this age is a clear sign of trouble.

Indeed, the reason that middle school is such a hellish time for so many people—and one way that it plays a key role in addiction risk—is that the beginning of puberty is when young humans begin to clearly delineate their first status hierarchies. This is why many people can remember the exact pecking order of their junior high school long after they have forgotten most of the official curriculum.

Emotion powers learning: it marks what's important for survival and reproduction, which is what matters in biology. (This is also why it's easier to remember dull material when you associate it with silly sexual innuendo, for example.) In adolescence, emotional life is especially driven by a desire for social success above all. The intensity of emotions around friends and relationships during the teen years burns these memories into our brains. The fact that this is also a time when the brain is undergoing a major restructuring also adds salience. Both the physiology of these brain changes and the psychological enhancement of memories it produces help sear everything that we experience at this time into our core selves—including experiences of drug use that can lead to addiction.

Evolutionary factors probably also play into the importance of social status for teens. The awakening of sexuality is also the awakening of sexual competition; adolescence is a time when, presumably, our ancestors would have begun the process of attempting to demonstrate fitness in order to secure the best mates. The early pubescent brain is mature enough to understand status and try to maximize it—but the means of securing clout and rank are mainly still more animal than human. Physical dominance, boldness, and visible traits like attractiveness are critical. Wealth also matters, though it becomes much more important among adults.

And unfortunately, more than ever before, research is finding that status stress—whether simply social or socioeconomic—is a critical determinant of health, including problems like addiction, obesity, and even heart disease and cancer. That's because the brain's stress system is reliant on connections with other people in order to function properly. From infancy, when we literally cannot turn off or down our stress hormones without parental nurturing, through romantic attachments, which wire our stress response to be calmed primarily by the affection of our partners rather than our parents, we basically need other people to keep severe stress at bay. Other people are also a source of stress, however. And this is where status and hierarchy come in, particularly when kids first truly encounter it in middle school.

Two major lines of research provide groundbreaking insight into the depth of the connection among status, stress, and health. One was conducted with troupes of baboons in Africa; the other followed thousands of British civil servants. The Whitehall Studies (named after the headquarters of the British state bureaucracy) were led by Sir Michael Marmot, who was knighted as a result of this research. The first wave included some 18,000 men. Marmot and his colleagues found that, contrary to popular ideas about "executive stress" and heart attacks felling the boss, it's generally far less healthy—and far more stressful—on the bottom. At age 40 to 64, low-ranked clerks were *three times* more likely to die of any cause compared to those at the top of the heap. The correlations were graded: each step up the ladder lowered risk, each step down raised it. Gradients favoring those with high status were found in heart disease, stroke, diabetes, obesity, addictions, infectious disease, and some cancers.

Since the study was conducted in the United Kingdom, which has a national health care system, the differences weren't linked to access to health care. And they weren't simply caused by worse habits, such as smoking, in the poorer men: only one third of the gradient between top and bottom could be explained by higher smoking rates, when researchers took other factors like childhood social status into account. In fact, even the smokers had a gradient, such that the higher a man's rank in the system, the less risk of death he had from smoking. A second Whitehall study included women and had similar findings, though for women, their husband's status influenced their own and, therefore, affected their health.

The same overall findings held true in Robert Sapolsky's studies of baboons. Baboons are social primates that have a strict hierarchy but do not have access to alcohol, drugs, junk food, or cigarettes. Nonetheless, top baboons have lower "bad" cholesterol, lower blood pressure, and better functioning immune systems—and again, the lower-ranked animals are at greater risk for all types of early death. Stress linked to low status harms health by raising blood pressure and cholesterol and altering immune response—all of which affect risk for all types of disease. High levels of stress hormones can also damage certain brain cells, which increases risk of depression and probably other mental illnesses as well.

While addiction can certainly strike people of any class, the risk is higher at the bottom because of the way low social or economic status increases stress. However, social connections themselves can actually mitigate the effects of low socioeconomic status: if you have a strong social network, no matter how poor you are, the negative effects of stress are reduced. But if you are bullied and berated and made constantly aware of just how far down the social scale you are, they are magnified.

To make matters worse, bullying during a sensitive developmental period like adolescence can alter the stress system for life. It's important to note here that not all stress is bad—indeed, a small to moderate amount of stress is necessary for learning because it differentiates novel experiences from familiar ones. Stress is only bad if it is overwhelming and makes you feel helpless, which, unfortunately, is exactly what bullying is intended to do. While challenging but manageable levels of stress build brain systems—like gradually increasing the weights you lift to train muscle—massive, crushing doses cause injury. If the brain is overwhelmed by stress in adolescence when key circuitry is developing, it can affect the way it reacts throughout later life.

One main pathway to these changes is seen in the link between depression and status stress. First, studies find that depression is especially strongly linked to high levels of stress hormones. In the brain, high levels of these hormones can result in an excess of the neurotransmitter glutamate, which is particularly damaging to brain cells in a region important for memory and emotion called the hippocampus. Depression tends to shrink this region; all long-lasting effective methods of treating depression, from electroconvulsive therapy (ECT) to medications (and, presumably, talk therapy, though this can't be studied in rats and we can't cut open human brains to check), lead to improved neural growth in the area. Severe stress in children, in fact, is more likely to lead to depression or other diagnoses than it is to cause diagnosable PTSD.

Evolutionary reasons may also link depression to bullying and status stress. Psychologists have long observed that depressive behavior in humans looks a lot like submission behavior in other animals. Physically, depressed people shrink back—just like low-ranked animals. They tend to be timid and compliant, lacking energy and drive. They often show the high levels of stress hormones that are seen in creatures at the bottom of animal hierarchies. And raising levels of the neurotransmitter serotonin with antidepressant medication, in fact, can boost the social status of animals in dominance hierarchies, while lowering it can cause status loss. Studies consistently find links between low socioeconomic status in humans and depression, with rates doubled among those at the bottom. (Although of course depression could also lead to unemployment and, therefore, low socioeconomic status, studies that follow children over time find that poverty tends to precede depression, rather than depression tending to cause poverty.)

What this means is that depression may have originated as an adaptation aimed at getting creatures to give up and submit when that would be in the best interest of their long-term survival and reproductive goals. When challenging the dominant ape is simply going to produce harassment and potentially deadly physical attacks, it is far wiser to lay low, for example. The suite of behaviors that accompanies depression may have evolved initially to get animals to yield so they can live to fight another day. And this could be why bullying and social rejection are such big triggers for it: bullying is basically an expression of dominance.

When being bullied triggers the physiology of submission, depression can set in. Initially, this is self-protective in terms of making the victim take action to avoid or appease the bullies. But if this social withdrawal and appeasement persists, exacerbates the bullying, or occurs in the wrong situation, it becomes harmful and counterproductive. And the hopelessness that results can be especially toxic during a period of development when children are faced with finding ways to cope with status stress that will influence how they do so for the rest of their lives.

Research is only just beginning to illustrate how much bullying affects mental and physical health. For example, a recent study of some 4,300 students showed that although just 6% of tenth graders who were never bullied ranked in the bottom 10% for their age in both mental and physical health, 45% of those who were bullied in fifth grade and were still being bullied had the worst health as tenth graders.

The psychiatric risks are particularly striking. In a study of young adults in North Carolina who'd been followed from age 9 to 16 and assessed for bullying involvement, the risk for depression tripled and the risk for anxiety disorders quintupled in victims. For those who were both bullies and victims of bullying, the results were even worse—their risk for depression was eight times greater and their risk for anxiety disorders was five times higher than in those who had no involvement in bullying. Another study of the same group found that when they grew up, victims and people who both bullied others and were victimized themselves were at greater risk of losing jobs, of being unable to manage financially, and of having overall poor health, compared to uninvolved children.

To be clear, these risks aren't associated with a single incident or short-term episode of bullying (what kids tend to call "drama" rather than bullying). They're linked with being overwhelmingly rejected by peers and scapegoated for years. And those who are targeted in this way tend to have preexisting problems that attract bullies to us, according to research. This is not to justify such behavior or to claim at all that the victims are somehow to blame. Instead, it is to point out that the harm associated with bullying interacts with children's own vulnerabilities in a way that makes the cycle hard to break. Just as choosing to take drugs early is a marker for risk of addiction, being chronically bullied is a sign of multiple health risks that escalate over time as they interact.

Having traits like ADHD, oversensitivity, difficulty regulating emotions, and depressive tendencies and then being bullied makes it much more likely that people will seek to escape from their feelings. If such kids do find drugs, the same traits that increased their risk for social exclusion will then increase their risk of learning addiction, which will now be compounded by both the physiological changes of adolescence and the distress of feeling rejected.

In my case, when I graduated from Greenwood Lake Middle School, I went on to a much larger high school. There, I finally found friends who shared my intellectual interests and didn't see me as completely alien or reject me. But I also discovered drugs.

EIGHT

Transitive Nightfall

The magical bower turns into . . . neon dust . . . pointillist particles

for sure now. Golden particles, brilliant forest-green particles, each

one picking up the light.

—TOM WOLFE, *THE ELECTRIC KOOL-AID ACID TEST*

I'LL NEVER FORGET THE FIRST TIME I saw Amy. She was sitting on the floor in the hallway of Brooklyn's Edward R. Murrow High School, a specialized career school for broadcast journalism. That year I was visiting Murrow once a week to work on a documentary about TV news for my upstate high school's gifted program. Becoming the next Edward R. Murrow—or my idol of the time, Walter Cronkite, the most trusted man in America—was my obsession du jour. Television journalism would be the last of my teenage special interests, before my focus narrowed in on drugs. I was in eleventh grade, about to make a long, strange trip from proto-yuppie to neo-hippie.

Amy had long red hair that fell in perfect, shiny waves around her shoulders, obscuring her face. She had porcelain skin with a sprinkling of freckles and big green eyes. I thought she was the most beautiful person I had ever seen.

I desperately wanted to befriend her; even more than that, I wanted to be her. A guitarist and singer, she cast an alluring spell on both the boys and the girls around her.

When I first glimpsed her, Amy was hiding her face behind her hair. Clutched between her knees was a container of Rush, a nasty inhalant drug (probably amyl nitrate) that you could buy at head shops, along with bongs and black lights. Being in Brooklyn and away from my "real" classmates gave me extra social courage. My friend Angelica introduced us and Amy and I began a conversation. She invited me to stay with her for the weekend the next time I came down to Brooklyn.

That Friday, we walked to Amy's house after school with Angelica, who had some hash that she offered to share. I didn't even know what hash was. I worried it might be an opiate, and thus potentially physically addictive. The others laughed at my ignorance and explained that it was just a type of concentrated weed. I was relieved; after all, I had already decided to try pot.

I watched excitedly as Amy took a chunk of sweet-smelling brown gunk and placed it into a small wooden pipe. Because it was so sticky, it took a little while for her to get it to light, while Angelica drew on the pipe. But soon, the bowl was glowing, and I could smell a rich, musky smoke. They each took a hit, then passed it to me. I wasn't even sure how to inhale it but managed to tentatively pull some smoke into my lungs. (Unlike the vast majority of people who eventually become drug addicts, I had never been a cigarette smoker. I'd tried once in eighth grade and found it awful.)

Soon, I felt something change, quite subtly, like a shift in the wind. Unlike many marijuana users, who find their initial experience disorienting rather than enjoyable, I didn't have to learn to like my altered sensations. I knew what to expect. Indeed, I felt I'd found exactly what I'd been seeking ever since I started reading books like Tom Wolfe's *The Electric Kool-Aid Acid Test* in ninth grade.

When I started high school, I'd begun a typical adolescent spiritual journey, spurred on by my still pressing fear of death. While studying for my bat mitzvah and struggling with questions about faith raised by the Holocaust, I also learned about Buddhism in a freshman history class focused on Asia and Africa. I was fascinated. The notion of reincarnation and the idea that we are essentially all parts of a God that has split into a million selves to tell itself a story appealed to me tremendously. I learned that Maia, or, as the Indians spell it, Maya, signified the real world, which is considered the world of illusion, the tale that God tells. Never mind that my mom had picked my name out of a sequel to *Mary Poppins* where Maia, one of the seven sister stars or Pleiades, comes down and talks to the Banks children. I thought that it was no coincidence that I had a name of such Buddhist significance.

I read everything I could find about Buddhism, and then, naturally, about the 1960s, a time that then seemed mythical to me, when a new youth culture blended Eastern religion, psychedelics, politics, and vibrant music into a movement that really had changed the world. I tore through Herman Hesse's *Steppenwolf* and *Siddhartha,* much of Timothy Leary, Ram Dass's *Be Here Now,* Aldous Huxley's *The Doors of Perception,* and a lot of Alan Watts. These books and ideas changed me. They offered a potential antidote to my fears that my character was fixed and irremediably selfish, as well as a more comforting way of dealing with my fear of death.

It made much more sense to me that souls would be recycled, that we would be part of God, than for us to be some sort of mortal creations of an external God. To try to attain enlightenment, I began to practice yoga and meditation on my own. And, to seek what seemed a faster route to spiritual experience, I soon wanted to try psychedelic drugs. The only problem was the '60s were over, and tuning in, turning on, and dropping out was out of fashion. It was the early 1980s and the former flower children were gearing up for success on Wall Street. I didn't know how to get drugs until I met Amy.

When we smoked that first lump of hash, I felt that our friendship was fated. My vision blurred slightly. Colors and sounds were suddenly more vivid—not disturbingly intense, just better and more compelling. Amy's Indian-print fabrics, rock posters, and tie-dyed wall hangings became spellbinding. I felt on the verge of laughter. The world was not so deadly serious anymore. Everything felt lighter. My worries seemed to float away, and because of this, I felt much more accepted.

Instead of being an unapproachable goddess, Amy became someone like me, someone with whom I was sharing a mystical experience. The music on the stereo—probably Cream's *Disraeli Gears*—was fuller, more complex and encompassing. I felt I had arrived, like I belonged for the first time in my life. "I've been waiting so long / To be where I'm going / In the sunshine of your looooooo-uu-ove."

Back home, at my own high school, I had always felt pressure to conform and a sense that, inside, I was different. True, I fit in better there than I had in junior high, but I always feared that my friends didn't really like me and were just tolerating me. With Amy, however, I felt understood. When we met new people, she'd often tell them that we were sisters—and this seemed to ratify the depth and permanence of our bond. She saw me as blood. I thought the drugs were the factor that made the crucial difference—the missing element that allowed me to open up to real friendship without constant anxiety about its authenticity. It took an artificial sensation for me to feel my relationships were real.

Getting high broke down the wall of fear that kept me from reaching others or accepting myself; it seemed to solve my lifelong problem of being both wildly curious about new experience and terrified of it. However, my immediate love for drugs also made me intimately aware of their power. I dove into the literature on my new obsession and decided that there were two classes of drugs: the psychedelics (including pot), which enhanced growth and awareness, and the powders (coke and heroin), along with alcohol, which shut that down, and were about mere pleasure and escape.

I thought I'd be okay if I stuck to the "growth" drugs: using for enlightenment was permissible, but using for escape or fun might lead to addiction. I was already scared enough simply dealing with all the changes in my mind and body as I moved through adolescence.

THE TEENAGE BRAIN is not just an immature adult brain. It is undergoing a transition comparable in magnitude only to its explosive development in the first few years of life. This stage is critical for understanding addiction: most learning disorders appear during a specific time during development, and addiction is no exception. Just as autism never appears suddenly in adults and schizophrenia is rare before puberty, addiction is overwhelmingly a disorder of late adolescence and early adulthood.

As noted earlier, 90% of all addictions begin during adolescence. While the odds that people will become alcoholic are nearly 50/50 if they start drinking regularly at 14, they are only 9% if they start after age 21. Part of the story here is that people who start at the youngest ages tend to have risk genes and/or trauma histories that push early use. But both the weaknesses and the strengths of the teenage brain and the way puberty affects it are also critical factors in how most addiction occurs

A child's elementary school years are a relatively quiescent time in terms of brain development. However, before that, between birth and the end of kindergarten, the brain has already grown to a stunning 95% of its adult size. If you consider the difference in body size and shape between a typical adult and a six-year-old, the fact that the first grader has an almost adult-sized brain in a body that is less than half of its adult size is remarkable.

Moreover, moving from childhood to adolescence isn't simply a linear catch-up process where the brain changes slightly while the body matures. Instead, from puberty until around age 25, the brain undergoes a remodeling that is almost as extensive as the period of rapid change in the first 5 years. Importantly, both adolescence and infancy are marked not only by growth but also by "pruning"—a selective reduction in both the number of connections between cells and in the number of cells themselves. This influences how easily the brain can learn—whether it is studying calculus or repeatedly taking cocaine.

The pruning process can be extreme. During early childhood, the brain grows billions of connections and then prunes away nearly half of them, with the timing and proportion of cells pruned dependent on the region. In adolescence, gray matter, which is made up of the cell bodies of neurons themselves, actually shrinks significantly, most notably in the prefrontal cortex, which does not complete its development until around 25. (Incidentally, this healthy shrinkage needs to be kept in mind when evaluating studies of drugs and the teen brain: smaller or shrinking structures aren't always a sign of pathology, but are sometimes a sign of efficiency.) Meanwhile, new "insulation," or myelin, also grows on the nerves that connect increasing numbers of circuits, maximizing their speed and refining other signaling qualities. This produces growth in what's known as white matter.

During development, the pruning of brain overgrowth is just as important as the addition of new cells and faster connections. While we tend to think that more brain cells is always better, failure to remove some excess cells can actually lead to disabilities. One of the most consistent findings in autism, for example, is early brain overgrowth, with some regions being "hyperconnected" or simply having too many cells and too many links between them.

One recent study of brain tissue from the temporal lobes of autistic and typical teens found that while typical adolescents had 41% fewer synapses than typical toddlers, in the autistic teens, the percent reduction was only 16%, compared to younger children with autism. Though this excess might have some potentially positive effects—like improved memory or enhanced perception—it may also lead to sensory overload or to "overlearning," where a person gets rapidly locked into a behavior that becomes repetitive and difficult to change. And both overload and overlearning might predispose people to compulsive behaviors, including addictions: overload by creating a desire to escape, and overlearning by rapidly creating habits.

Still, in general, it's fair to say that brains work best when their connections are optimized. Too many cells or connections can cause nearly as many problems as too few, and miswired links to the wrong places can curtail communication and signaling. Adolescence is when a great deal of circuit optimization normally takes place. Because the brain is in flux, it is particularly vulnerable during this period: what is learned during the teen years will shape both the brain itself and the psychological coping skills people rely on for the rest of their lives.

Critically, the optimization process isn't evenly spread all over the brain. It moves slowly, from back to front, starting with the most vital and primitive emotional and regulatory areas and ending in the most complex and uniquely human regions. Of course, most of the brain's basic circuitry, for functions like seeing, hearing, and moving, has already been fine-tuned by puberty. What remains to be calibrated are the motivational systems, which sit mainly in the center of the brain and must be prepared to face the challenge of finding a mate and reproduction.

Successful adolescent development requires a drive for novelty and social contact with same-age peers to pull teens away from family and toward friends and potential partners. During the teens, the motivational systems must also learn to handle the hormones that make these years famously difficult. Unfortunately, the very last brain areas to develop are those that modulate our feelings and desires, the circuits that allow us to think critically, plan wisely, and master our urges and ourselves. As neuroscientist Robert Sapolsky put it, in the teen brain, the systems that drive a desire to take on the world are "already going at full speed, while the frontal cortex is still trying to make sense of the assembly instructions."

At 17, then, when I met Amy and started taking drugs, I had a brain primed to seek novelty and adventure, a brain evolved to push me away from the familiar, safe world of family and pull me into one dominated by the peers who would historically have mattered most in determining the course of my adult life. The systems that underlie motivation—dependent primarily on the neurotransmitter dopamine in a particular circuit in the midbrain—play a nasty trick on prepubescent kids. A change in dopamine levels draws them away from home and routine and makes the new, exciting, and dangerous seem incredibly alluring.

As the dopamine circuits rewire themselves, the previously reliable pleasures of childhood play start to grow old. What you love becomes boring and loses its flavor. Sugar itself tastes less sweet and feels less fun. (This may be why candy somehow never tastes as good as it did when you were a kid.) While these reward regions are maturing, the process makes pleasure both more attractive and harder to get. This is because during adolescence, new experiences may spur a much bigger burst of dopamine than they do in childhood.

Even in the normal course of events, thanks to these changes in neurotransmitters, most adolescents get at least a taste of anhedonia—the dull, utterly joyless dread that seems to be created by an attenuation of dopamine signaling. Prolonged periods of anhedonia are one of the most unbearable symptoms of depression and of withdrawal in addiction. The propensity toward this deep pleasurelessness in adolescence is why teenagers, despite their frequent mood swings and operatic dramas, so often also complain of boredom and ennui. Few adults ever experience quite the intensity of boredom that a teenager does. Nor does the lure and feel of a thrill ever seem so enticing.

Changes in dopamine signaling during adolescence attract teens toward new, dangerous, exciting challenges to replace the lost thrills of childhood— and to slay boredom. These changes heighten interest in social life and sexuality by altering the way the brain responds to signals about risk, reward, and punishment. Imaging studies show, for example, that the brains of adults and teens react pretty much identically if you give them the opportunity to seek moderately rewarding experiences. Like adults, teens are attracted, but not potentially fatally so.

But offer the prospect of a *huge* reward and the teens' response is greatly magnified, compared to the adults'. The prize looms large, far out of proportion to thoughts of the future and potential risks. If adolescents think that speeding, smoking, or cutting class will win social approval—the biggest prize of all at this age—this can blot out most considerations of crashes, cancer, or college. Furthermore, if you merely offer a paltry pleasure, it may not be seen as rewarding at all to a teenager. In fact, the teen brain responds to a small reward as though it's an insult or a punishment, which may explain some of the ingratitude and sarcasm teens display when adults praise them, feed them, and ferry them around to their activities. A motivational system that is prone to boredom and places such high value on big rewards is bound to prompt at least some hair-raising behavior. And for many adolescents, drugs seem not only like an easy way to get pleasure but, more important, a path to popularity.

Interestingly, teens aren't cognitively clueless about risk: if you ask them to estimate their odds of getting HIV from sex or becoming addicted to cocaine, they will actually *overestimate* them substantially. This doesn't stop them from being reckless, however, in part because their concern is overcome by their focus on big rewards. Unlike adults, they typically haven't yet experienced the emotional fallout from a disastrous decision, which creates a healthy fear of repeating such an error. Teens also often see themselves as invulnerable—and both this and the risk taking can probably be traced to the various effects of different dopamine levels.

Dopamine levels, of course, are also affected by drugs. Any pleasurable drug or experience will directly or indirectly influence the dopamine-driven circuitry of motivation—otherwise the experience wouldn't be desirable or exciting (more about this in the next chapter). And since the brain's motivational systems are being optimized during adolescence, it's not surprising that exposure to potentially addictive drugs might be the most risky at this time. Both chemically and psychologically, the impact of altering this system while it is trying to set itself up for adulthood is enhanced. Chemically, drugs alter dopamine levels, which may change the way the developing system calibrates itself, making it more likely to become reliant on this artificial stimulation. Psychologically, if you always use drugs to cope at a time when you are supposed to be learning healthy ways to manage social fears, you don't develop better alternatives. This, too, makes you more likely to get hooked.

It's important to note that, although it often seems to parents that the bizarre qualities of teenage behavior are unfortunate "side effects" of their brains' unripe nature, their persistence across history and in multiple cultures suggests that they are an adaptive feature, not a bug. Teens need to become more independent and to learn to manage the risks that are present in their environment; they can't do this without getting out there and learning by experience. Keeping them completely sheltered from risk and choice doesn't help: brain development is "experience dependent," meaning that it requires particular input at the right time in order to proceed appropriately. Locking kids up "till their brains are grown" will simply either delay or distort maturation. Of course, this doesn't mean that letting adolescents go wild is a good idea either—merely that some degree of adolescent risk taking seems to be a necessary part of normal human development. No matter what adults do, teenagers will try to evade them, just like their parents did when they were young.

Indeed, the very same plasticity that makes the teen brain vulnerable to bad influences also makes it flexible and able to learn rapidly in a way that it will never be again. While curiosity and boldness can be dangerous, they also lead to some of humanity's greatest achievements, from the exploration of new lands to the creation of new ideas. It's not coincidental that many of the most revolutionary discoveries and scientific advances in human history have been made by people under 30.

In teens and young adults, unmyelinated nerve cells that haven't yet been insulated don't transmit information as quickly or efficiently as myelinated ones do. However, once this insulation is laid down, neurons also can't change as much or learn as easily. A brain can be either highly responsive or packed with knowledge: expertise comes at the cost of flexibility, to some extent. Openness to learning is also openness to risk, for better or worse. And this means that a brain primed for learning is also a brain primed for addiction.

THE CHANGES I underwent as a teenager weren't just chemical, of course. As in childhood, they were continually shaped by culture, by context, and by my own reaction to it. I attended high school between 1979 and 1983—exactly the time when adolescent alcohol and other drug use in America hit an all-time high. In 1981, for example, a whopping two thirds of twelfth graders reported having used an illegal drug, the highest rate ever recorded; today it is 50%. When I was a senior myself in 1982, 14% of teens reported having tried a hallucinogen like LSD or mescaline at least once; now, the figure stands at 8% (although, like the other drug trends, it has bounced around with no relationship to ever-rising antidrug funding and law enforcement).

Basically, I grew up in a culture where all types of drug use were quite common, if not necessarily accepted. Although the era of the three-martini lunch was winding down, the drinking age was still 18 in New York, and one third of the adult population smoked cigarettes. My upstate high school actually had a smoking lounge for *students* and some kids sneakily smoked pot there as well (I was not one of them; I still cared about academic achievement and feared getting in trouble). Student drug testing and "zero tolerance" had not yet begun to damage the futures of teens like me.

At the same time, the backlash against the '60s and the turn toward political conservatism were already well under way. With Ronald Reagan's election in 1980, American culture seemed to be violently repudiating hippie ideals like equality for women and minorities—and especially, drug-fueled spiritual quests and the search for a less commercially focused community. Ambition, individualism, consumerism, and the veneration of the market now firmly replaced utopian dreams of communal, equitable living and "back to the land." Selling out was back in.

My own journey mirrored these tensions and extreme swings. When I'd started ninth grade, I was still the uptight, career-obsessed outsider I'd been in junior high. Once a week, I helped produce, write, and coanchor my school's local public access cable TV show; I participated in at least five extracurricular activities and even convinced Walter Cronkite himself to let me interview him.

But all my early success didn't make me as popular as I'd hoped it would. I didn't get that bragging about my achievements was off-putting; I thought it would make people like me. I still thought that I had to be somehow superior in order to just be seen as okay; I didn't understand that this made me seem like a snob. And so, I began to think that success might not offer the social benefits I wanted. Making one of my many switches from one extreme to another, I zigzagged from compulsive careerism to psychedelic mysticism. Via Amy, I was drawn toward the only remaining vital psychedelic community of the 1980s: the one surrounding the Grateful Dead.

AMY AND I couldn't wait to go to our first Dead show, which we attended in 1982 at Madison Square Garden. The Dead's music became the sound track to our acid trips, just as it had been for Ken Kesey's merry pranksters and millions of others since. It changed my life, in both positive and negative ways.

If writing about music is like dancing about architecture, writing about psychedelics and the music that enhances them is yet more absurd. Still, what architecture I soon found the Dead could conjure up on some sublime nights. Spirals and whorls; geodesic domes; the most intricate tile work in the algorithms of the holiest Moroccan mosques; fractals, recursive trails in the finest colors, which echoed out into infinite, starry space. The sound filled it all in; made cheap, glistening plastic beads into the finest diamonds; gave life to dazzling moments the way children animate toys. In those instants, it was timeless and I felt like I was at the white-hot center of every psychedelic experience, beyond words but craving expression.

The rhythms and riffs wove patterns into every crevice of the night. There was a glimpse of utopia, a place where everyone did belong and no one was lost or lonely. High, you could talk about these earnest yearnings. If rejected or mis-understood, you could disown them as mere drug talk; if accepted, you could gather them in like nestlings, knowing you had found your people.

Tripping taught me perspective taking in an extremely visceral way. When you see how profoundly a drug can change your own perceptions, it becomes almost impossible not to understand that other people's perspectives may vary just as widely. This may be one way these substances increase empathy. Through psychedelics, I learned a whole new way to see the world.

Because hallucinogens so obviously altered my perspective, I became aware that it was only one limited view, one shaped by chemicals and people around me. It could be changed! Who are we if a few molecules can alter our senses and beings so enormously? What does this mean for mind and matter? Considering these ideas inspired me, rather than doing harm.

Indeed, having these experiences helped me move away from my "fixed" worldview, which framed my social problems as ineradicable and my character as inalterably bad. I began to feel like I was part of something larger and, there-fore, not so left out or self-centered. The sensation of tripping also mirrored aspects of the way I felt as a child. Things were wonderful, awesome, and vivid, but also threatening, looming, and loud. I wanted order and predictability, and needed to develop some way of sorting the data, some way to make sense of the world that could protect me from its sharpness and messiness. At first, I found this in my friendship with Amy and our exploration of psychedelic philosophy and spirituality and the community around the Grateful Dead. But I couldn't hold on to it.

When many teenagers start using drugs, they do so to proclaim their inde-pendence and demonstrate rebellion. Often, the drugs are taken at least as much for their symbolic value as evidence of separation from their parents as they are to achieve a particular high. That was not the case for me: oddly, one of the in-sights I had on acid was that it wouldn't actually hurt me to call home and let my mom and dad know where I was when they asked me to or to do chores like washing the dishes without whining about it. I could give this to them with-out losing myself, I discovered.

Unlike teens who seek to rebel, in fact, I became less resistant and more helpful when I started getting high. Marijuana and psychedelics helped me un-derstand the reciprocity of relationships in a way that I hadn't previously comprehended. These drugs made me see why being kind really mattered—and that who you are is not simply a result of inborn predispositions unfolding but a function of what you do and choose. Although there almost certainly are other, safer ways to learn this, I did so through drugs. Cultures that have historically used these drugs in sacred ceremonial contexts like rites of passage may have had real wisdom that we now have lost.

Unfortunately, I also learned some incorrect lessons from my psychedelic experiences. I learned how powerfully positive drugs could be—but not how harmful. I learned how exaggerated antidrug propaganda was—but not where the true risks lie. I also learned that drugs were an excellent way to solve my social problems. All of this soon made me vulnerable to the substances I had originally—and correctly—pegged as dangerously addictive.

NINE

On Dope and Dopamine

Cocaine turns you into a new man—and the first thing the new

man wants is more cocaine.

—ANONYMOUS

I WAS SITTING ON THE EDGE of Jerry Garcia's bed in a nondescript and surprisingly unluxurious hotel room in New Haven when he offered me a line of cocaine. I was 17; it was 1982. He was sitting in an armchair, only inches away. His large bearded face wore his iconic enigmatic smile. I'd seen it earlier from a far greater distance, as he played onstage at the Coliseum, wringing euphoria from his guitar strings. He wore his trademark dark-colored pocket-front T-shirt.

Until that day, I'd resisted coke, citing the hippie code of avoiding "white powders" and sticking to pot and psychedelics because these drugs didn't let you flee yourself. As I'd discovered to my regret, if you take mushrooms or acid when you feel low, they do not enhance your mood, but instead exacerbate your anxiety and distress. If only I had stuck to the "no needles or white powders" rule, I might have lived an entirely different life. But even the Dead themselves hadn't been able to successfully maintain the antipowder stance.

"Cocaine has some very weird karma behind it," Garcia told me in his famously nasal speaking voice, which was so incongruous in his bearish body. I nodded, but snorted the line anyway. Some of that karma would soon become mine.

I had only taken coke one time before I did so with Jerry—and that was just a few hours earlier in the hotel room of one of the band's technical staff. As with both psychedelics and marijuana, I had a feeling of heightened sensation, but with coke, things seemed more real, more crystallized and sharp, not less. It might seem odd that any drug would be able to affect your sense of reality itself, but if you think about it, the brain must have a way to chemically distinguish between dream and reality and between fantasy and fact. Cocaine is only one of many drugs that can interfere with your sense of realness and tune it up or down.

On coke, I felt a smug sense of superiority, of being in on something, of being powerful and utterly desirable. I felt intellectual clarity, sophistication and strength, not the giggly childishness of pot. My heart rate elevated, and I felt a quickening, a sense that everything was moving faster and becoming more exciting. If acid made you feel like you'd discovered the secrets of philosophy, cocaine made it seem like you were able to conquer the world.

My boyfriend, Ethan, and his friends were among the Dead's large coterie of dealers. It was through him that I'd gotten backstage passes to that night's show and then afterward found myself spending a few hours with my idol. I'd met Ethan a few months earlier on the commuter bus I took to the city to see Amy and work on my gifted project; I'd sat down next to him because I thought he was cute and we'd bonded in a discussion about the Dead.

Not surprisingly, the fact that I was doing lines with Jerry himself made me even more ecstatic. Even without the cocaine, just spending five minutes with him would have been a peak experience for any Dead fan. But getting to hang out for a few hours, discussing the etymology and Buddhist echoes of my first name with the man whose music I worshipped, was blissful. As the Dead song goes, I felt "dizzy with possibility." Many people say that no experience of a drug ever matches the first one. But my cocaine initiation was especially peerless, even though nothing more than a few hours of stimulant-fueled conversation ensued that night.

WHAT WAS HAPPENING to me chemically as I started taking cocaine? Given that I was in the minority of people who do get hooked, what changed in me that isn't altered in those who do not? To understand why learning matters in addiction, it's crucial to look at how a vulnerable brain reacts when exposed to a drug like cocaine in a pattern of use that is potentially addictive. Here, there's no avoiding dopamine, the chemical that has become synonymous with pleasure and is often described as the root of all addiction.

Dopamine is a much misunderstood neurotransmitter. Found in less than 1 million neurons—a tiny proportion of the 86 billion such cells in the brain—it has an outsized impact. In pop science, dopamine is best known as the "pleasure" or "reward" chemical that puts the hook into addictive drugs and fuels craving (it doesn't hurt that it has the word "dope" built right in). But seeing dopamine only as the brain's way of producing pleasure oversimplifies and misrepresents its role. If the brain worked as simply as using one neurotransmitter for one distinct psychological function, it wouldn't be complicated enough to allow us to be capable of forming such questions.

The real connections between dopamine and pleasure, desire, and memory show how learning actually shapes addiction and why addiction is, in fact, a learning disorder. To understand this process, we need to explore how pleasure came to be understood in the brain and what dopamine has to do with it. The story begins in the early 1950s, in Canada. James Olds was a postdoctoral student working with neuroscientist Peter Milner at McGill University in Montreal. In 1954, the two discovered what would soon become known as the brain's "pleasure center."

They found that if you implant an electrode in a brain area now known as the nucleus accumbens and send a current through it, rats seem to enjoy the stimulation so much that they will hit a lever to repeat the experience up to 2,000 times an hour. Research conducted around the same time on human epilepsy patients also suggested that stimulation of this region during surgery produces a feeling of exhilaration. Psychiatrist Eric Hollander has observed patients receiving such stimulation in today's more advanced operations for severe OCD. He says, "They all of a sudden feel great. The anxiety goes away. They want to do things. It's just like a stimulant." Indeed, patients implanted with electrodes incorrectly placed in these areas will sometimes behave exactly like people with addictions, hitting the stimulation button thousands of times in an hour and neglecting their hygiene and personal lives.

By the late '70s, neuroscientist Roy Wise of Concordia University in Montreal and his colleagues had shown that dopamine is the neurotransmitter responsible for the sensation produced by this stimulation. Wise, who is now at the National Institute on Drug Abuse, first proposed what has become the dominant theory about dopamine and addiction. He claimed that dopamine in the circuitry linked to the nucleus accumbens is the currency of pleasure, the chemical signature of bliss. This region was almost certainly activated in my brain the day I first took cocaine with Jerry.

Wise's hypothesis was based on rat research. His group had found that while rodents will ordinarily work to earn injections of cocaine or amphetamine, the animals stop doing so if given high doses of antipsychotic drugs, which block dopamine receptors. Human experience with antipsychotics also suggests that they can dampen pleasure; common side effects include apathy and anhedonia.

But other research soon complicated the straightforward interpretation of dopamine as the neurotransmitter that represents pleasure. For one, some dopamine neurons in the so-called pleasure center have been found to fire more in response to punishment and distress than to reward. That doesn't make sense if dopamine in this area only signifies pleasure. Second, many of the brain's dopamine neurons are involved in the control of movement, not emotion, and are located in circuitry that is distinct from the areas that register pleasure. That's why Parkinson's disease, which destroys dopamine cells starting in a region primarily involved in motor control, is known mainly as a movement disorder and why antipsychotics that block dopamine can cause Parkinson's-like movement disorders.

Surprisingly, despite these anatomical distinctions, understanding the role of dopamine in Parkinson's may also shed light on its role in addiction. To start, the symptoms of Parkinson's aren't limited to shaking, rigidity, and difficulty with walking. There are also psychological and emotional effects: depression and amotivation are sadly common experiences throughout the progression of the illness. The entire dopamine system can be affected.

Moreover, it's not always so easy to distinguish precisely between problems with movement and problems with motivation. The linguistic link between the two may be more than just metaphorical, and the anatomical distinctions between regions involved in movement and in being moved emotionally may be less important than they once seemed. The connections between dopamine in addiction and dopamine in Parkinson's may have profound implications for the relationship between mind and brain. They suggest that dopamine may be required for the fundamental expression of drives and desires—in other words, free will.

Neurologist Oliver Sacks was among the first to explore the philosophical and psychological implications of this aspect of dopamine, connecting it to will and willingness in his 1973 account of patients with a severe form of Parkinson's caused by a 1920s encephalitis outbreak. In his book *Awakenings,* he wrote:

> The first qualities of Parkinsonism which were ever described were those of festination (hurry) and pulsion (push). Festination consists of an acceleration (and with this, an abbreviation) of steps, movements, words and even thoughts—it conveys a sense of impatience, impetuosity, and alacrity as if the patient were very pressed for time; and in some patients it goes along with a feeling of urgency and impatience, although others, at it were, find themselves hurried against their will.

Here, dopamine dysfunction is creating not just excess movement—but also excess and aberrant motivation. It is increasing drive and want. This could happen as dopamine cells become dysfunctional and perhaps overcompensate at times, as they get damaged by the disease.

But Parkinson's patients can also have the opposite problem, which is difficulty initiating or sustaining motion—or even desire to move. Before the development of L-DOPA, a precursor to dopamine that can help patients with Parkinson's recover some function, many of the people Sacks treated had found themselves literally frozen in particular positions, unable to move without assistance for years. This paralysis is consistent with a very low level of available dopamine. By reducing the availability of dopamine in the brain, Parkinson's can, frighteningly, affect not only what its victims are able to physically do but also what they are psychologically able to want or will.

Sacks describes one type of Parkinsonian "akinesia"—or lack of activity—this way:

> [S]ome of them would sit for hours, not only motionless, but apparently without any impulse to move: they were seemingly content to do nothing and they lacked the "will" to enter upon or continue

any course of activity, although they might move quite well if the stimulus or the command or the request to move came from another person—from the outside.

This sudden return to movement could be dramatic: a historical case that Sacks describes involved a man who had seemed completely paralyzed but who once "leapt from his wheelchair" to save someone he saw drowning in the ocean. After completing the rescue, he was returned to his rigid, statue-like state. Similarly, one of Sacks's own patients could juggle, but only if thrown objects and kept completely uninterrupted—if she dropped a ball or was otherwise distracted, she would once again become motionless. She wasn't able to instigate movement and could often only do so if faced with an external pressure—like having a ball thrown to her.

This suggests that dopamine is involved not only in movement but also in the will or even the desire to move. Parkinson's patients who are "frozen" may be physically capable of moving, but they don't seem to have the chemical capacity to *want* to do so. Without dopamine, they need external direction to motivate them. Although it's possible that dopamine in motor regions signals will or intention, and in other circuits it signifies pleasure, there is actually a simpler explanation, which also illuminates addiction.

To see why, we first need to parse pleasures more finely. Research now suggests that there are at least two distinct varieties of pleasure, which are chemically and psychologically quite different in terms of their effects on motivation. These types were originally characterized by psychiatrist Donald Klein as the "pleasures of the hunt" and the "pleasures of the feast." To understand what goes wrong in addiction—and the role dopamine plays in addictive learning—these experiences of pleasure must be understood as distinct.

As the phrase suggests, the pleasures of the hunt are the thrill of the chase: excitement, desire, stimulation, intent, a sense of power and confidence in being able to seek and get what you want. In contrast, the pleasures of the feast are those of satisfaction, comfort, relaxation, attainment, and sedation. Put another way: the pleasures of the hunt are those of lust and sexual longing and the pleasures of the feast are orgasm and afterglow. Or as a drug user, you might distinguish between stimulants like cocaine and methamphetamine, which imitate the hunt—and depressants like heroin, which mimic the feast. In fact, Klein originally made the distinction based on his observations of differences in the experiences of people with cocaine addiction and those who preferred heroin.

The University of Michigan's Kent Berridge and Terry Robinson discovered that dopamine is primarily involved in just one of these pleasures. In Roy Wise's original theory, dopamine accounts for all types of pleasure, and the reason that drugs are addictive is that they elevate dopamine so far out of its natural range that this artificially inflated intensity drives irresistible craving. But Berridge and Robinson's work suggests that only one type of pleasure gets elevated—and that this creates a different type of problem. To me, their perspective characterizes both the lived experience of addiction and the existing data far more accurately.

In Berridge and Robinson's view, pleasure is divided into "wanting" (hunt) and "liking" (feast). As we'll see, the distinction is especially important in addiction because each type has a different influence on learning. Like many discoveries in science, this one was made when the researchers were trying to understand why an experiment didn't work out as they'd predicted. They'd used a chemical that selectively destroys dopamine cells in the nucleus accumbens of rats to eliminate their "pleasure centers." Not surprisingly, after these key dopamine cells were eliminated, the rodents became so amotivated that if the researchers hadn't manually fed them, they would have starved to death.

"They wouldn't want to eat. They wouldn't want to drink," Berridge says. "We'd have to artificially nurse them and artificially feed them, the way you would in a hospital intensive care ward." The rats behaved as though they had extremely severe Parkinson's, which they essentially did. Destroying their dopamine cells had taken away their motivation, leaving them with no desire or will to do anything at all, even what was necessary for survival.

At the time that they did this experiment, Berridge and his colleagues thought that dopamine was necessary for any type of pleasure. They agreed with Wise, whose view had become the conventional wisdom. So, they expected that the rats wouldn't be able to enjoy their meals: if dopamine was required for pleasure, the animals shouldn't be capable of liking even the most sumptuous food, since they had virtually no dopamine. Without dopamine, they shouldn't like or be pleased by anything, in fact.

That was not the case, however. Thankfully for people who have Parkinson's or must contend with side effects of dopamine-blocking drugs, even having extremely low levels of dopamine in the nucleus accumbens doesn't eliminate all types of positive emotion. The rats may not have wanted their food, but when they got it, they still liked it.

Some readers may now be wondering how it's even possible to tell if a rat "likes" something or takes pleasure in it. Berridge had a clever way of finding out—at least, for pleasure related to food. He knew that normal rats lick their lips when they are given sweet foods, which they prefer, but not bitter ones, which they avoid. He had predicted that the dopamine-deprived animals would be as apathetic and "meh" about eating sugar as they had been about seeking meals. But, no, when sweets were fed to them, the rats licked their chops.

At first, of course, because their work contradicted the dogma that dopamine is necessary for pleasure, the researchers thought they had made some type of mistake. "I was flabbergasted and thought, 'What could this mean? How could it be that there's all this evidence out there that dopamine is pleasure, but when we look at the direct hedonic impact of a sensory pleasure, like sweet, it looks totally normal?' " Berridge says.

However, repeating the experiments and expanding on them led the researchers to a different conclusion. While dopamine *is* involved in motivation or the pleasures of the hunt, that's not the only way we can feel good. Dopamine *is not* necessary, it seems, for enjoying sweetness, comfort, satiation, and calmness—research suggests that these pleasures are more strongly linked to the brain's natural opioids, or heroin-like chemicals, instead of to dopamine. And this has implications for the broader understanding of addiction.

To understand why, we have to look more closely at the implications of the "dopamine is pleasure" perspective and what they say about learning in addiction. Roy Wise's "dopamine" or "anhedonia hypothesis of addiction" suggests that by artificially elevating dopamine levels far beyond their normal physiological range, drugs like cocaine cause such intense pleasure that they compel people to seek to repeat the experience, far more than natural rewards like sex or sugar could ever do. This creates craving for more. Unfortunately, however, the brain seeks to stay in a specific range of chemical balance, which is critical for survival. As a result, it has processes that kick in to normalize neurotransmitter levels when chemicals like dopamine get too elevated. This means that with drugs, the amount needed to attain euphoria will increase as the brain tries to restore normalcy.

Over time, these "opponent processes" cause tolerance, meaning that more of the drug is needed to get to the ecstatic peaks and that ordinary pleasures increasingly become paler as well, due to the increased threshold for dopamine action. Once an addict has elevated tolerance, in order to simply experience a normal range of feelings—one that includes good moods, joy, contentment—the drug must be taken, because otherwise the opponent processes keep dopamine levels too low. Addictive learning here is driven simply by the desire to feel okay. This has been the basic neuroscientific paradigm of addiction since the 1980s.

This position makes sense if dopamine is seen as simply "the pleasure chemical" and if addicted people are motivated primarily by a need to avoid the dread and ennui of low-dopamine states that come with withdrawal. The definition also fits with what neuroscientist George Koob, who heads the National Institute on Alcohol Abuse and Alcoholism, has labeled "pharmacological Calvinism." This is the sad and frequently common experience that when you get high now, it seems as though you are "borrowing" pleasure from your future, from an account with a finite and unreplenishable supply. Pharmacological Calvinism means that there's no free lunch: using up extra dopamine today by taking drugs will require payback in low mood tomorrow. The bliss of drinking will be followed by the hammering of the hangover. In this view, unearned pleasure is actually impossible in the long run. Any high will be balanced by some type of low.

Berridge and Robinson's research, however, suggests that the reality is somewhat more complex. If dopamine is what creates the sense of pleasure, animals shouldn't be able to enjoy food without it. Yet they do. The theory these scientists developed to resolve this problem—known as the "incentive salience" model—also explains an infuriating aspect of stimulant drug problems that simply isn't captured by other theories. This experience is so characteristic of stimulant addiction that it appears in some form in virtually every detailed account of it I've ever heard or read. Here are a few cocaine addicts' descriptions of what it feels like:

> "I can remember many, many times driving down to the projects telling myself 'You don't want to do this! You don't want to do this!' But I'd do it anyway."

> "[M]y body's saying no and my mind's saying no, but . . . we started all over again. I didn't need it, I didn't want it . . . it's like some kind of molecular thing in my cells would go for it, you know. I felt like a fucking robot."

> "I used to smoke some [cocaine] that wasn't good, feel sick and want some more. That's totally fucking crazy. The point that is best learned from the whole experience is the craziness, the completely illogical short-circuiting of the normal human mental process that takes place in obsessive addiction."

Here's the problem as it appeared in my story. At the end of my addiction, when I was shooting cocaine dozens of times a day, I wasn't able to enjoy it—even if I took incredibly high doses that should have been far more than enough to raise dopamine, despite tolerance. I lived with my cocaine connection and was generally able to get as much as I wanted. If cocaine elevates dopamine and I felt lousy because my brain lacked enough cocaine to elevate my dopamine sufficiently, more coke should have done the trick.

Instead, what would happen was this. I'd tell myself that I didn't want to shoot coke because I knew it would make me anxious and paranoid. I knew from the center of my cognitive brain that this was true: in the summer of 1988, I repeated the experiment hundreds of times and at least 99% of that time, every shot of cocaine produced fear, distress, and a severe feeling of discomfort, sometimes even an overwhelming fear of death. And yet I continued to inject coke, dozens of times a day. I did so each time with an overwhelming emotional sureness that I wanted nothing more than a shot, accompanied by an equally firm intellectual knowledge that if I had one, it would suck. I had come to truly, madly, deeply want a drug that I equally truly, madly, deeply did not like and, in fact, detested. But I couldn't learn to stop myself before I took it.

Berridge and Robinson's theory of "incentive salience" far better explains what happened to me than does the idea that I was simply suffering from a lack of dopamine and attempting to right the imbalance that tolerance and increasing doses had created. Their argument is that dopamine produces desire, not satisfaction—"wanting" but not "liking." In this view, elevating dopamine with drugs like cocaine leads to escalating desire—not escalating pleasure. Of course, this doesn't mean that dopamine plays no role in pleasure; the pleasures of the hunt can, after all, be quite enjoyable. Instead, it suggests that the wanting type of pleasure is not the whole story. As anyone who has ever suffered from sexual frustration knows, only desire that is likely to be sated is enjoyable. If satisfaction is not likely to be forthcoming, that very same yearning can become agonizing, not at all fun. You can wind up wanting more and more of something (or someone) that you like less and less.

These different types of pleasure are also important in understanding the role of learning in addiction, because "wanting" is critical to learning while "liking" is less so. If you are curious, feel capable, and have a goal in mind—i.e., you want something—you will be highly motivated to learn whatever will help you achieve it, and those lessons will really sink in. In contrast, being calm, content, and satisfied is less directly motivating. "Liking" can obviously make an experience memorable, but if you don't also *want* to repeat it, it won't change your behavior.

In addiction, this means that because being addicted escalates wanting more than liking, the drug experience gets deeply carved into your memory. Anything you can associate with achieving a drug high, you will. As a result, when you try to quit, everything from a spoon (you could use it to prepare drugs) to a street (this is where the dealer lives!) to stress (when I feel like this, I need drugs) can come to drive craving. Desire fuels learning, whether it is normal learning or the pathological "overlearning" that occurs in addiction. You learn what interests you with ease because desire motivates. In contrast, it's far more difficult to learn something you don't want to understand or care to comprehend.

Berridge and Robinson's research also helps resolve another paradox: If dopamine signifies pleasure, then the brain should become less and less responsive to it as tolerance to a drug develops. But while tolerance clearly does occur, the opposite result is also seen in the brain. As I took cocaine, paranoia began to set in at *lower and lower* doses—not higher ones. The summer of 1988, it also took increasingly less drug to achieve the state of heart-pounding anxiety and mortal dread that I experienced so frequently. Neuroscientist Marc Lewis described his experience of this effect in his addiction memoir this way: "I kept pumping [cocaine] into my vein, this non-sterile solution, until my reeling consciousness, nausea, racing heart, and bloated capillaries told me that death was near. Later that night, I begged myself to stop. . . . But the urge would not relent."

Such an effect is antithetical to tolerance; it's known as sensitization. And Berridge and Robinson were able to demonstrate that, at least in animals, addictive drugs like cocaine make the dopamine system more sensitive in some ways. Craving—but not pleasure—increases. If dopamine represented pleasure, then sensitization would make the drug actually feel better and better—and at lower and lower doses. This is not what people with addiction experience. If it worked this way, taking drugs would be a rational and readily affordable choice. Instead, the sensitization makes addicts feel worse at lower doses.

The role of sensitization and tolerance in addiction also points to another aspect of the importance of learning in addiction. Both processes are critical parts of normal learning and motivation: if you aren't sensitive to novelty and you don't get tolerant to the familiar, creating memories is very difficult, since it's hard to tell what you already know from what you don't know and what you need to know from what is irrelevant. (Note: tolerance is sometimes called habituation when nondrug experiences are involved; other researchers use the terms synonymously, which is what I will do here.)

Moreover, sensitization and tolerance are also fundamental properties of the nervous system on a cellular level. They are critical to implicit learning, which alters behavior by shaping our emotions in relation to our experiences. This affects how "good" or "bad" something seems and, ultimately, how much we want or like it. If you associate the smell of the ocean with wonderful childhood memories, for example, you'll probably enjoy the beach and frequently head for the shore—but if your seaside memories are of being bullied and excluded, the salty scent may not be at all pleasant to you and you may come to prefer mountain vacations. Sensitization and tolerance are part of the way the brain tags situations as safe and inviting or fearful and threatening.

Sensitization occurs when an extreme or painful signal is amplified. It can be seen even in primitive organisms like sea slugs, which respond to noxious experiences like pain or stress by becoming more reactive to cues that predict it. Eric Kandel, in fact, won the 2000 Nobel Prize in Physiology or Medicine for research on sea slugs that revealed the molecular underpinnings of memories, including that of sensitization.

This is what he found. Normally, if their tails are touched, sea slugs will retract their siphons, which are projections that they use to remove waste and excess water. But Kandel showed that this reflexive response becomes sensitized if the stimulation is painful, like a shock. In other words, a slug will pull back its siphon more quickly the next time it's touched after being shocked, even if the second touch is lighter and not uncomfortable. Kandel went on to elucidate the mechanisms in synapses and cells that cause this signal amplification and that represent the biological basis of this type of memory.

Similarly, the startle response in both humans and other animals can become sensitized by extreme stress; this is why even New Yorkers without PTSD were especially jumpy when they heard loud noises in the weeks that followed 9/11. PTSD itself, in fact, may be an extreme type of sensitization learning. Although human memory is obviously far more complex than that of sea slugs, it's striking that addiction can interfere with mechanisms involved in learning that occurs in an implicit, unconscious way, associating experiences with safety or danger, like sensitization.

In contrast, habituation or tolerance is the opposite type of learning. Rather than an exaggeration, habituation involves an attenuation of signaling in response to experiences that are predictable and safe. Kandel could demonstrate habituation in sea slugs, too: a small, nonpainful touch that would cause siphon retraction at first would cease to do so over time as the slug realized it wasn't dangerous. Again, this type of tolerance isn't just something that happens in addiction. It is the process through which novelty gets old and the unfamiliar becomes ordinary. It's responsible for what's known as the "hedonic treadmill," where what once was thrilling and new—whether a sexual partner, an activity, or an intellectual challenge—can become boringly familiar with repetition.

While both sensitization and tolerance seem like a drag—they amplify fear and pain while attenuating pleasure and joy—they clearly aid survival because of the way they motivate learning. Without habituation, that background hum or other distraction can't be overcome and a fearful challenge can't be mastered and made manageable. And if everything always seemed new and exciting and we didn't habituate to it, the world would be overwhelming and we wouldn't be able to focus on what's genuinely changing.

Without sensitization, on the other hand, our attention wouldn't be turned toward the experiences that are real threats and we wouldn't be able to respond to them effectively. It might take far too long for us to react to a potentially life-threatening situation. A balance between both of these processes is what we need to be able to cope with the familiar and the unfamiliar, the challenge and the routine. And that balance is changed during addiction.

Addiction is a learning disorder in part because of the way it affects both the habituation and sensitization processes and also skews them. "Wanting" for a drug is sensitized and attention is drawn to drug-related cues far out of proportion to their value. "Liking," however, falls prey to tolerance, meaning that the joy leaches out of the drug experience, and even other pleasures become muted. Learning is also the key to addiction because by simply varying the timing, regularity, and dosage of a drug—in other words, the factors that affect the experience we remember—the neurochemical and psychological effects can be altered dramatically. Large doses, given irregularly and unpredictably in varied environments, tend to produce sensitization; small, regular, predictable doses always taken in the same place and at the same time produce tolerance.

I'll discuss more about why the patterning of addictive experience matters in the next chapter, but it's important to note here that sensitization and tolerance are not just automatic properties of certain drugs. Different routes of administration, different dosing schedules, and changes in environment can make the same drug produce either sensitization or tolerance to particular effects.

Interestingly, autism also seems to affect both of these processes—amping up some types of sensitization and lowering some forms of tolerance. For example, research on autistic children finds a sensitized startle response and a reduction in habituation by the amygdala, a brain area that processes fear and other emotions. While greater sensitization could be linked both to increased sensitivity to experiences like loud sounds and to a rise in related fear, reduced habituation by the amygdala would further escalate this problem by making it harder to learn that something new is safe. This combination might mean that for autistic people, ordinary experience could sometimes be traumatic enough to produce PTSD-like responses—and that those who have access to drugs might be at extra risk for addiction.

In my case, one of the reasons I loved heroin was its ability to simply turn the volume of my emotions and senses down. Most opioid addicts appreciate this, whether the intensity of their emotions comes from something like autism or depression or whether it comes from traumatic experience. As Lou Reed's famous lyric describes it, "When the smack begins to flow / I really don't care anymore." Nearly universally, opioids are described by people who love them with words that evoke being nurtured and free of need: warm, safe, loved, comfortable.

Curiously, unlike with cocaine addiction, that soothing property didn't fade away: if I could get enough heroin, the drug always worked. The heroin high never turned on me the way cocaine did, probably because pharmacologically, opioids primarily affect "liking" and replicate the physiology and satiety of feeling loved. It makes sense that a drug that produces comfort and satisfaction once obtained would produce less craving than one that primarily creates a desire-based high. While extended sexual desire without satisfaction becomes torturous, no one minds lingering afterglow. Heroin addiction, therefore, isn't as much a condition of sensitized "wanting" as cocaine addiction is. Because opioids act directly on the natural systems that mediate "liking," the high remains at least somewhat satisfying, if not as great as it initially seemed. However, even in opioid addiction, there is certainly some sensitization of wanting and diminution of liking; it's just not as pronounced.

With cocaine, it's possible—although it takes a seriously massive and potentially deadly amount of repetition—to learn that it really, truly does not work anymore, no matter how badly you think you want it. With heroin, however, the dissociation of wanting and liking is not as obvious. I think this is why, while I sometimes say that I reserve the right to take opioids again in my 90s, I no longer feel a desire to ever return to cocaine. My emotional knowledge of its consequences has finally caught up with my previously ineffectual cognitive sense that it will only bring pain.

But that didn't happen until I was 23, the time at which my brain was likely completing the development of the cognitive control regions. And that development may have been what allowed me to finally recognize that the consequences of continued cocaine use would not be good, regardless of how wonderful it was that first time with Jerry.

TEN

Set and Setting

A pleasure too often repeated produces numbness; it's no more felt as a pleasure.

—ALDOUS HUXLEY, *POINT COUNTER POINT*

IN 1983, I STARTED COLLEGE AT Columbia on a hot, humid, late-summer day. I was assigned to a dorm called McBain Hall on West 113 Street. I was both excited—and terrified. Somehow, I'd managed to maintain my grades throughout my high school drug adventures, even scoring a National Merit Scholarship after I'd stupidly taken LSD the day before taking the PSATs. I felt like my whole life had been preparation for the moment when I would finally start college; now that it was here, I swung between anticipation and apprehension. As I entered the lobby, I discovered that the elevator was broken. My room was on the sixth floor. My parents and I had to carry all of my things up the stairs. The new outfit I'd bought was sticky with sweat by the time we'd finished; there went my idea of making a good first impression on my new classmates.

When we were done, I looked around my new home. It seemed sterile. It was an impersonal oblong box with two twin beds and some wooden furniture. The only window overlooked an airshaft, offering a view of the shaded windows of other dorms. The showers and bathroom, which were shared by dozens of people, were around the far corner. My assigned room seemed too small for one person, let alone two. I hated the prospect of having so little privacy—not because I was snobby, but because of my compulsions and my need for long periods of solitude in a safe place. Freshmen had no choice, however.

My family prepared to leave, and I lost it. I didn't want to stay here. I was overcome by anxiety, I didn't know anyone, I had such difficulty with friends. My mother saw the tears welling up in my eyes and couldn't bear it either. "No use postponing the inevitable," she said, and gave me a tight hug and a kiss good-bye. After they left, I sat on my bed, distraught. I didn't know what to do.

As was my usual pattern, academics were easier for me than socializing. In fact, I loved the classes and the intellectual stimulation. I took a difficult physiological psychology course so I could learn more about how drugs worked on the brain; I was the only freshman enrolled in it. I made the dean's list. I would do well meeting people and talking about ideas but somehow couldn't translate that into sustained friendship—or, at any rate, couldn't feel truly connected and secure with the friends I actually did make. To make matters worse, I went upstate every weekend to see my Deadhead high school boyfriend, Ethan, absenting myself from the most important social events on campus.

The difficulty of adjusting to the changes that come with early adulthood is a common theme in many stories of addiction, whether people come to it via depression, other mental illnesses, economic hardship, trauma, or Asperger's. Creating an entirely new network of social support while coping with new academic or employment challenges is hard for everyone—but it's especially difficult for those who are in some way wired atypically.

And as I started college, I had no idea that I was already beginning to become addicted to cocaine. My vision of the choices I had about my drug use was becoming occluded. But I wasn't aware of the narrowing process. It's hard to tell when one starts losing control over a certain behavior, when the line between what you feel like you "want" and what you think you "need" becomes blurred. However, I do know that at Columbia, my attitude toward cocaine underwent major changes. The valuation systems in my brain were shifting.

At first, coke had been a treat—to celebrate a special occasion, like meeting Jerry Garcia. I also saw it as an aphrodisiac that elevated sex to a higher, deliciously exalted plane. It made me feel glamorous and special and valued. But after I got to Columbia, the time that I wasn't high came to seem depressing and drab. I began to ask Ethan to buy it, and I was greatly disappointed when we didn't partake. Without noticing, I began to put the drug before the relationship—and to secretly pine for cocaine, rather than for him. There was a blindness to my behavior and a lack of consciousness about my own motivations and desires. I would tell myself I didn't care about coke, then become upset and irritable when he showed up without any. I began to feel increasingly socially isolated, overly dependent on a man who insisted he didn't love me and didn't want a long-term relationship and yet sure that I would be unable to find a more suitable boyfriend at school.

Further, while I'd been acutely aware of social rank since junior high, it was only when I got to Columbia that I truly became conscious of class and its insidious effects on social life. In high school, I'd pretty much been surrounded by middle-class kids just like me. I'd had some experience with working-class and poor classmates but little where I was the one with the low socioeconomic status. Now, though, I was exposed to the upper class for the first time. I had no idea how to fit in to this world of designer clothes, prep schools, and entitled, cynical attitudes.

I had a constant sense that I was walking around without knowing that I had toilet paper on my shoe or chocolate at the corners of my mouth; something about me was always not quite right. I felt as though everyone else had read and absorbed the etiquette book except me. I had no idea that this was, at least in part, related to class differences—and I now think that colleges should help students recognize and explicitly address the challenges of socializing with people who have radically different relationships to money, in order to ease the transition and prevent or at least minimize alcohol and other drug problems.

But back then I had no clue how to be or at least feel more acceptable. What I did know, however, was that drugs were a valuable social currency. And that was how, in short order, I began to sell coke. The drug had become extremely, glamorously popular: celebrities wore tiny silver coke-spoon necklaces, and the hit comedy *Saturday Night Live* didn't bother to disguise its coke-fueled humor. It was the early '80s and greed was good; the hippies had completed their transition to 30-something-era yuppies. A drug that makes you feel bold, confident, and on top suited the age perfectly.

It was also the heaviest recorded period of illegal drug use by young adults in American history. While many people think college drug use peaked in the '60s or '70s, in fact, my generation, Generation X, proportionally took the most drugs. Among people around my age now starting to hit their 50s, for example, just under half report having tried cocaine at least once—a remarkable percentage. In 1983, the year I started college, nearly one quarter of college students reported having tried coke at some point; the latest figure for undergrads today is just five percent. And at Columbia, in New York City, which was then pretty much the world capital of cocaine use, rates almost certainly were much higher.

Of course, back then coke had the reputation of being relatively harmless. Among the people I knew, the idea that cocaine was a "hard" drug like heroin seemed as ludicrous as the reefer madness that surrounded pot. The attempt to scare kids away from drugs by claiming that marijuana was a "gateway" to the hard stuff actually helped make it into one. When you discovered that you had no difficulty controlling your use of marijuana or psychedelics, you were much less likely to believe the harsh truth about cocaine. We thought it was just more fear mongering.

Given all of this, becoming a dealer seemed like the answer to all of my problems. It offered instant status, friends, and money, not to mention a constant supply. Ethan suggested that I try selling and he started out by fronting me an eighth of an ounce—3.5 grams—for $250. I'd break it up into half and quarter grams and sell it for $100 per gram. I didn't cut it and usually kept a half gram for myself. At first, I'd make a quick $50 and have coke for the weekend, which I would happily share with friends.

Soon, though, I had a thriving business and was buying and selling much more. My cost dropped when I bought in quantity, but I kept my prices the same. Within a few months, I was making hundreds of dollars a week, which seemed like a huge amount of money at the time. Since coke was little different from pot to me, selling it seemed no worse than bartending: I didn't think dealing was wrong because I didn't think using was wrong. Surely the law would soon catch up with public opinion and widespread use of the stuff. Dealing also provided a convenient identity; it literally gave me something to bring to the party. As long as I had coke, I knew that I was wanted. It's hard to feel left out when everyone is waiting for you to arrive.

Further, for someone who finds unstructured social situations uncomfortable, selling, preparing, and/or giving away the drug gave me something specific and ritualized to do. I loved making "bindles" out of shiny magazine paper, cutting out squares, folding the corners, tapping the powder in from the tiny hopper of a miniature scale or a Deering grinder, then sealing them up. Cutting lines with a razor on a mirror—especially chopping up rocks that sparkled like shale or pearl—was also wonderfully satisfying for someone with obsessive tendencies.

By my sophomore year, I had it down to a science. Inside my new, single dorm room's standard-issue wooden desk, in the drawer normally used for small office supplies, I kept a mirror and a little black plastic gram scale. When people would visit to buy or use, I'd slide it open and cut out lines or weigh what I was selling and package it right there. If I needed to instantly hide what I was doing, I could shut the drawer and nothing drug-related would be visible.

Now I could dine out, dress fashionably, and get behind the velvet ropes. I could pay for my friends who otherwise couldn't afford to go out to clubs. I got a beeper and began selling not just to other students but to people on Wall Street and in the fashion industry. Most nights, I would chat animatedly while I weighed people's purchases, ask them if they liked it ground up, and offer free lines while they waited. If I wanted to go out, I'd make deliveries to parties and night spots. Sometimes, if I was paranoid about having dozens of people streaming in and out of my room, I would do everything by delivery and would go from apartment to dorm to apartment, greeted eagerly wherever I went. If I had any social discomfort, I could look at the beeper on my hip, make a business excuse, and leave.

At this time, the dance club Area was the place to be seen. Set up like an art gallery, with decorative themes, artwork, and sets that changed every month, it was frequented by Andy Warhol, Madonna, Sting, JFK Jr., Debbie Harry, and other '80s celebrities at the height of their fame. One night I saw Robin Williams, amid the models and club kids; I also spotted the professor who'd taught my psychopharmacology class. Outside the club, hundreds of people were silently begging the snooty doormen to choose them to gain admittance; I was sometimes among them when my club connections fell through. Inside, the place was awash in cocaine. The women's bathroom had become coed in order to facilitate use—and men would lead women they'd picked up in there to turn them on to cocaine. Between trips to the ladies' room with my friends, I'd dance the cocaine energy out of my bones.

Before spring or winter break, on New Year's Eve and at exam times, I'd sometimes make thousands of dollars a day. But, of course, my own use was escalating as well.

IN ADDICTION, IT'S not just the type of drugs you take that matters or your reasons for taking them. The dosages, the timing of those doses, and even the place where you get high can make all the difference in terms of how your behavior will be affected. Although it might seem like dose pattern, culture, and environment should have little impact on a chemical taken in a sufficient amount to produce a psychoactive effect, in fact, these apparently extraneous factors can be the difference between addiction and casual use and even, sometimes, between life and death.

When I'd first taken psychedelics in high school, I read Timothy Leary. The Harvard professor and acid prophet emphasized the importance of what he called "set and setting" as key influences on drug experiences. Set is a person's frame of mind, mood, expectations, and the prevalent cultural ideas about a drug; setting is the physical and social environment in which a substance is taken. The effects of set and setting are most florid with psychedelics.

Taking LSD when feeling happy and safe with friends at a Grateful Dead show is a far different experience from, say, being unwittingly dosed by CIA agents and then interrogated, as happened in what were later determined to be illegal experiments on civilians in the 1950s. No drug takes effect without context; psychedelics in particular exaggerate people's mood and predilections in interaction with their social and cultural surroundings and expectations. That's why the CIA soon thought that acid would allow them to terrorize and weaken enemy agents through "mind control"—while the hippies thought that dosing people would sow free love and bring about world peace.

But set and setting don't just affect psychedelics. They influence the experience of all drugs and can actually play a role, along with dosage and dose schedule, in whether addiction develops or not. The way cultural pressures, individual expectations, and other aspects of people's state of mind affect the amount of control people have over their drug use was first explored in Norman Zinberg's classic, *Drug, Set, and Setting*. The book examined controlled use of marijuana, heroin, and psychedelics in the absence of addiction and was among the first research to show that nonaddictive drug use was even possible.

Zinberg found people who took heroin only on weekends for decades without negative consequences; he also did the first study showing that although nearly half of American soldiers in Vietnam took drugs like opium and heroin, 88% of those who had been addicted while overseas did not become readdicted upon returning home. Other researchers later confirmed that even among those who used heroin a few times after coming back, when they were out of the war zone, the overwhelming majority did not return to addiction, to the great surprise and relief of the military leadership. Indeed, those who relapsed—around 1%—were mainly those who had had drug problems before the war.

So why do set and setting matter so much? Many factors are involved, but one critical piece involves the basic patterns of learning about rewards, which were first elucidated by behaviorists like B. F. Skinner in the 1950s. Although I didn't know it, when I took cocaine, my set and setting were unfortunately highly apt to produce addiction, in part because I was able to take varying doses irregularly and increase my access to the drug over time.

Skinner observed how rewards and punishments affect behavior and how changing their delivery and timing can be used to promote desired actions and reduce undesirable ones. This early work shed important light on why some patterns of reward *by themselves* can lead to addictions, even without drugs. The fact that learning is the key driver of these behaviors shows how fundamental it is to the development of addiction.

The most important addictive learning pattern—technically known as intermittent reinforcement—was discovered by Skinner himself, in a classic case of scientific serendipity. In 1956, the behaviorist pioneer was studying learning in rats. One weekend, he realized he was running low on the food pellets he used to reward the rodents when they had learned to press levers correctly. Since he had to make these rat treats himself and he didn't want to interrupt the experiment to make more, he decided to give the animals their rewards less frequently to stretch out what remained of his supply.

To his surprise, rather than reducing the odds that the rodents would press the lever, this variable reinforcement actually made them respond more. It also made them persist in trying for far longer periods of time after he'd stopped providing rewards. Later studies found that the most effective schedule of reinforcement to hook animals on a behavior pattern was the most unpredictable: if pigeons, for example, get rewarded only 50% of the time for doing a task (basically, completely randomly), they will perform the required behavior more frequently and will be more resistant to changing it than if they are always rewarded or are rewarded in a more predictable way. Humans behave quite similarly.

In fact, some researchers argue that brains evolved as prediction machines—so even animals like pigeons seek patterns in the environment that will help them maximize the time and minimize the energy spent obtaining important resources. As a result, our brains experience pattern finding in and of itself rewarding—and being unable to figure out the link between behavior and reward or punishment is maddening.

Many of humanity's greatest creative and scientific achievements have probably been driven by this tendency, by our desire to seek order in chaos and feel joy when we find a new way of making sense. One of the greatest pleasures of music, for example, is interplay between predictable and pleasant patterns and moments of unexpected but harmonious surprise. These joys, like those of drugs, play out, in part, through dopamine: the neurotransmitter is released in "wanting" and motivation circuits when you make an accurate prediction of a pattern.

And so, once you've learned that, say, a G chord will follow a C in a piece of music, the highest levels of dopamine will be released when you hear C, and a slightly smaller dose will follow to an appropriately timed G. When you listen to a song you like, then, part of the pleasure comes from your brain accurately predicting the next notes and beats—and then rewarding itself when it gets it right. When musicians change things up and play a variation that you didn't predict, this, too, can be pleasant because after you hear it, you realize how it fits into the pattern, and this "clicks," too, elaborating your prediction algorithm. But if a bad note or chord follows, dopamine can actually drop, signaling that the predicted pattern has not appeared and expressing the brain's displeasure at both the inaccuracy of its forecast and at any discord.

Figuring out these patterns is one reason people often like to listen to the same tunes over and over; they can then discern how they occur in different voices and instruments. Moreover, the love of patterns and the ability to take pleasure in detecting and predicting them can lead to success in areas as varied as the arts, medicine, science, programming, and engineering. It's also why puzzles and games remain popular entertainment and why mystery stories are so compelling.

Gambling addiction can also be seen as one of the clearest expressions of this phenomenon. In pure games of chance, winning occurs completely at random—and there's no doubt that some people are prone to becoming compulsively engaged with such games. Just as with heroin or cocaine addiction, gambling addicts will risk their relationships, homes, jobs, and freedom simply to continue to play. Internet games—or even just social media like Facebook and Twitter—are also characterized by both intermittent reinforcement and addictive qualities, as anyone who has ever been annoyed by someone else's refusal to disengage from these media or by their own inability to stop procrastinating by watching cat videos can attest. In extreme cases, gaming and Internet addictions have resulted in disruptions in relationships and work that are every bit as serious as drug addictions.

But there's no drug in these situations to neurochemically alter self-control, no substance to "hijack" the brain's reward system or directly elevate dopamine levels outside their normal range. Simply by creating an unpredictable pattern of highs and lows, gambling and other behaviors can become addictive—and the fact that this occurs without a drug offers insight into what happens with psychoactive substances and why addiction risk exists at all. A pattern-seeking brain is prone to getting fooled by random rewards that only appear linked with behavior; attempting to find structure in intermittent reinforcement can get us stuck looking for an order that doesn't exist.

In fact, like many compulsive gamblers, Skinner's first intermittently reinforced pigeons even developed "superstitions"—behaviors like spinning around or other particular motions that they associated with times when food rewards had actually been delivered. These actions didn't make the random rewards any more likely, of course—but like a gambler's lucky charm, they probably gave the animals a sense that they could affect the outcome, even though it was an illusory one.

All of this is why, when substances are involved, the pattern of use can be every bit as important as the drug's pharmacology—and why brain changes linked to simply taking a drug don't create a "disease" of addiction. As behavioral addictions prove, these brain changes can occur even without drugs. And with drugs, they don't occur automatically. The pattern of use, the users' prior history and brain wiring, and the cultural context of use all matter.

Irregular and varied dosing—another element of unpredictability—are also important in producing sensitization and tolerance to drugs. As we've seen, sensitization escalates desire but not satisfaction—"wanting" but not "liking." Tolerance, in contrast, does not have this effect: it basically nullifies the "high" if dose isn't varied and escalating. Both processes can feed a cycle that only worsens the problem. The timing, setting, and consistency of use can actually change sensitization and tolerance so significantly that one dosing regimen of a particular drug will produce the exact opposite effect with another.

This is one reason why medical use of drugs carries a far lower risk for addiction than recreational use does, even though, on the face of it, it seems bizarre that factors such as having a legitimate prescription and taking specific doses at particular times and in particular places would alter addiction potential.

Addictiveness seems like it should be a fundamental and invariable pharmacological property of a specific drug—but in fact, it is hugely affected by environmental factors. Even in animal experiments, the pattern of drug availability, the dosing, and the timing of use matters; you can get very different rates of addiction with the same drug by varying these. The implications of these facts are critically important to understanding addiction and creating better drug policy.

For example, the role of dosing patterns and their effects on tolerance is important not just in illustrating why addiction is a learning disorder but in using that knowledge to treat it. Widespread misunderstanding about these phenomena, in fact, contributes significantly to stigma against maintenance treatments of heroin and prescription pain reliever addictions, using opioids like methadone, Suboxone (buprenorphine), or even heroin itself. Maintenance treatments are the only therapies that can lower mortality by 75%—something that would be considered a miraculous success in any other type of treatment but addiction care. The effects of timing and dose can be seen particularly clearly with opioid addictions and they are critical in understanding why they are often so deadly.

On the surface, maintenance treatment does look like merely "replacing one addiction with another." Indeed, giving methadone or Suboxone instead of heroin does actually involve shifting use from one type of opioid to another. But what's different about maintenance is the pattern of use: in active addiction, the timing, setting, and dosing of opioids maximizes sensitization, whereas during maintenance, the pattern of use maximizes tolerance. For street users, the irregularity of supply and timing means that drug wanting is sensitized: taking large, irregular doses is the best way to increase "wanting" via sensitization. Since most addicts tend to use as much as they can at any given time, this means that they will swing from periods of being high to those of withdrawal, depending on how much money they have, what they have to do to get that money, and how reliable their dealers are.

Consequently, although they will develop some tolerance to "liking," it will not be complete because of the irregularity of the dosing. Even if the occasions when the drug produces euphoria become less common, they still do occur, providing intermittent reinforcement. In contrast, if you provide a steady, pure, stable supply, tolerance will predominate: the more steady and regular the dosing is, the more tolerance will build. Very rapidly, an opioid user on maintenance will no longer experience a high because he or she will be on a stable plateau of tolerance.

Indeed, if the person is stabilized on a high enough methadone dose, even if they try to take additional opioids "on top," no high will ensue. The end result is that on a dose that could kill a user with no tolerance, stable maintenance patients can work, drive, and be otherwise unimpaired, simply because their dose is not varied and is taken regularly and consistently. Tolerance also means that maintenance does not block emotional growth or automatically make people distant in relationships. In order to use opioids to escape emotion, you need to get "high," and stable maintenance patients are too tolerant to do so. Unfortunately, because people don't understand these basic facts about the role of timing and dosing in addiction, maintenance patients are stigmatized as being constantly "high" and "not really in recovery," even though tolerance means that this is inaccurate.

Importantly, the learned relationships between a regular time and place and the use of a drug is also sometimes a matter of life and death. Tolerance in general is generated by routine. Taking the same dose of the same drug in the same place maximizes tolerance by strongly linking the experience of being in that specific situation to the physiological process that generates habituation. As a result, when routine changes, so can tolerance.

For decades, in fact, users and physicians have reported cases where people take the same dose of the same drug that they have previously used uneventfully, but suddenly, mysteriously overdose on it. Often, it later turns out that these folks took the drug in a new environment or social setting, not as part of their typical habit. Sometimes, a new source of stress is involved—for example, a lost job or breakup. In these cases, the unconscious cues that normally activate tolerance don't do so—and without tolerance, the normal dose becomes an overdose. It seems that sometimes, context alone can change a safe dose to a lethal dose. And this effect can even be seen in rats: if they are made tolerant to a high dose in one place and then given the same dose in a new environment, 50% will actually die of overdose.

Of course, the type of drug use—whether it is a stimulant or depressant, for example—is not entirely irrelevant. Stimulants, because of their direct effects on dopamine, seem to produce far more severe sensitization of wanting and diminution of liking; opioids, because they act on opioid receptors that directly mediate liking, create a less extreme dissociation between the two. And this is why "stimulant maintenance" treatment like using Ritalin to treat cocaine or methamphetamine addiction does not seem to be as successful as opioid maintenance. It is much harder to create tolerance with a drug that imitates the pleasures of desire than it is with one that simulates the experience of satisfaction. While these drugs may be useful to treat people whose addiction is linked to untreated ADHD, so far the data are not as promising as for opioid maintenance.

Alcohol, incidentally, also does not work for maintenance. Unlike opioid tolerance, alcohol tolerance doesn't completely eliminate impairment, even if the drug is given in steady, regular doses. That means that even with high levels of tolerance, heavy alcohol users on a consistent dose are still significantly impaired. Contrary to the claims of those who criticize opioid maintenance as "just replacing vodka with gin," the difference is that opioids produce complete tolerance in steady-state dosing, while alcohol does not.

And while you might think that social conditions and environment only matter in the lives of humans, even rats turn out to show surprisingly large differences in their susceptibility to addiction based on nondrug factors in their lives. The best example of this is a groundbreaking series of rat experiments conducted in Canada in 1977. Having read research showing that rats will starve themselves to death if given free access to drugs like cocaine and heroin, Bruce Alexander of Simon Fraser University wondered whether there might be more to it than drug access alone.

Along with his colleagues Barry Beyerstein, Robert Coambs, and Patricia Hadaway, Alexander decided to see whether environmental and social conditions had any effect. Like humans, rats are highly gregarious creatures that don't do well under conditions of solitary confinement. Might they take drugs excessively in the classic experiments not because the drugs are so good—but because the rest of their lives are so bad? Humans kept in total social and sensory isolation rapidly begin hallucinating and show all types of abnormal behavior. A recent review of the literature found that in every single study, after mere days of involuntary isolation, psychiatric symptoms invariably appeared in at least some participants. It's not hard to imagine why someone might rationally decide to take potentially fatal doses of drugs if they were caged alone with no apparent hope of escape and no alternative pleasures. But in more normal environments, human addiction that is life-threatening is far more rare—even when drugs are widely available.

To find out whether social setting matters for rats, Alexander's group decided to create a virtual rat paradise—the ultimate enriched environment. They created a 29-square-foot enclosure—more than 200 times the size of a typical cage—with plenty of space to roam, lots of hidey holes for nesting, and, of course, lots of other rats. "Rat Park" also included walls painted with forest scenes, wood chip bedding, and intriguing objects to play with and exercise on, as well as abundant food and water. The lucky rats chosen to live in Rat Park would be compared to those who lived alone in standard, bare cages.

During the experiment, the researchers offered both groups of rats access to morphine-spiked water, which was sweetened enough to overcome the bitter taste that would normally make rats avoid such a beverage. They also had access to plain water. But while the caged rats lapped up the morphine draught, the Rat Park residents mainly refused. Under some conditions, the caged rats took 19 times more morphine than the rats that had companions and space to explore. Even when all of the rats were made physically dependent on morphine and taught that drinking it could relieve withdrawal symptoms, the Rat Park rats resisted it. The caged rats drank eight times more morphine than the Rat Park rodents in this part of the experiment.

When given options, rats, like humans, tend not to slip into addictions. This is why, no matter how many times the media panics over a new drug that's "more addictive than heroin" where "one hit gets you hooked for life," the vast majority of people who use it don't become addicted. People with decent jobs, strong relationships, and good mental health rarely give that all up for intoxicating drugs; instead, drugs are powerful primarily when the rest of your life is broken.

Further, the dose itself, not just its variance, also matters and this interacts with the effects of the environment. In another set of experiments by different researchers, for example, rats were placed in a new environment and offered moderate doses of cocaine. The doses were enough to cause a high, but not an overwhelming one. In the strange place, the researchers found, rats took more of the drug than they did when the same doses were available in their home cage. When the rodents were offered large doses that produced a robust high, however, they went for them equally avidly at home or in a new place. Large doses seem to be so rewarding that animals seek them out whether or not they are accompanied by other types of new experiences—but moderate doses just don't seem that interesting to rats that are safe at home. Although a few early experiments failed to replicate the Rat Park effect, most did so.

And many other studies have since found that social contact and enriched environments have major effects on animal drug use. For example, allowing mother rats access to their babies reduces cocaine use (showing that even in rats, the drug doesn't abolish maternal instincts, as some have claimed about crack). Male prairie voles (a type of field mice) that have established a monogamous relationship take less amphetamine, a phenomenon also seen in human males who are married. And providing a running wheel reduces cocaine use by 22% in male rats and a whopping 71% in females.

Social rejection and defeat also appear to have a large influence—in, as you might predict, the opposite direction. For example, rats take half as much time to learn to self-administer cocaine if they have first been beaten in a fight by a more dominant rat than they do if they haven't been previously defeated. They also take much more cocaine if given the opportunity to binge after losing fights.

Interestingly, offering large amounts of sugar also cuts cocaine use. A 2007 study found that 94% of rats preferred either saccharine or sugar to an intravenous injection of cocaine. This finding, of course, prompted dozens of variants on "Sugar is more addictive than crack" headlines, which, you may now realize, are actually not warranted by this kind of research. Addictiveness is incredibly dependent on context: in some situations, rats clearly do prefer cocaine to food, otherwise those early experiments in which some starved themselves to death by binging on cocaine would never have worked or been replicated.

Of course, politicians were busy demonizing crack when that early research was published. They used these studies to argue that cocaine is more powerful and dangerous than any "natural" reward like food or sex. Now, with the obesity epidemic, the new enemy is sugar, and so experiments that make sugar look more dangerous are being spotlighted. The tautological nature of arguing that drugs are addictive because they activate regions designed to attract us to food and sex while claiming that sugar is addictive because it activates the "same areas as drugs" is never explored. Drugs work because they act like sugar or sex do in the brain; sex and sugar are not potentially addictive because they act on some specially designed "drug area." Brains did not evolve in order to allow us to become addicted.

Indeed, we get addicted only in certain contexts, which makes all the difference in terms of whether any given human will eat normal amounts of sugar, binge on it, take cocaine occasionally, or become a full-on crack addict. By itself, nothing is addictive; drugs can only be addictive in the context of set, setting, dose, dosing pattern, and numerous other personal, biological, and cultural variables. Addiction isn't just taking drugs. It is a pattern of learned behavior. It only develops when vulnerable people interact with potentially addictive experiences at the wrong time, in the wrong places, and in the wrong pattern for them. It is a learning disorder because this combination of factors intersects to produce harmful and destructive behavior that is difficult to stop.

IN MY CASE, unfortunately, the chaotic pattern of my cocaine use was pretty well a recipe for addiction, and some part of me recognized this years before I ever sought help. This was visible in the way my relationship with psychedelic drugs changed. The more coke I did, the less I wanted to trip. During my freshman year, I would occasionally take LSD, standing on the roof of the East Campus dorm, the city spread out below me like a wildly lit pinball machine, the evening sky orange from the city's lights. The world seemed senseless: the spurts of traffic moving up the streets, the pulses of red and green lights, the buildings in their various shapes and heights—all random.

Unlike when I tripped in high school with Amy, I felt little enthusiasm or wonder—and when I did experience these emotions, I tried to hide it because I thought I would seem childlike or unsophisticated to my Ivy League classmates. LSD had stopped making me feel communion. The experience became hollow. This emptiness also made me feel guilty because it reminded me of the spiritual ideas I hadn't even realized I'd abandoned. My psychedelic use dwindled, then stopped entirely. But, one day a friend wanted me to trip with her, and worn down by her persistence, I agreed. It was my worst psychedelic experience.

I freaked out. I cried for hours, lost in dark and tangled hallucinations. I was inarticulate with pain. I saw that I was a drug user—nothing more. I knew that taking psychedelics again would make me confront my attachment to cocaine. But I didn't want these revelations; I didn't want to look at how far I'd moved away from my ideals about using drugs as tools for growth rather than escape. If I could have accepted these insights, it might have been a turning point for me. In fact, because hallucinogens prompt such revelations quite frequently, many believe they may be useful in therapy for addiction and quite a few studies now suggest that there is real promise in this idea. Unlike addictive drugs, psychedelics don't produce compulsive behavior; they are too exhausting for most people to take daily and since they make bad moods worse, use rarely escalates over time. So if I'd returned to tripping, it wouldn't have meant simply switching addictions. But I didn't have anyone to show me other ways to cope. I didn't know how to handle social life without a chemical escape route. And so, I never tripped again.

Needless to say, I also did not stop taking cocaine. Not surprisingly, the more I used, the more I sold. Since I wasn't exactly your most subtle dealer, I eventually got onto the radar of the campus authorities. By the summer of 1985, the end of my sophomore year, I was called in to meet with the deans. I was so clueless that I thought I was meeting them about my grades when I got the letter scheduling the appointment. After a hearing that left me so hysterical that I was immediately brought to a psychiatrist, I was told I would be suspended for a year.

This is the way private colleges tend to deal with drug problems: they try not to involve the police and, ideally, they refer to treatment. It's the way addiction is treated when "one of us" has a problem—and, when done right, it is far more effective than punishment. In my case, though I don't understand why, I was simply told to leave and then return, presumably, to sin no more. Despite being caught dealing and despite daily cocaine use, I couldn't accept that I was addicted—and bizarrely, the psychiatrist I was taken to see because the administrators feared I might kill myself after the hearing didn't diagnose me, either. I easily convinced him that I so feared death that suicide was not an issue. I wasn't even given a follow-up appointment.

My parents took all of this on board unhappily—but they, too, somehow didn't realize I needed serious help. I often wonder if the school had referred me to a better psychiatrist or mandated that I seek treatment, if I would have gotten the help I needed sooner. But instead, I quickly got myself a plausible summer job as an assistant to a film producer and told everyone I was "taking a year off" to explore the movie business. Inside, however, I was gutted. All the self-hatred that lay just beneath the surface, all the insecurity that had been magnified by my ongoing social difficulties now turned into frank depression, although this was hidden by my worsening addiction.

I thought that I had ruined my life and that Columbia would actually not readmit me, no matter what they now said about a temporary suspension. I didn't see myself as having been exposed as a drug dealer; I saw myself as having been revealed as the evil, undeserving person I really was. All of this played out as I chased yet another inappropriate man and added heroin to my cocaine habit.

It was then that I began to learn the relationship between addiction and love.

ELEVEN

Love and Addiction

Love will make you do things that you know is wrong.

—BILLIE HOLIDAY

THE CARLTON ARMS HOTEL SITS ON 25th Street and Third Avenue in Manhattan, at the end of what is now a traffic-free block in the expanded campus of City University's Baruch College. But it's a relic of a much different time. In the '80s, it was home to "madmen, junkies, comedians, ex-cons, pushers and hookers, transvestites, drunks and nuts of all kinds," according to its former manager, Ed Ryan. Artists like Brian Damage, known for installations at '80s clubs like Danceteria and Studio 54, decorated each room in signature eye-popping style, often in garish and psychedelic colors.

Today Herman the one-eyed junkie, Alabama Bob (who preferred coke), and the rest of the itinerant oddballs I knew from the time it was a single-room occupancy and welfare hotel are gone; the other similarly low-rent residences that once dominated the neighborhood have been torn down or renovated, often becoming dorms or luxury apartments. Although the Carlton is still relatively cheap, a room there now costs at least double the $35 that someone paid for the room where I first took heroin in 1985.

It was not long after I'd been suspended from Columbia that I began studying up on heroin, as I had with almost every other drug I tried before I took it. I read *Naked Lunch*. I listened to the Velvet Underground over and over: "Heroin" . . . "I'm Waiting for the Man" . . . "Sister Ray." Lou Reed's craggy voice detailed the dire consequences of drugs and needles in the lyrics, while his tone and music betrayed their ecstasies and his love for injecting. "It's my wife / It's my life. . . ." I'd always said I'd never try heroin because I knew I'd love it; but now that I wasn't in school and my life was basically over, I began to think that I had nothing to lose.

I'd come to the Carlton looking for Matt, with whom I was then in a tortured relationship. Originally my cocaine connection, Matt was also the older brother of my high school boyfriend, Ethan. When Ethan had decided to quit cocaine—and me, during my freshman year at Columbia—he'd given me his brother's number so I wouldn't lose both my business and my boyfriend at the same time.

Matt and I became involved about a year later. At the time, he was "on a break" with his girlfriend, Susan. I was single. When we'd first met during my freshman year of college, Matt lived in a cozy apartment in the village on Lafayette Street, surrounded by his collection of hundreds of jazz, rock, funk, and psychedelic records, books, and, especially, comic books, or as they were starting to be called, graphic novels. At the time, he was the hub of a humming underground drug network, selling everything from quarter grams to quarter kilos to people ranging from the Grateful Dead and jazz musicians to movie industry tech people, sound engineers, college students, businesswomen, and Wall Street traders. Tall, thin, with brown hair, a moustache and mischievous eyes, he was well versed in the things I then loved: psychedelic music and drugs. The whole scene fascinated me.

But, although I didn't realize it, by the time we started seeing each other, Matt was beginning a decline into addiction. He couldn't decide whether he wanted me or Susan. And I was still afraid that I was unlovable. None of my boyfriends had ever said the L word to me—indeed, like Ethan, they had typically told me over and over that they did not want a commitment and that I was basically not wife material. At least, though, I could get Matt to sleep with me.

Because of the tenuousness of our romantic relationship, however, I tended to use my drug connection with him not only to buy but to monitor him via the phone pager we used to communicate for deals. Usually, when I beeped him, I really did need to arrange buys—but I was also trying to keep tabs on him, to try and prevent him from disappearing to "secretly" binge. That would leave me unable to supply my customers. Of course, I also didn't want him to have any chance to see Susan.

I was as obsessive and driven about my relationship as I was with cocaine and had been with my other prior intense interests. And I was every bit as hooked on my 1980s-era pager, which was planted on my hip as soon as I awoke, as I would become later with e-mail and Twitter. Like cocaine itself, Matt was intermittently reinforcing—his unpredictability and chaotic charm kept me guessing, desperately trying to find the hidden pattern that would make him love me. It was both that relationship and the cocaine that took me to the Carlton that day.

By this point, Matt and I were living together in an apartment near Columbia on 113th Street. We'd moved from snorting cocaine to smoking it; this was just before crack hit the front pages. In fact, long before crack became the subject of massive hysteria, thousands of New Yorkers had already begun smoking freebase. In 1984, on the stretch of Broadway between 96th Street and Columbia's main gate at 116th Street, there were dozens of bodegas featuring glass crack pipes, metal mesh screens, and miniature butane torches, sometimes hidden behind the counter, but more often in plain sight, even in the window. Other neighborhoods had similar concentrations of such stores and would soon have crack dealers on many corners.

By 1986, networks and newspapers seemed to be marketing crack with their hyperbole about it being the scariest, strongest, and most addictive drug ever. Although that wouldn't be a selling point to ordinary people, to those who seek extreme experience—that is, many of the people at highest risk of addiction— it signals that the drug must offer a powerful rush and is the most forbidden fruit, both of which are highly attractive to them. During the run-up to the presidential election of 1988 around 1,000 stories about crack and cocaine appeared in the major newsmagazines and national papers; on NBC alone, an incredible 15 hours of airtime were devoted to the story in the seven months before November. But by then, many people in our crowd had already been making it themselves for years from baking soda, cocaine, and water. We didn't think freebase made you into a monster; however, we sure knew it could make things get weird.

That's why I preferred freebasing at home. The high was all-consuming; at least the media had gotten that part right. Once you started, interruptions like answering the phone or the door or, worst of all, going outside to get something felt unbearable. Once you were high, any unpredictable element in the environment was terrifying. Even music was tricky: the popularity of the bland New Age label Windham Hill in the '80s probably owed a lot to its usefulness as unobjectionable wallpaper sound during freebase binges. Shadowfax was my band of choice for this purpose.

And if snorting cocaine had provided a taste of intermittent reinforcement, freebase was far worse. Sometimes, you could melt it with the torch into a perfect swirl of sweet, chemically redolent white smoke that would curl enticingly in the round bulb of the pipe, then blast you into the stratosphere, at least for a few minutes. Other times, you'd chase the rock with the flame and evoke paranoia or just numbness and craving for one more, just one more, just one more and that's it. Nights often dissolved into days in the thrall of "one more." The worst part was running out and searching desperately for any tiny scrap of drug you might have missed, smoking paint chips or other debris picked from the floor or carpet, just in case it might be freebase.

Unlike me, Matt liked to freebase cocaine away from our apartment, at various fleabag hotels, the filthier, the better, it seemed. He preferred smoking coke with people who were far more addicted than he was; ironically, at this point in my addiction, that actually ruled me out. I was usually still able to end binges to meet commitments, if barely. One time, just to show how much control he had over his use, Matt tossed a few rocks of freebase out the window at the Carlton—the equivalent of a rich man throwing around hundred-dollar bills in the presence of beggars. The people he was smoking with weren't proud; they simply tied some bedsheets together and climbed down into a courtyard to attempt to retrieve them, explaining to a neighbor they were searching for "precious stones." Matt didn't lower himself—but then, he didn't have to. He could always get more.

When I couldn't find him that afternoon in 1985—morning for us, as we rarely rose before 11—I had somehow learned that he'd been with Susan. After paging him endlessly, using the code "911" after my callback number to illustrate the business-related urgency of the matter and repeatedly getting no reply, I decided to track him down myself. I called the Carlton's communal phone—it was in the hallway and hotel staff tended to know who was around, sometimes even rousing themselves grouchily to call people to the phone—and found out he was there.

Just before I arrived, he had sold some coke to a very strung-out junkie couple whom I knew only as Pablo and Gigi. She had long, straight, dirty-blond hair, big round glasses, and many small scabs dotting her arms and legs that she explained by saying, "I pick because I do coke; I do coke because I pick." He had long dark hair with sunken brown eyes and was slender and only slightly less ravaged.

They sold heroin to support their own habits and, by some miracle, had just gotten hold of two ounces of China White smack. Since Matt was probably carrying at least an ounce of cocaine with him, that $35 SRO room, with its unstable furniture, punk décor, and crooked floors, was at the time home to many thousands of dollars' worth of drugs. Had we been caught, under the Rockefeller laws everyone would probably have been looking at 15 to life.

Consequently, the last thing anyone wanted was a screaming fight that could potentially draw unwanted attention from management or worse. And so, when I began shrilly expressing my displeasure with Matt about Susan at the top of my lungs, either Pablo or Gigi quickly cut out a line of heroin and offered it to me.

I didn't think twice. In my rage I did it. In one motion, I snorted two lines as if I'd been sniffing heroin all my life. Suddenly, the fury was gone, replaced by perfect, blissful silence. I didn't care about Susan anymore; I didn't even care about Matt anymore. I just didn't care. I needed nothing and no one. It was complete satisfaction. All desires extinguished. Instant nirvana. The physical sensation was like being hit hard with something infinitely soft, warm, comforting, enveloping. Every molecule of my body felt nurtured. I was home. Although I felt nauseous and ran to the bathroom, I didn't even throw up.

Heroin gave me the comfort that all the other drugs had only teased me with, a sense of well-being that ended all worries. The high wasn't delicate; it couldn't detour into darkness like marijuana, psychedelics, and even cocaine might. Soaring on heroin, I felt safe, wrapped in a cozy, protective blanket. While many people find this sense of cushioning disconcerting—or even unpleasantly numb—for me, it felt like I had finally found the insulation I had always needed. For what seemed like the first time ever, I felt truly safe and loved.

EVER SINCE ADDICTION was first described, people have compared it to love. Before compulsive drug use was seen as a disease, it was the original sin of loving too much. As noted earlier, addictions were historically seen as dangerous "passions" for particular substances. Poets and songwriters have also linked these ideas in metaphor forever—from the author of what may be the world's oldest song that survives in writing, an Egyptian love song inscribed on a 4,300-year-old tomb, containing the lyrics, "I love and admire your beauty / I am under it," to Roxy Music's "Love Is the Drug." However, it was not until 1975, when Stanton Peele and Archie Brodsky published the groundbreaking *Love and Addiction,* that the two were given a thorough side-by-side psychological examination. Point by point, the authors illustrated how unhealthy relationships—whether with drugs or with people—share the same fundamental qualities.

For one, nearly every behavior seen in addiction is also found in romantic love. There's an obsession with the qualities and particulars of the beloved; there's craving if the object of the addiction is unavailable; and, in some cases, people engage in extreme, uncharacteristic, and even immoral behavior to ensure access. Withdrawal prompts anxiety and fear; only the drug or loved one can relieve these pains. Both conditions profoundly alter people's priorities.

Importantly, like addiction, misguided love is a problem of learning. In love, people learn powerful associations between their lovers and nearly everything about them and around them; in addiction, these connections are made with the drug. Soon, relevant cues like sights, sounds, and smells will spur relapses into obsessive behavior. For example, a distraught lover may dial his partner when he hears "their" song; a visit to the Carlton Arms by one of its formerly addicted denizens may spur longing for cocaine. Stress, too, often leads to both pharmaceutical and romantic relapse.

In both love and addiction, the stress relief system has become wired to the object of the addiction—you need the drug or the person to feel at ease, in the same way that young children need their parents. In addition, both romantic obsession and addiction rarely appear before adolescence; both are shaped by life's developmental stages. But to really understand just how intimately love and addiction are linked—and what this shows about addiction as a learning disorder—you need to look inside the brain.

At around the same time *Love and Addiction* was published in the '70s, researchers led by C. Sue Carter at the University of Illinois were starting to unravel the neurochemistry of what is known in animals as "pair bonding." During these same years, in Baltimore, Candace Pert and Solomon Snyder would become the first scientists to isolate a receptor responsible for the effects of heroin and similar drugs. Pert and Snyder's work would ultimately lead to the discovery of the brain's natural opioids, the endorphins and enkephalins, which are important not only in addiction but also in love. The chemistry of love and addiction turn out to be startlingly similar—and both are intimately connected to learning and memory.

Oddly, our understanding of how humans bond chemically began with research on the sex lives of two types of obscure field mice. One kind, the prairie vole, belongs to the select five percent of mammal species that are monogamous, meaning that they form long-term sexual and child-rearing bonds with members of the opposite sex (though, as in humans, this bonding doesn't rule out sexual infidelity). Another type, known as the montane vole, hooks up but never settles down. Montane voles mate promiscuously and the males don't parent. When Carter and her colleagues realized that the key difference between these animals was their mating patterns, they recognized that studying their brains could potentially reveal how monogamy is represented anatomically.

And after the researchers peeked inside the voles' brains, these anatomical distinctions became clear. The two species' dopamine systems are wired quite differently. In female prairie voles, the pleasure and desire circuitry contain large numbers of receptors for a hormone called oxytocin. In male prairie voles, this circuitry has many receptors for both oxytocin and another hormone called vasopressin. But in montane voles, it's another story. Both male and female montane voles have far fewer of these respective receptors in the relevant regions.

In terms of behavior, this matters tremendously. As Carter and her colleagues discovered, oxytocin is critical to the social lives of mammals. "It's taking over parts of the nervous system and putting information into them about a sense of safety and trust," she told me. Without oxytocin, mice cannot tell friends or family from strangers—and mothers do not learn to nurture their young. Also, the particular distribution of oxytocin and vasopressin receptors in the pleasure systems of prairie voles is what makes them monogamous. The specific geography of these receptors allows the memory of a special partner to be wired into a vole's brain, making him or her "the one and only." This happens during mating, linking the scent of the partner with the pleasure of sex and the comforts of home. Later, when that partner is present, the stress system will be calmed and dopamine and opioid levels rise. In contrast, when the partner is absent, stress rises and withdrawal symptoms ensue. While some prairie voles still "cheat" with other partners, they don't typically leave their "spouses" for their "lovers" or "mistresses."

Montane voles, in contrast, are not wired for monogamy. Sex feels good to them, of course, but the particular partner doesn't matter. Montane voles don't

have enough receptors for oxytocin or vasopressin in their pleasure regions—
so they never link the memory of a specific mate with the joy of sex. For a montane vole, any attractive member of the opposite sex will do. Only sexual
novelty, not familiarity, brings pleasure in this species.

Humans seem to be wired more like prairie voles. We form pair bonds, but
we can also enjoy sexual variety. The relevance of the oxytocin and vasopressin systems for human bonds can be seen in genetics: studies show that differences in genes for oxytocin and vasopressin receptors play a role in the way
people handle relationships and in conditions that affect social skills, like autism. For example, some research suggests that a variation in the vasopressin
receptor gene in men—similar, but not identical to, the one that affects monogamy in male voles—is associated with a 50% reduction in the likelihood of
marriage and poorer quality marital relationships in those who do marry.

In both men and women, oxytocin is copiously released at orgasm. It also
spikes during childbirth and nursing, causing uterine contractions during labor
and, later, the "letdown" of breast milk. At the same time, it helps bind parents
to their particular babies. Consequently, oxytocin has become known as the
"love hormone" or "cuddle chemical." (Vasopressin is far less well researched
and understood, but it is critical to bonding in male mammals and also seems
to spur aggression against competitors or intruders who might harm young or
mate with females.)

Oxytocin apparently teaches us who is friendly or at least familiar—and
who is not. Chemically, it seems to help wire the memories of our loved ones
into the programming of our pleasure centers. Unfortunately, oxytocin can apparently also do the same for the memory of drugs. In drug addictions, rather
than associating a person with stress relief and pleasure, those connections are
made with the drug. Intriguingly, in fact, a few small studies suggest that oxytocin may relieve withdrawal symptoms in heroin and alcohol addictions—
perhaps as a result of this wiring and the link between love and addiction.

Oxytocin on its own doesn't seem to cause pleasure or desire, however. At
least when given as a nasal spray (the best method we now have that doesn't
require brain surgery), it's indistinguishable from placebo. But it does subtly alter behavior. For example, studies have shown that oxytocin elevates trust and
helps autistic people to more accurately detect other people's emotions. It's not
entirely benign, though: while oxytocin strengthens bonds and encourages al-
truism among "us," it also elevates hostility toward "them," with studies showing that it can increase racism or other types of discrimination based on who

we see as part of our in-group and who we see as outsiders. Oxytocin makes social signals stronger and more memorable, but it doesn't necessarily make them feel good.

Instead, the pleasure of love, sex, and social connection comes from our old friends dopamine and opioids. Oxytocin is part of what allows us to connect happiness with particular people, parent to child, friend to friend, and lover to lover. The happiness itself, however—and the comfort, relaxation, and warmth we feel in love and with our loved ones—comes at least in part from our endorphins and enkephalins. In contrast, the desire for and motivation to be with those we love is probably more dopamine driven: studies find that blocking certain dopamine receptors prevents prairie voles from forming partner preferences, for example. As we saw earlier, "wanting" is more linked to dopamine, while opioids seem to be involved in both wanting and liking. By linking the release of these substances to the presence of our loved ones, oxytocin makes us want and like them, and it may do the same for drugs.

The way oxytocin wires social connections is highly context dependent, however. This makes evolutionary sense. Contra Freud, it's not likely that a species of mammals that found its own parents to be their first choice in sex partners would thrive. Such interbreeding would rapidly lead to genetically defective offspring. However, the father of psychoanalysis wasn't totally wrong, either: the chemistry of romantic and parental bonds is quite similar. Preferences formed in childhood do affect romantic predilections in animals and humans as well.

For example, adult male rats who were stimulated with a paintbrush to simulate maternal licking while exposed to a lemon scent during infancy ejaculate more quickly (presumably meaning that they are more excited) when paired with similarly scented females. Childhood play with peers also affects sexual arousal later—female rats that played with either almond- or lemon-scented female playmates as pups later preferred to mate with males with the familiar scent, rather than the unfamiliar one. The specifics of how oxytocin and other chemicals vary when making sexual, parental, and other social bonds are not known, though clearly these bonds are not identical.

In addition, the type of nurturing a child receives enormously affects the development of the bonding systems. As noted earlier, having a highly affectionate and responsive mother turns on different suites of genes, compared to being raised indifferently. Not surprisingly, neglect and trauma make social connection more difficult. These changes, too, are mediated by oxytocin, vasopressin, opioids, and dopamine. They affect not only the way the next generation parents its own offspring but how they relate to others more generally. For instance, women with borderline personality disorder—a condition marked by extreme emotional reactivity and clinginess alternating with hatred or cold rejection—respond to oxytocin by becoming *less* trusting in experiments involving cooperation, and this is related to their levels of childhood neglect and sensitivity to rejection. Both chemistry and environment play a role in how you learn to love and who you learn to love.

During infancy, oxytocin focuses your brain on remembering the characteristics of the people who raise you and linking these cues with stress relief, even if your caregivers are inconsistent or cruel. Consequently, if you grow up in a violent and/or neglectful home, this may skew your reaction to later romantic bonds. In essence, oxytocin teaches you what to expect from a partner. If your parents are warm and reliable, you learn to expect that in romantic relationships. On the other hand, if you learn that affection is barbed, you may find it hard to recognize love in a healthier context; in fact, you may gravitate toward brutal or uncaring partners. You may also compulsively seek drugs because you discover that they provide a sense of being truly loved that you can't find elsewhere. If your bonding systems are miswired for any reason—genetic, environmental, or both—you also may not be able to feel the love other people actually do have for you and may also look for relief in drugs.

Indeed, because oxytocin wiring depends on both genes and the environment, it varies incredibly widely. This complexity is magnified because it differs not just between people but changes within individuals as they form relationships over the course of development. Since oxytocin wiring is basically designed to addict us to each other, it also plays a critical role in addictions. Like love, addiction is learned in light of a particular developmental context; your childhood affects your risk for addiction in part because of how it affects the way you experience love. This means that each addiction is as individual and particular as each love, making the experience of addiction and the road to recovery immensely variable. Moreover, to love, you typically have to persist despite negative consequences—as Shakespeare put it, the course of true love never did run smooth. It is a rare relationship that never requires compromise or perseverance.

Love really is a drug—or rather, it's the template for addictive behavior.

* * *

GIVEN MY LIFELONG difficulties with relationships, it's hardly surprising that I wasn't exactly lucky in love, at least early in my life. During my addiction and recovery, this was made even more difficult by the codependency movement of the 1980s. This movement, which grew along with Al-Anon, the 12-step program for family members of alcoholics, was based on flawed psychology and wound up promoting harmful ideas that persist in addiction treatment and policy to this day. It recognized the link between love and addiction—but in a peculiar and ultimately damaging way.

The idea of codependency itself—that some people are overly dependent on their partners and try to escape their own issues by trying to solve other people's problems—is relatively uncontroversial. The stereotype of the wife of an alcoholic who makes excuses for him while trying to convince him to quit does reflect some degree of reality. Such people clearly exist, they are frequently involved with addicts who provide lots of distracting problems, and it can be therapeutic to recognize and address the way that the "controlling" behavior that can result can be counterproductive.

But the codependency movement stretched this idea to dangerous extremes. Because addiction was defined as a disease, codependency became one as well. No one, however, was ever able to come up with a diagnostic tool that reliably distinguished between "codependents" and those without the disorder. Moreover, the problem of codependency was soon combined with the idea of "tough love," which diagnosed nearly any caring behavior toward people with addiction as "enabling" their drug use to continue. As a result, codependency counselors and Al-Anon members recommended withholding love and material support. Add this to an individualistic culture where any type of dependence on others is seen as weakness and you have a recipe for pathologizing normal human needs while increasing the pain and stigma associated with addiction. Indeed, at the peak of codependency's cultural currency in the early '90s, some psychologists declared that 94% of all relationships were dysfunctional, and a popular cartoon showed a convention for "Adult Children of Normal Parents" as being sparsely attended.

In my case, even before I became addicted, I was already at sea with the ambiguous rules of mating and dating, and I foundered. I didn't understand how to flirt, for example; I saw it almost as a form of lying because it didn't reveal your true intentions. As I preferred to approach everything directly, I would instead throw myself at men I liked, with typically disastrous outcomes— hilarious only in retrospect. My emotional intensity, belief that intelligence should be seen as an asset (sadly, often untrue for women, particularly young ones), combined with my directness tended to terrify guys, who would rapidly retreat into a "Let's just be friends" or "Let's just have casual sex" stance. I can look back on my adolescent and young adult self with compassion now, but at the time, each rejection only compounded my negative view of myself.

This strategy, combined with my tendency to, shall we say, persist despite negative consequences, did not serve me well. From my first lover onward, it meant that I tended to wind up with men who were "just not that into" me—often guys who would explicitly say that they didn't love me. I would tolerate almost any type of misbehavior, glad simply for some form of sex and affection, and these pairings would stretch into years-long relationships. To make matters worse, I was so desperate for connection that I couldn't allow myself to see any of it as unhealthy.

Indeed, I came of age romantically in what was probably the least romantic time in American history: the late '70s and early '80s, when the idealism of the '60s had diminished into the sleaze of the "let it all hang out" era. This was the time when divorce peaked—and when leaving your wife or husband and your kids to "find yourself" was not only not stigmatized but idealized. The ethos was totally confusing to teenagers, especially girls. We saw grown men lusting after fifteen-year-old Brooke Shields as she purred, "Do you know what comes between me and my Calvins? Nothing." We were told we should seek our own pleasure, while the Knack sang "Good Girls Don't." We had none of today's awareness of sexual harassment and date rape—catcalls and men and boys trying to have their way with you, no matter what you wanted, were simply part of life. Or, you were supposed to always want it, to not have "hang-ups." There was no template for "dating"—the word itself was old-fashioned and seemed a relic of the *Happy Days* '50s. Sex was everywhere—but love seemed like some type of delusion, at best.

It's not surprising, in this context, that Peele and Brodsky's *Love and Addiction,* with its strong claim that love was a form of addiction, became a bestseller in 1975. The authors' aim was to destigmatize addiction by comparing it to the most natural, healthy emotion of all: love. But given the cultural era in which it appeared, instead of making addiction seem less pathological, it wound up making love seem more so.

The book was eagerly seized upon by the founders of the codependence movement, who saw it as endorsing their ideas about the sickness of people who had relationships with addicts or were raised by them. By 1991, Robin Norwood, author of *Women Who Love Too Much* (1985), had called it "required reading" for every such woman. Melody Beattie, who wrote *Codependent No More* (1986), was another key figure in the movement, which grew as the drug panic of the '80s and '90s raised fear about everyone touched by addiction. Being obsessed with your lover soon became a sign that you had the addictive disease of codependence; wanting to spend all your time with a new partner was abnormal, not healthy. Any addiction-like behavior during love was a flashing danger sign; you were supposed to end relationships with any obsessive qualities to nip codependence in the bud.

Because it was rooted in 12-step programs, the codependency movement was also deeply committed to the idea that addiction is a disease. And if addiction was an illness, then sick codependent love had to be a medical disorder, too. The role of learning and culture and how they interact with biology and psychology was ignored. "It is a sad irony for us that our work contributed to labeling of yet more 'diseases' over which people are 'powerless,'" Peele and Brodsky wrote in a 1991 preface to an edition of *Love and Addiction*, printed when codependency was all the rage. While they had wanted to show that normal love could go awry in a compulsive and life-contracting way—just as drug use can—instead, their work was interpreted to mean that all relationships were mere addictions and most love was delusional and self-centered.

This logic fit in with the highly individualized zeitgeist of the 1970s and '80s in America. Psychology, having rejected evolutionary ideas about human behavior because they were tainted by associations with racism, sexism, and eugenics, saw individuals as completely self-sufficient. Biology was irrelevant. You didn't need anyone else to be happy; you just needed to actualize yourself. As some feminists put it, a woman needs a man like a fish needs a bicycle—and as the self-help saying has it, "you can't be loved until you can learn to love yourself." While some of these ideas were a necessary corrective to inaccurate biological determinism, they went way too far.

Human biology doesn't dictate lifelong gender roles, nor does it illustrate any meaningful racial differences (there are far greater differences within "races" than between them, genetically and otherwise, for one). But biology does make all of us an ineluctably social species. We now know that we are fundamentally interdependent, psychologically and even physically. A baby, for example, literally needs to be held and cuddled for his stress system to be properly regulated; without repeated, loving care by the same few people, infants are at high risk for lifelong psychiatric and behavioral problems. Before this was widely known, in fact, one in three infants raised by rotating staff in orphanages died—essentially, from lack of individualized love. This was only discovered in the 1940s, when Rene Spitz, a psychoanalyst, compared infants raised by nurses who rarely touched them in sterile hospitals to babies raised by their own mothers, who were prisoners but were allowed to keep their babies. All of the prisoners' children thrived—but one third of the poor babies kept in a supposedly safer hospital died, and many were profoundly delayed in developing language. Without the obsessive, physical love of parents, either biological or adoptive, infants quite literally fail to grow and physically waste away.

And while a romantic relationship isn't necessary for health, having at least some close relationships is. Research finds that loneliness can be as dangerous to health as smoking and more harmful than obesity, in fact. The more and higher-quality relationships a person has, the more mentally and physically healthy they tend to be—and that's not just because people want to be friends with those who are healthy. Improving relational health improves health in general, for both children and adults.

When I was addicted in the late '80s, however, human interdependence was not understood as well or as emphasized in psychology. Passionate love was decried as addiction and any need for others was suspect, especially in someone who was already an addict. Back then, any relationships an addicted person had were presumptively labeled "codependent"—even if the person was in recovery. The idea that a person with an addiction could be genuinely loved for herself was mocked.

Although research rapidly showed that there is actually no way to scientifically distinguish between people with "codependence" and everyone else, the idea continued to spread and is still taken seriously by a distressingly large percentage of professionals in the addictions field today. In fact, *Codependent No More* remains a bestseller among books about addiction. While it's certainly true that some people behave addictively in their relationships, it's hard to say more than that about it with any data firmer than anecdote. There is no "codependent" personality, no "brain disease" of codependence, no predictable course of any such "disorder." A "diagnosis" of codependence is about as scientific as a horoscope—and far less entertaining.

For me, at least, it did great harm. Once I started recovery, I was told that basically *anyone* who was involved with me was merely playing out codependence. That meant there was no way of distinguishing between good relationships and bad ones. And so, since I knew I needed love, I ignored signs that any of my relationships were unhealthy. I figured scraps were better than nothing at all. I failed to realize that the same thing that distinguishes addiction from passionate interest also divides unhealthy love from that which is the highest experience of humanity. That is, love is real when it expands and enhances your life—and troubling and problematic when it contracts or impairs it. Whether you love a person, a drug, or an intellectual interest, if it is spurring creativity, connection, and kindness, it's not an addiction—but if it's making you isolated, dull, and mean, it is.

Contrast this with the "codependence" account, where obsessive, passionate love can't be "real" love—and Romeo and Juliet had a disease. More sensible, I think, is to see it like this: obsessions can get out of hand, but love is inherently obsessive and needs to be that way to keep us bound to each other. When you are in love, in fact, the levels of serotonin in your brain fall and are comparable to those seen in people with OCD. Having low levels of this transmitter isn't always bad, though. It is healthy and normal to be obsessed when you meet a potential life partner. If this love isn't reciprocated or the person is abusive, that's another story—but obsession itself doesn't make love an addiction.

If all-encompassing passion is itself pathological, true love is a disease. To me, this seems not only demeaning and dehumanizing but silly. Dismissing one of the greatest human sources of meaning and joy—and yes, pain and loss, too—as illness is no way to help people. And what it says to those of us with addictions is that we are the unlovable monsters we feared we were. Not helpful.

I haven't even yet mentioned the feminist critique of codependence, which is also important and accurate. On top of all the other problems, labeling caring as codependence particularly pathologizes behavior that is typically associated with women. It makes us the problem for "loving too much" rather than recognizing human interdependence and normal relational needs. Indeed, calling the official diagnosis for addiction "substance dependence," which was the label for the condition in the editions of the DSM published during codependency's heyday, subtly implies the same thing. "Dependence" itself is pathologized, when, as we've seen, dependence isn't the real problem in addiction: compulsive and destructive behavior is. (And indeed, the DSM-5, published in 2013, recognizes this, replacing "dependence" with "moderate to severe substance use disorder.")

Further, saying that a woman who tries to help her alcoholic husband by hiding his drinking from his boss has a "disease" because what would really be better is for him to face the consequences is not only inaccurate but incoherent. Trying to help someone you love, even ineffectually, is admirable, not sick. In fact, as we shall see, the idea that being kind and "enabling" addicts does more harm than good is, itself, damaging. Healthy relationships are essential to recovery: while love isn't always all you need, without it, few people get better. Love can't always cure addiction—but lack of it or inability to perceive it often helps cause addiction. Compassion is part of the cure, not the disease. Our societal belief that toughness is what works instead is a huge part of why our drug policy is so disastrously inept and harmful.

Of course, determining which relationships are healthy and which aren't, what qualifies as too much passionate involvement and what is appropriate, and figuring out whether someone is enhancing rather than diminishing your life isn't always easy. Human relationships are messy and the trade-offs aren't always clear. When your friends hate your new partner, it's hard to tell if they are picking up on her social awkwardness or seeing a troubling flaw that you have missed. Similarly, determining whether your obsession with rock climbing is healthy or dangerously escapist isn't always simple. Figuring out whether the risks outweigh the benefits—with any type of passionate engagement—can be tricky.

Obviously, these determinations will ultimately also be highly subjective, which is part of why addiction is so difficult to define and address. Addiction and love are both deeply culturally constructed. This means not that they aren't real but, rather, that they can't be defined in some generic context, that they must be understood where and when they occur specifically. The exact same type and level of drug use may be healthy for one person in one place at one time—and unhealthy for another or even for the same person in a different situation. A glass of wine a day may be healthy in middle age, for instance, but not if you can't stop when you take medication that shouldn't be mixed with alcohol. Similarly, the same type of passionate attachment that is healthy in one relationship may be harmful in another.

And all attachment, of course, is molded by and requires learning. Critically, the learning that occurs in addiction or love is distinct from the way, say, you learn a fact about history or science. The role of oxytocin, dopamine, and opioids in wiring future cravings to past memories of our passions means that we learn love and addiction much more permanently than we do things we care less about. Part of the function of emotions themselves is to carve important experiences into memory, so learning love or addiction is visceral. These experiences aren't just stored like other memories; the changes are deeper, longer lasting, and more pronounced. That's is why it's a lot easier to remember your first love than it is to remember the specific material you were taught in school the year you met him. Love and addiction change who we are and what we value—not just what we know.

AFTER MY SPECTACULAR first experience with heroin, it wasn't long before I became addicted to it. At first I made a half-hearted attempt to limit my use to weekends, but it wasn't long before those "weekends" began to meet in the middle. At the time, I worked for an independent film producer whose office was on West 23rd Street; I'd go across town to the Carlton on my lunch hour, score from Pablo and Gigi, and return to work in a much better mood. Since I wasn't doing anything more complicated than filing and running errands, it didn't seem to interfere with my performance. However, my boss was notoriously difficult, typically going through an assistant a month. I lasted three and couldn't tell you whether my heroin use hurt or helped or made no difference whatsoever.

It did make me less difficult—or at least less clingy—in my relationship with Matt, however. I now had a much more reliable source of the opioid safety and comfort that I'd sought in him; that made me need him less. Heroin also provided a whole new subculture to explore. In fact, the first time I experienced withdrawal symptoms, I was actually somewhat excited, because I'd read so much about what it was like and wondered how I'd feel going through it myself. The first time I started to become shaky and sweaty when I didn't have heroin, I was oddly proud, feeling as though I was now part of an exclusive club, like Lou Reed and William S. Burroughs. In fact, the first time I kicked the withdrawal symptoms, it really wasn't worse than having a moderate flu—and so when I started using again, I thought it would always be that easy. Instead, it got tougher each time—both physically and mentally. And I soon decided that if I was going to be a proper junkie, I'd really have to start injecting it, rather than snorting it.

TWELVE

Risky Business

THE MAN WHO GAVE ME MY first injection was skilled at finding small veins. He had been a veterinary technician at one point, putting cats and dogs "to sleep" as painlessly as possible. But I had to work hard to convince Keith to "give me my wings," as losing one's IV drug virginity is known among the cult of the spike. The moral responsibility for such an introduction can weigh heavily— and because of the sexual overtones of the act, Keith wanted my boyfriend's permission before he would proceed. As a feminist, of course, this outraged me. But as a fellow drug user, I also completely understood the desire to avoid pissing off an important connection.

It was winter, in late 1985 or early 1986. I was 20 and still trying to escape my despair over my suspension from Columbia, often with what appears to me now as reckless bravado. Matt and I were sharing a one-bedroom apartment on 49th Street between Sixth and Seventh. It was in one of the very few residential buildings in an area dominated by the art deco towers of Rockefeller Center and the looming Time-Life and Exxon skyscrapers.

That day, I sat on our double futon bed as Keith prepared a dose of cocaine; heroin injecting would come later. Matt stood in the corner, near the scale where we weighed out the coke. Keith skillfully slid the thin diabetic needle under the skin of my forearm. It went in easy, like it was made to, and there was sudden red in the barrel. I was startled by the nonchalance of it. For half an instant I felt as though I was out of my body, watching him inject someone else—but then as he pushed the plunger home I was jolted by the familiar taste of cocaine, cool and chemical. A metallic taste appeared on my tongue as if by magic. My mind was rapidly overcome by a crystalline euphoria, a bliss that was surprisingly satisfying and completely free of craving. I felt perfect. The tiny red mark that remained in the crook of my arm was a badge, demonstrating that I had earned my wings.

And so began my romance with the needle, one that would soon see me shooting up dozens of times a day. Because of cocaine's effects on "wanting" rather than "liking," injecting it soon became the most compulsive experience I have ever had; it was only the first few times that I shot up that I didn't experience overwhelming craving for more within minutes or less. Both injecting powder and smoking crack (which cannot be injected because it is not water soluble) are far more intense and drive far more desire for rapid repetition than snorting. It's never again like it is at the start.

Once I'd broken the injection barrier, of course, I wanted to shoot heroin, too. Within weeks, I had discovered what became my drug of choice: the speedball, which is a shot containing a mixture of both, injected simultaneously. Mystifyingly, even though they both went in at once, the cocaine would always hit before the heroin, blasting its presence with that weird taste and a heartpounding feeling of triumph and power. Then the heroin would flood in, calming, soothing, and peaceful.

I gave myself over to it completely, soon engaging in rash and potentially deadly behavior that I now cringe to consider. During the Grateful Dead's spring tour of 1986, for example, not long after I'd had that first shot of cocaine, I paid for a suite at an expensive hotel in Philadelphia. The first night they played, I went to—and enjoyed, as usual—the show. But the second night, even though I had backstage passes due to our ongoing connection with the band, I didn't even venture out to hear the music I'd supposedly traveled from New York to see. Drugs had overcome all of my most passionate interests. I stayed in my hotel.

Some of my friends had large quantities of heroin; I had the cocaine. Because I didn't yet know how to inject myself, I had others do it for me—in this case, a long-haired guy named Chris and another young man who went by the street name of Ignatz. Wanting another shot when both were busy, I eventually went into the suite's large, marbled bathroom with a needle. Amazingly to me, I easily found a vein, and that was the last day I ever relied on other people to inject me.

Ignatz, a blond, charming junkie who could and repeatedly did convince Matt and me to front him drugs when we knew the odds of him ever paying us back were minuscule, figures in another appalling and risky binge I conducted around this time. As a trade for cocaine, I'd obtained a large quantity of the opioid painkiller Dilaudid—a drug that is about twice as potent as heroin. To see what it was like, I began shooting it mixed with both heroin and cocaine, which added an odd but not unpleasant feeling of pins and needles rushing through my body to the exhilarating rush of the cocaine and the tidal wave of calm from the heroin. And I just kept doing it.

After hours of these extremely strong injections, I apparently overdosed. I began humping the bed and yelling "oh babies" and then ceasing to breathe in between these furious bouts of activity. Ignatz, who was sitting in our living room while Matt and I were in the bedroom, quite logically figured we were having sex when he heard these strange cries. But after a much longer time than an ordinary sexual encounter would last, Matt emerged from the bedroom—leaving Ignatz quite curious about what was going on.

Soon, they both became concerned, trying to shake me and ensure that I stayed conscious. Matt later told me that they thought I might be having some type of seizure. I remember little of this beyond shooting the drugs and having a vague sense of a compulsion to move repeatedly. But then, I woke up in tears with a sense of hovering outside my own body in the bathroom, in complete despair because I thought I was dead. Eventually, I must have passed out. However, I woke the next day, apparently unscathed other than being completely unable to recall anything between the bathroom experience and the last shot I took.

Looking back, I can't believe how lucky I was. Either of those incidents could easily have resulted in a fatal overdose or profound brain damage. Any injection for which I used someone else's needle could have infected me with HIV, hepatitis C, or other potentially deadly diseases. My life would soon be altered by meeting someone who taught me to protect myself from blood-borne disease—and that would spur the activism that also helped fuel my early recovery.

But before that, this young woman who spent her life terrified of tame experiences like roller coasters and even driving, who had no interest in riskier activities like skydiving or mountaineering, and who was utterly petrified by the idea of mortality took overwhelming risks with drugs—with virtually no thought or consideration of her own safety. Just writing this now fills me with shame and terror. I still can't understand my own behavior. My actions, however, do seem completely in line with what we now know about how the adolescent brain places an oversized emphasis on the value of rewards with little real recognition of risk. And the way it learns to negotiate risk behavior is a critical part of how addiction is learned.

CORNELL'S VALERIE REYNA studies how young people's risk decisions can go awry—and her conclusions are counterintuitive. Her research suggests that the major reason that teens and young adults are unreasonable about risk is not that they are too emotional when they consider it—but rather, that they are too rational. Although my own behavior during my late teens and early 20s now seems completely irrational to me, her work helps make sense of it.

As noted earlier, studies find that adolescents often significantly overestimate their odds of bad outcomes from activities like sex and drug use. For instance, when asked to estimate the odds that a teenage girl who is sexually active will become infected with HIV (in a study conducted in the '90s, when AIDS was not as treatable as it is now), the average guess was 60%. The actual risk was less than 1% for an American teen in most parts of the country.

But even absurd overestimations like this don't deter youth. And that's not because they don't consider them. Instead, there are two important factors. First, young people do tend to weigh immediate benefits more heavily: the visible prospect of pleasure literally looms larger in their minds than anything else that might happen later. Second, adolescents get lost in deliberation when they do consider negative consequences—and being out in the weeds doesn't tend to spur good judgment. Interestingly, this isn't a problem limited to teens and risk decisions; it's a difficulty seen with any type of inexperience or lack of expertise. If you haven't encountered a similar problem before, you don't know which factors matter most.

That's because of the way the brain learns to process information. When you first learn any process, you have to consider what you do carefully, think through every step deliberately, and monitor yourself closely. But once you become experienced—whether with dancing, decision making, drug taking, or calculus—your thinking becomes much more automatic.

Your brain ultimately calculates the "gist" of the data or behavior and offloads its processing to less conscious and, ironically, more emotional brain regions. This is why "overthinking" can interfere with athletic or artistic performance. Once you know what you're doing, your expertise lies not primarily in conscious consideration, but rather in what your brain and body are now able to do *without* thought. Research on expert decision making among doctors, for example, shows that the best physicians actually consider fewer variables when they make good choices—their gut tells them what to ignore. But this "gut" has to be trained by years of experience of making choices by using data.

As a result of the learning that creates these emotional algorithms, when adults think about risk taking, they tend to automatically get a bad feeling that immediately says, "No way!" Their brains have had years of experience with making choices and can now highlight the worst potential negative consequences rapidly and unthinkingly by creating emotions, not simple thoughts. Emotions, in fact, are probably best described, as the neuroscientist Antonio Damasio pointed out, as decision-making algorithms honed by eons of evolution. The emotions we feel now are the ones that prompted our ancestors to make the decisions that increased their odds of survival and reproduction. From fear and pain to love and desire, our feelings were built to guide our actions, and they incorporate the results of experience over time.

These emotional algorithms, of course, are largely unconscious. But like much about the brain, they require experience for development, and those that help us make good choices about risk need training. I can tell that mine are in pretty good working order now, in fact, by the shudder that runs through me as I try to write these scenes and consider all the harm that could have befallen me and those around me back then. In contrast, teens and young adults haven't developed this rapid gut-level calculus. Instead, they "rationally" and deliberately think through the odds of success in things like playing Russian roulette, drinking Drano, or setting their hair on fire. One fascinating study in which adolescents were asked to consider whether such absurd and dangerous acts were a good idea found that it took teens a full sixth of a second longer to say no than it did adults. That seems like a tiny amount of time, but it is long in terms of much that goes on in the brain. And in that sixth of a second lies a world of experience, one that no one has been able to induce faster than it takes teens to actually grow up—if they survive their inevitable bad choices.

Ironically, the same kind of process is the essence of addiction. Drug taking starts as a rational, conscious choice and through repetition becomes an automatic, unconsciously motivated behavior. Addicted people, unfortunately, seem to offload their ongoing choices about taking drugs to the now "expert" systems that handle unthinking actions, just as musicians no longer have to think about the mechanics of producing the notes they want to play. But the unique property that addiction has as a learning disorder is that, unlike playing music or learning math, addiction changes the values that govern decision making—in favor of getting high.

WHEN I THINK about how insanely I behaved at the height of my addiction, I can see how strongly my choices were skewed. Of course, there were all types of factors that went into it: my own fears and pain, my social difficulties and beliefs about being unlovable, my suspension from Columbia, and, yes, a desire for pleasure. Part of me sometimes felt as though if I wasn't going to be able to be extremely good, I might as well be extremely bad. The desire for immediate relief obliterated any consideration of long-term outcomes, even deadly ones—and I certainly did rationalize my relationship to risk by thinking I was smart enough to take steps to minimize it. (And that wasn't simply fantasy: doing things like using clean needles when I wasn't ready to stop taking drugs actually did save my life.)

The drugs themselves and the process of addiction, moving from freely choosing to get high to feeling more and more compelled to do so, were clearly also important. Before I was able to get enough experience to make risk judgments in an automatic and sensible way, my brain was already being influenced by another set of automatic processing algorithms built through frequent drug taking. These changes increased my perception of the value of drug experience, while lowering the weight given to other considerations. Although intellectually, I still cherished the idea of future success in love and work, emotionally, it became increasingly hard to make the choices that would have maximized the odds of a productive and connected future. I consistently chose what felt better *now*.

One economic theory of addiction, in fact, suggests that the condition can be explained almost completely by the variance in values placed on present and future between addicts and others. Originally proposed by psychiatrist and behavioral economist George Ainslie, this theory suggests that addictive behavior occurs when people repeatedly choose pleasure now without worrying about future pain. And it's certainly true that addicted people do consistently overvalue current pleasures—and that they do so while continually underrating those that could be better in the future if gratification were delayed.

Ainslie's theory also nicely accounts for why poverty, chaos, and trauma would increase addiction risk. In such situations, it is actually rational to be tightly focused on the present, because your own experience suggests that a positive future may never come. If the world is unpredictable and people are unreliable, an available reward now *is* more valuable than an uncertain gain later. In these cases, children who eat the one marshmallow in front of them rather than waiting for two later are actually making the right choice, given the environmental constraints they typically face. One study, for instance, found that when children were given the option to have one treat now or two treats later by either experimenters who had previously kept their promises or those who had not, they sensibly scarfed down the first sweet if their prior experience was that the researcher was flaky. Being present focused and prioritizing whatever good experience you can actually get right now is certainly part of what can lead to addiction—but it can also be an effective survival strategy in an uncertain world.

However, I don't think being too present focused is the whole story. In my case, for instance, while I certainly did make some choices that risked my entire future for passing pleasure and relief, it wasn't as though I was *unable* to plan or delay. During my addiction, for much of the time, I was able to attend college, study for tests, and write papers—meeting deadlines as well as or even better than some of my fellow students. But as I became more addicted, I also became *less and less capable* of making good decisions—even though I clearly retained some ability to plan and delay. What's tricky about addiction is that choices become less freely made. They aren't entirely automatic, though—even in the worst parts of it.

This leaves room for people to argue extremes. People with addictions typically don't shoot up when the police are watching; we often work quite hard to hide our behavior. We also specifically plan things like drug deals by taking steps to avoid detection. Consequently, some claim that we are choosing entirely freely, since we clearly are weighing at least some factors and their potential downsides into our choice process. In contrast, because we also do things that are quite incredibly self-destructive—like shooting dozens of speedballs without considering overdose risk—others argue our brains are "hijacked" and our actions entirely determined.

This makes the question of addiction and choice profoundly vexed. To me, a good analogy for the way will is hampered—but not eliminated entirely—is as follows: Consider a prisoner who is locked in a cell that contains a completely hidden trapdoor to an escape route. On the surface, there is no way out: the bars are strong and placed tightly together, the walls are stone, and the door is securely locked. The window is out of reach, and too small and barred, besides.

If the prisoner does not know that the possibility of escape exists via the trapdoor, she is not "free" to choose it—even though another prisoner who does have that information can easily liberate himself. By analogy, while addicted, there *are* alternative behaviors available to you—and you sometimes even recognize that they do exist—but you simply can't enact them or believe with enough conviction that they will genuinely help to power yourself through the necessary changes.

This situation is maddening to people who love someone with addiction: you watch them destroying themselves in ways that seem unimaginably selfish and stupid, but despite all efforts, they don't stop and often can't see the hurt that they cause. In my case, depression and self-hating beliefs (and perhaps Asperger's) made it difficult for me to believe anyone genuinely cared for me. I literally couldn't take in love that was actually present, no matter how clearly expressed. It didn't reach me, perhaps because depression had lowered my ability to feel the pleasure of relationships.

I know this was part of how my father's depression worked. When I was young, he was constantly critical of my work and behavior, and I felt that no matter what I did, it "wasn't good enough." I'd show him a test or an essay that I'd gotten an A on; he'd focus on the areas where it could be made even better. I'd play the piano or draw a picture and he'd always point out some remaining flaw. If I tried to apologize for some childish infraction, it always seemed that his response was not entirely forgiving—no matter what I did to try to make him feel better, it wasn't enough.

I didn't realize this was not just perfectionism; it was also linked to the fact that *he* couldn't experience much pleasure. In reality, it wasn't that my efforts were always poor, it was that he often couldn't feel good or genuinely proud, no matter how good the people and world around him were. Unaware of this deficit, he often blamed his poor moods on flaws in the work or in those who couldn't figure out how to please him. Exacting standards like this can spur achievement, of course, but the insatiability of anhedonia can create neediness that drives people away or makes them miserable, too.

Similarly, during my own depression and addiction, the love that my parents, friends, and siblings did have for me was actually there. My father, in fact, relentlessly reached out to me, refusing to give up even when I reacted in anger. When I was at my worst, from 1986 onward, he'd still call and visit, even when I tried to dodge him. During that same time, I'd spend hours crying on the phone to my mother, trying to get some kind of reassurance from her and some sense that I'd be able to find my way again. But I couldn't take love in, couldn't feel any of the acceptance or the comfort normally given in relationships—except, sometimes, through drugs. And taking them was exactly what the people who loved me wanted most for me to stop doing. I loved my parents—and they loved me—but I still couldn't quit.

> *A mother's love for her child is like nothing else in the world. It*
>
> *knows no law, no pity, it dares all things and crushes down*
>
> *remorselessly all that stand in its path.*
>
> —Agatha Christie

While the love of a child for a parent is obviously deeply engrained and inseparable at first from the utter dependence infants have on adults for their survival, the love of a parent for a child requires more understanding. Newborns are noisy, demanding, often stinky, and, to many people, not even particularly cute. Not only are they unable to feed or care for themselves, they can't even smile or laugh—and the experience of birth for most mothers is hardly pleasant. That doesn't even begin to get into the terrible twos—let alone the teens. What could possibly induce parents to feed, change, and nurture these painfully delivered needy beings for the decades it takes for them to become fully independent?

The answer, of course, is the same reward, motivation, and pleasure systems that drive romantic love and addiction. Indeed, the origins of the dopamine, oxytocin, vasopressin, and opioid systems in parenting are seen earlier in evolutionary history than they are for monogamy. Before there was pair bonding and marriage, there was parenting. The ability to form romantic relationships has similar chemistry to that of the parent-child bond not to keep psychoanalysts employed but because evolution rarely builds new systems when it can co-opt existing ones.

While metaphorically, as noted earlier, the connection between romantic love and addiction is blatantly obvious, parenting, even more than coupling, requires the core addictive feature of compulsive behavior despite negative consequences. Virtually all parents who are honest with themselves have had moments when they regret having had children—and studies of people's daily levels of happiness find that parents actually have less of it than nonparents. This includes lower marital satisfaction, reduced overall life contentment, and less of an overall sense of mental well-being. Surprisingly, these reductions in quality of life are not only seen during the toughest times when children are at their neediest; studies find it to be true throughout the adult lifespan of parents. Either something is missing in our measures of happiness (which is quite possible) or children often don't bring the pleasure that they seem to promise.

To be sure, it's also abundantly true that children bring many times of pure joy. As my own mother once noted, you rarely laugh more or hear more laughter than when you spend your days with small children. But just like in addiction, these good times in parenting are interspersed between much more trying experiences, often widely. Indeed, you probably never spend more time crying or witnessing tears and screaming than with little ones, either. Whether the difference in happiness between parents and nonparents is real or whether it stems from an inability to measure the real contentment found in the deep meaning of parenting remains widely debated. Whatever the data ultimately show, however, there's no doubt that having kids can be punishing and does require persistence despite that.

OF COURSE, TO those who see addiction as the lowest of immoral behavior and parenting as an ideal of selflessness, comparing the two is outright offensive. And this is true precisely because the comparison can't account for either the profound social meaning of parenting or the emptiness of many drug-induced pleasures. In fact, in the early '70s, when Jaak Panksepp, who discovered the role of endogenous opioids in the maternal/offspring bond, wrote his first paper on this topic, it was rejected by the most prestigious journals for exactly that reason. He recalled in an interview that the editor of *Science* told him that the subject was "too hot to handle." The journal didn't want to go out on a limb and compare mother love to what were seen as the filthy habits of despicable addicts. "If it was wrong, they didn't want to be responsible," he said.

Today, however, at least among those who study the chemistry of social bonds, the data have become so strong that the issue is seen as settled science. Exploring the learning involved in parenting offers yet more insight on the learning processes of addiction and their compulsive nature.

Even normal parenting, for example, involves a touch of OCD. New parents notoriously become obsessed with the safety of their children and with ways of protecting them. Nearly every new mom and dad can recount a story of checking on a sleeping baby, often repeatedly, just to be sure he or she is breathing. New parents often chastise their own parents for their apparently lax handling of infants, while the grandparents laugh knowingly about how they did the same to their own parents, with completely different but equally essential ritualized safety protocols.

In fact, women who suffer from clinical OCD itself tend to get worse during pregnancy and the early postpartum period, suggesting similar brain mechanisms for the behavior. And pregnancy causes the initial onset of OCD in a full 10% of women with the condition. Some data suggest a connection to the brain's serotonin systems, as is the case with romantic love as well. In new love, early parenting, depression, and OCD, serotonin activity seems to be reduced. And that probably occurs because basically, some degree of obsession is normal and healthy in sustaining critical relationships, like those of couples and those between parent and child.

Another crucial aspect of learning in early parenthood, of course, is learning to bond with your particular baby. Contrary to popular belief, parental love is not always "instant" and the capacity for bonding is not destroyed if the baby isn't immediately placed on the mother's belly or into the father's arms. While many parents do experience overwhelming feelings of love when their children are first born, at least as many do not have the "love at first sight" experience—and often feel guilty about this. But just like in addiction, it takes repeated exposure and repeated engagement for any type of love to become engrained and wired into the brain. Although some people may feel overwhelming desire to repeat an addictive experience immediately after trying it—just as some parents do feel the flood of love immediately upon seeing their newborns and some couples feel love at first sight—nothing is actually "instantly addictive" because learning takes actual, not just planned or wanted, repetition.

And babies, thankfully, come equipped with features that make them addictive to adults. The survival of the human species depends on it. Although they often look like otherworldly aliens with conical heads and smushed faces in photographs, when held in real life, new babies are eerily attractive. Evolution has kitted them out with a set of features for which our language is remarkably inept and which we trivialize without being aware of their life-sustaining importance. What we call "cuteness" and respond to most commonly with the nondefined verbal response of "Awwww" is, in fact, an evolutionary adaptation that primes caregiving. Although newborns are not at maximum cuteness—this begins at about six weeks, when they reward frustrated and exhausted parents with a winning (and, not coincidentally, intermittent) smile—they have the beginnings of all of its features. These include relatively large eyes, small body size, high-pitched voices, and endearingly clumsy behavior.

The similarity of these features across species is why we find most animal babies adorable. Simply seeing a baby smile or watching a puppy or kitten scamper playfully can bring immense joy. Even though culturally, we tend to diminish and dismiss such pleasures as cheap and sentimental, the reality is that they are the foundation for the human ability to form lasting connections, first with our parents, then with our friends, relatives, lovers, and, perhaps, children and grandchildren. The fact that addictions can be built on the same system is not an insult to parents or to the meaning of love—but a testament to their strength and power.

THIRTEEN

Busted

Before you can break out of prison, you must realize you are locked up.

—ANONYMOUS

I OPENED THE DOOR WITH A needle in my arm. Seven plainclothes narcotics cops burst in, five burly men and two women, all shouting. I hastily finished my shot and threw the works down, attempting to be discreet about it. I had been expecting my friend Lina, who should have been returning with money for the cocaine Matt and I had just fronted her. I was also suffering from a painful ear infection, which is how I'd obtained the drug I was shooting. It was Demerol, a narcotic I'd been prescribed by the Columbia Health Service. I must have been quite ill: the doctor had prescribed me an opiate as well as antibiotics, even though I'd told her that I had a history of heroin use.

Of course, I wasn't supposed to be injecting the Demerol. In fact, I'd actually managed up until exactly that point to abstain from drugs almost entirely for a few months, in hopes of being able to be readmitted to college after my "year off." Now, I was clearly off the wagon and life was about to get exponentially worse. My idea that I had recovered and could safely use drugs occasionally was about to be definitively falsified.

Before that awful day in September 1986, reining in my drug use had seemed relatively easy, at least theoretically. I wasn't exactly a fan of heroin withdrawal— I'd gone cold turkey about four times at this point—but dope sickness did not keep me from quitting during my rare attempts at it. Instead, I always got in trouble a few weeks after stopping, when I felt well again and thought, "Just one will be okay." This time, though, the brief period of abstinence that I'd now violated with the Demerol had been created by an even worse experience—one that I actually thought had solved my drug problem.

Through sharing needles, I'd contracted hepatitis A earlier that summer. (Typically, people get B or C via injection; somehow, I got A, which is usually spread by bad seafood.) My hep A infection, which is normally less severe than the other types, soon made me so sick that even healthy food seemed poisonous— let alone drugs. Broccoli, for example, was indigestible, and if I tried to eat even the tiniest bit of fat, I'd become overwhelmingly nauseous. One pizza craving ended disastrously in the bathroom.

In fact, the way I'd found out that I was sick was by discovering that heroin didn't work to salve what I'd assumed were withdrawal symptoms. And taking more heroin—from a batch that I could see worked fine for everyone else— astonishingly made me feel even worse. I got terrified. Bloody urine and gray feces soon had me headed for the ER. I was so sick that anything I ingested by any means only made me feel toxic and enervated. In such a state, not taking drugs, even while surrounded by them, was relatively easy.

So easy, in fact, that after I left the hospital, I thought I'd been cured, my problem solved. I still didn't understand that ending addiction wasn't just about making it through withdrawal. I also didn't know that I was almost certain to relapse since I hadn't learned alternative ways of coping and was still living in a drug-filled environment. I continued to believe that addiction was primarily driven by physical dependence. Since I was free of that, I thought I was well.

Now it was less than a week into my first semester back at Columbia after my suspension. I'd been allowed to return because I had convinced not only myself but also the school officials that I no longer had a drug problem, thanks to the hepatitis. I was not alone in my belief that getting through withdrawal was all that was needed.

In my essay seeking readmission, I'd written about my illness and recovery from it—and about my genuine desire to study and learn. For the most part, I was surprisingly open: the school knew that I'd left due to a cocaine problem and I wrote about how I'd then moved on to heroin before "recovering" via my liver disease. I really thought I was making a new start. But I didn't mention that I was still living with and basically working for a coke dealer. I wasn't quite sure what I planned to do about that.

Looking back now, I am shamed and horrified by the entire sequence of events, which took place in our apartment on 49th Street, near Radio City Music Hall and Rockefeller Center, just off Sixth Avenue. It exemplifies both the sheer mindlessness of addictive behavior—and of the way we deal with it as a society. I had no idea what to do when I opened the door and realized that it wasn't our customer, Lina.

Her "friends" turned out to be Long Island–based narcs who had been set on her by a high school chum who needed to bust someone to avoid prison. They had been desperate to meet me when Lina was arranging the sale. I had refused. To protect yourself as a dealer, avoiding selling to people you don't know is generally a good policy. However, I was recorded on a phone call earlier that day coordinating the meeting for the deal. By declining to leave the apartment to make the sale, I ultimately spared myself an additional charge of selling directly to the cops. But Lina was charged with selling, which is actually a more serious charge than I wound up facing, even though I was far more involved with drugs.

Lina was a naive NYU sophomore from Nassau County in Long Island; she had dyed black hair and a few piercings, but she was hardly tough or sophisticated. Her friend from home had been busted in Nassau County. To get a lighter sentence, he needed to find another dealer to implicate. Lina, unfortunately, became his target. I knew her through a crowd that I had gone clubbing with at '80s hotspots like Area and The Tunnel, in happier times before taking drugs had taken over my life. Ironically, Lina wasn't a dealer or even a regular user— let alone an addict. She occasionally did coke and was making this sale as what she thought was a favor for an old friend. I didn't know it at the time, but while the police were storming my apartment, she'd already been arrested and was being held downstairs in a van.

Immediately after they arrived, two of the officers took me into the hallway outside my apartment. I was now high, still feverish and completely dazed; I was also terrified. Their guns were prominent and visible to me in their holsters. They promised that if I signed the form that they shoved in my face, they wouldn't arrest me. Stupidly, I complied. To this day, I still don't understand exactly why: it must have been some combination of fear, fever, intoxication, and perhaps my ongoing Aspie tendency to take what other people say at face value. Aside from selling drugs, it's probably the single most idiotic thing I ever did. Of course, the police were lying to me; if I had been thinking at all clearly, this should have been obvious. The document turned out to be permission to search. They had no warrant. If I hadn't signed, there might never have been an attempt to prosecute me.

The narcs rushed into the bedroom. There, they found Matt, who was sitting in his underwear weighing coke on the scale. Nearby was obviously a large quantity, at least a kilo. That was not typical: Matt was holding most of it for his connection, who had wanted not to have the weight in his possession in case of just such an event. Stashed nearby in a file cabinet was $17,500 in cash, most of it needed to pay the connect for the drugs. In my blindly compliant state, I showed the narcs exactly where it was.

As they searched, the cops stomped around, sneering at our messy apartment, with one woman saying sarcastically that it belonged in *Better Homes and Gardens*. Their behavior was so bizarre and over-the-top clichéd that the whole thing seemed even more unreal. One stocky man with a gun wore a Hard Rock Café T-shirt, which he probably thought was cool, but which clued-in club goers saw as a tourist trap. (It is strange the details you notice and the thoughts you recall from events that forever change your life.)

And it soon got even weirder. Matt had literally been caught red-handed, or I guess I should say white-handed, but they had no interest in him. When they slapped the handcuffs on me and dragged me off, he actually thought I'd been kidnapped by a gang posing as cops, since real police wouldn't throw away the marlin in favor of the minnows. He just sat there, dazed and confused. For my part, I was in shock, too. I remember being pulled into the elevator, walked by the cops past our doorman and into the street. For one second, while I stood on 49th Street, a strange feeling of relief and utter freedom flooded over me. The thing that I most dreaded would happen had already done so: I didn't need to worry about it anymore. Then the fear returned.

I would spend the next five years of my life dealing directly with the repercussions of what happened that day. My actual recovery would not begin for another two years—and my addiction only worsened after my arrest. While there are certainly legitimate policy arguments about the best way to deal with crimes like drug dealing, there's no doubt that the criminal justice system is ineffective and often actively counterproductive in dealing with addiction. My experience is just one of millions that shows why.

THROUGHOUT THIS BOOK, we've seen that addiction is not defined by dependence on a particular substance to function or by a desire to avoid withdrawal or by simply being obsessed with the object of the addiction. If it were merely any of the above, it might be possible, perhaps, to use punishment via the criminal justice system to fight it. If withdrawal was really the problem, hepatitis— or, indeed, a two-week stay in jail or somewhere that I would have no access to drugs—could actually have cured me.

Instead, addiction is defined by using a drug or activity in a compulsive manner *despite* negative consequences. And "negative consequences," of course, is simply a less morally charged phrase for a whole range of experiences that can be experienced as punishing; the terms are fundamentally synonymous. In other words, if punishment worked to fight addiction, the condition itself couldn't exist.

Think about it for a minute: addicted people continue taking drugs despite losing jobs, loved ones, their homes, families, children, dreams, even sometimes body parts. I continued after contracting a disease that made me feel as though I had been poisoned. I continued after being suspended from the school I'd spent most of my life dreaming of and working toward attending. I continued while facing the daily risk of overdose and AIDS—after I'd already nearly died from an overdose and contracted hepatitis. And I continued even when the cocaine made me feel paranoid, terrified, and as though I was about to die, even though the thing that most frightened me of all was death. While there are many experiences that are not common to all addictions, the compulsion to continue using no matter what is its essence.

In this light, the idea that other sorts of threats or painful experiences will stop addiction makes no sense. Addiction is an attempt to manage distress that becomes a learned and nearly automatic program. Adding increased distress doesn't override this programming; in fact, it tends to engage it even further. If learning were occurring normally during addiction, addicted people would soon learn not to take drugs because the consequences are so bad. The fact that they do not is the crux of the problem.

Moreover, a whole series of studies shows that the brain responses of many addicted people to reward and punishment are abnormal, regardless of what substance is involved. In one, about two thirds of people with substance addictions showed an elevated emotional response to the prospect of monetary gain—an overvaluing of reward. This group, however, responded normally to losses. For these addicted people, similar to what is seen in teens, there appears to be a heightening of desire for reward that may occlude consideration of future punishment. But more interestingly, the remaining third of the participants did not respond to punishment at all. Even after they'd learned that drawing cards from one particular deck resulted in more loss than gain, they continued to select cards from it, showing the characteristic trait of persistence despite punishment. Similarly, other studies have found reduced brain activation during punishment (typically monetary loss) in people addicted to cocaine and methamphetamine.

So why then do so many believe that addiction ends when people "hit bottom" and that criminalizing drug use helps people "bottom out"? Let's set aside for a moment questions about how to deal with drug dealers who don't have addictions and what level of punishment or consequences might be appropriate when selling is illegal. What I want to start to explore here is how punitive and moralistic treatment that claims to view addiction as a disease does not really do so and instead bolsters the law enforcement approach.

The problem begins with the shadow cast by our laws and their history. Indeed, to paraphrase geneticist Theodosius Dobzhansky on biology and evolution, nothing about addiction treatment and drug policy makes sense, except in light of history. To understand how we came to use punishment to "treat" a condition that is literally defined by its resistance to punishment, we have to return briefly to the history of ideas about addiction and how this influenced our laws related to drugs.

As noted in chapter 2, America's first drug laws were born in a climate of overt racism, during the Jim Crow years. The rhetoric used to win their passage was explicitly racist and supporters played on white men's fears of miscegenation and losing power. The concept of the fiendish "addict" used to advocate for the laws hewed closely to racist stereotypes.

This unfortunate use of drug policy in support of racism did not end with Prohibition; it simply went underground, reemerging in 1971 with Richard Nixon's declaration of war on drugs as part of the Republican Party's "Southern strategy." This strategy targeted southern Democrats who were disaffected from their party because of its support for civil rights laws. Expanded further by Ronald Reagan, the strategy used code words like "crime," "drugs," and "urban" to signal to racist voters that Republicans would "crack down" and be "tough" in dealing with black people. As Michelle Alexander points out in her bestseller *The New Jim Crow,* selective enforcement of harsh drug laws created a new—and apparently legal—way to segregate, control, and incarcerate black people.

But this is only one part of why America remains addicted to a punitive—and failed—drug policy.

FOURTEEN

The Problem with Bottom

Just when you think you've hit rock bottom, you realize you're

standing on another trapdoor.

—MARISHA PESSL

I SAW TEARS IN MY FATHER'S eyes when he bailed me out, even though he did his best to hide them. It was September 14, 1986, around midmorning. He looked like he'd aged years in the short time since I'd last seen him, a few weeks earlier. A heavy smoker—his only regular indulgence—he lit up a Marlboro as soon as we got into his latest beat-up used car. I could feel his pain when he hugged me.

I'd spent only three nights incarcerated, but it had seemed like forever. During that time, I'd been told by a jail doctor who treated me for the ear infection that I "didn't deserve" to go to Columbia; I'd also been surprised by the kindness shown me by inmates who were in far worse situations than I was. Meanwhile, my dad had been searching for a bail bondsman and trying to find me a lawyer.

After I was released, he drove me to the first diner we could find, where I wolfed down some pancakes and tried not to think about what had happened. As a vegetarian (yep, I was an IV-drug-using vegetarian), I hadn't been able to eat much in jail. Meals there consisted of things like bologna sandwiches on white bread, which I picked at. I knew better than to ask for anything else.

I felt wretched and was filled with shame. Everything in my life was wrong, everything. But knowing this didn't stop my craving; in fact, it enhanced it, even as I understood that this was completely insane. All my other obsessions had been replaced; all the intensity that had previously driven my special interests, from volcanoes to TV news to the Grateful Dead, now had its red-hot center on heroin. Like an obsessive stalker, I could think of nothing other than tracking and winning the object of my affections.

I wasn't physically dependent on opioids. I'd only had that one shot the day of my arrest after months of abstinence. But it didn't matter. I could picture the stash of Dilaudid pain pills in my apartment that I knew the cops hadn't found: it was in the same file cabinet where I'd kept the money they'd confiscated. I couldn't wait to get home and shoot up. All I could think about was that relief. The idea blotted out everything else.

My father took me to my aunt's elegant rent-controlled apartment on the Upper West Side, where my mom and Grandma Marge were conferring with her about my problems. It was like the world's most awkward family holiday party; no one knew what to say or do. If I had had cancer, they would have called their doctor friends at Sloan Kettering and Mount Sinai and mobilized social networks to find the absolute best care for me. Since it was addiction, they were too embarrassed. And even if they had sought expert guidance, it probably wouldn't have been of much use.

In fact, as recently as 2009, Tom McLellan, one of the leading academic researchers on addiction treatment and a former deputy drug czar, admitted that when he needed it, even he didn't know how to find evidence-based treatment for his own addicted son. While resources for accessing treatment that is scientifically supported are slowly being developed, in the 1980s, the idea wasn't even close to consideration, let alone implementation. I can look back now and think that it's odd that my family didn't take me right to a rehab, but at the time, even highly educated and medically literate people didn't know what to do. Addiction was unspeakable.

After an hour or so of excruciating small talk, I managed to convince my parents to let me go back to my apartment, promising, of course, that I wouldn't do what I—and probably everyone else—knew I was going to do anyway. I can't recall ever feeling more shame, distress, and humiliation. For most of my life, I'd been my family's pride and joy; they'd reveled in my academic success. But I'd thrown everything away—my first chance to go to Columbia and now, I presumed, my second one. I was an embarrassment; I could not control myself. If being broken down by despair was going to end my addiction, this is when it should have done so.

However, the disgrace of my arrest and incarceration did not push me to seek recovery; it only made the craving worse. Shame and guilt didn't provide any new tools that would allow me to change. Without a clue as to alternate ways of coping, I couldn't see any way out. I was just like the prisoner in the cell with the hidden trapdoor: with no hope for escape or information that would make it possible, I was as securely caged as someone in an escape-proof room. When I finally got home, I shot the drugs and began the worst two years of my addiction.

ALTHOUGH MY STORY has many uncommon elements, my response to incarceration and the threat of further punishment is not one of them. Indeed, the research shows clearly that our national commitment to a punitive approach is both deeply misguided and hard to change for multiple reasons. I'll explore the data on the failure of punishment—as well as an important but little-known reason for our persistence in using it. Because we do not understand addiction as a learning disorder—one that is actually defined by its resistance to punishment—these ideas are particularly pernicious.

First, from all types of studies, there's no doubt that incarceration itself doesn't stop addiction. A systematic review of the research on criminal recidivism (including drug crimes) overall found that in 11 of the 23 studies included, probation or other community-based sentences were *superior* to prison in cutting repeat crimes. Only two studies suggested positive results of incarceration—and the rest found no difference. Following addicted people over time and looking at how incarceration affects their odds of recovery offers another way to test whether punishment works. Here, incarceration once again fails. A study of over 1,300 injection drug users in Baltimore interviewed repeatedly between 1988 and 2000 found that people who had been incarcerated during that time were half as likely as those who were not to be among the 20% who successfully quit injecting during the study. A Canadian study of 1,600 IV drug users followed from 1996 through 2005 had similar results: recent incarceration cut the odds of recovery by nearly half.

The data on adolescents show even stronger evidence of harm. One study of over 100,000 American children arrested between 1990 and 2005—mainly for drug crimes or assault—found that those who received custodial sentences were *three times* more likely to be incarcerated as adults, regardless of the severity of the initial crime, as compared to those given alternative sentences in the community or to those who had their charges dropped. This means that for youth, prison is essentially three times worse than doing nothing at all.

The data here are particularly persuasive because this study was large and basically randomized, though not by the experimenters. The researchers compared youth who had faced harsh judges with those who were overseen by justices with a lenient sentencing history. The assignment of judges to juvenile cases is rarely linked to the nature or severity of their crimes—it's determined by essentially random factors like the day of the week the crime was committed and the judge's rotations, other cases, and calendars. This alone dramatically reduces the odds that the differences seen in the data were linked to tougher kids getting tougher sentences. Instead, it shows that tougher sentences make kids worse.

A Canadian study, which followed nearly 800 low-income youth from age ten into adulthood, found an even bigger effect. This research was able to include kids who committed crimes but didn't get caught (determined through interviews with them and their parents and teachers) as well as those who were arrested. It found that the adult arrest rates for people who had any contact with the juvenile justice system were *seven times* higher than for those engaged in a similar level of delinquency who weren't caught. And the odds of adult crime were more than *37 times* higher if the teen was actually locked up in a reform school or juvenile prison. (The researchers controlled for IQ and other obvious factors that might have influenced these results.)

In addition, there's another way to evaluate the effectiveness of incarceration, which is international comparisons. In 2014, the British government assigned experts to look for a relationship between the toughness of a country's drug policy and the rates of drug use. (Although use rates are obviously not synonymous with addiction rates, they do bear some relationship to them.) But no correlation was found—the same conclusion drawn by a 2008 multinational study published in the peer-reviewed journal *PLOS Medicine*. In fact, some of the countries with the toughest drug policies have the worst addiction problems.

According to a 2013 study, America, which leads the world with its incarceration rate, also tops the charts in marijuana addiction and cocaine addiction, suggesting again that the criminal justice system is not an effective way to reduce drug-related harm.

And what countries have the world's worst rates of heroin and opium addiction? Here, it's not the United States, although we do lead the world in painkiller misuse. But hard-line countries like Russia, Afghanistan, and Iran—some of which have the death penalty for drug offenses—have higher rates of illegal opioid misuse. A full 2%–3% of their populations have taken heroin or opium in the past year, compared to 0.55% for the United States as of 2012.

The data are even more stark when you contrast drug war spending and incarceration rates with addiction rates within the United States. Funding for the drug war, which is concentrated primarily on law enforcement, international interdiction, and supply-side efforts, went from $100 million a year in 1970 to more than $15 billion annually in 2010, increasing by a factor of over 31, even after accounting for inflation. During that same time, addiction rates either remained flat or rose. The best data come from large national population surveys on mental health.

The first one, known as the Epidemiological Catchment Area (ECA) study, was conducted between 1980 and 1985 to determine the prevalence of various mental health issues in the general population. Around 20,000 Americans participated. This study found that around 6.1% of citizens in the first half of the '80s had had some type of illegal drug problem at some point in their life, either the less severe "abuse" diagnosis then used by psychiatrists or the dependence diagnosis, which is equivalent to addiction. This basically provides a snapshot of addiction in America at a period when, according to the annual household and school surveys, illegal drug use was at an all-time high. (These yearly surveys are intended to track drug problems, but unfortunately, the earlier ones have no data on addiction rates, only rates of use, so they can't track long-term addiction trends.)

Fortunately, an even larger but similarly rigorous mental health survey that does include addiction data was conducted between 2001 and 2005, although it used a slightly different way of diagnosing addiction. In the time between the ECA and this survey, known as the National Epidemiological Survey on Alcohol and Related Conditions (NESARC), the incarceration rate in the United States more than quadrupled. Much of this increase was driven by drug arrests and drug crimes. So what happened to addiction? In 2001 through 2005, the prevalence of DSM-defined illegal drug abuse or dependence was 10.3%, suggesting that our trillion-dollar law enforcement spending spree may have actually increased addiction rates—or at any rate, didn't decrease them.

Given these dismal and clear statistics, why do so many people—including those running rehabs who claim to be experts—argue that retaining criminal penalties for drug use is the only way to avoid an apocalyptic plague of addiction? Why do so many contend that treatment cannot work unless it is backed by punishment, when we know punishment doesn't work? How can advocates claim that addiction is a disease like any other—and then contend that criminal sanctions are required as part of treatment, which is not true for any other disease?

The answers have more to do with the ongoing racist and moral framing of addiction in drug policy than it does with the effectiveness of incarceration. They are also linked to the still prevalent misconception that simply removing access to drugs by locking people up during withdrawal will solve the problem—even though, as we've seen, just detoxing without further support rarely leads to recovery.

BUT UNTIL PEOPLE understand exactly *why* punishment isn't a cure for addiction—and why we still secretly think that it is—it's hard to argue effectively for alternatives. Addiction frightens both affected families and the public, and addictive behavior frequently angers them, often for good reasons.

Even though I'm telling my own story, I spent a lot of time cringing while writing this and considering some of the things I did in the past. I am not proud of having been a dealer and I now find much of my behavior in my late teens and early 20s around drugs incomprehensible. I can't imagine how I could have taken the risks that I did; my thinking seems completely illogical and idiotic. Many adults when considering their own teenage follies or the choices made by their children similarly shake their heads in disbelief: while the adolescent brain may have some adaptive properties, its way of handling risk is not one of them in the modern world, and heavy drug use and addiction certainly exacerbate this. Yet there it is. And the fact that so many other young people—addicted or not—continue to behave as recklessly as I did clearly illustrates the futility of the tough approach.

So what does keep us so attached to punishment and so resistant to actually changing our practices? In addition to politics and racism, the key ideology that props up the punitive approach comes not from law enforcement but from treatment. Misinterpreted ideas taken from 12-step-based rehabs are integrated with the criminal justice system and are part of the belief system that quietly upholds current policies. They are deeply embedded not just in nearly every public and private system that addresses addiction but also in popular culture and conventional wisdom about it. And their support for a moral approach under the guise of medicine has taken us in the wrong direction.

IF YOU'VE EVER seen a story about celebrity addiction—or have any familiarity with 12-step programs and American pop culture—you've likely heard the term "rock bottom." This concept is at the root of our public justifications for using punishment to fight addiction. Without understanding the insidious effects of this idea, it will be difficult to move toward treating addiction as the learning disorder it is.

Conventional wisdom has it that people with addictions must "hit bottom" before they can recover—and that harsh and humiliating treatment facilitates this process, while "enabling" or being loving and kind is counterproductive. Even though these ideas are not supported by evidence, they are frequently used to justify punishment, cruelty, and abuse of people with addictions.

These notions originated in 12-step programs, which are now a required curriculum in at least 80% of American addiction treatment programs. Twelve-step language like "enabling" and "bottoming out" has entered the vernacular, and the idea that addiction is a "disease" of "powerlessness" is pervasive in rehab and beyond. It is even endorsed by the medical specialty group of doctors who work in addictions, the American Society of Addiction Medicine. To understand the ubiquity of these incorrect ideas, we need to look at their history and how the data actually contradict them. Although these ideas might have been far less troublesome if 12-step groups remained voluntary and were not integrated into coercive treatment and the criminal justice system—and although many people in 12-step programs themselves do not support them—unfortunately, they have spread in a way that has done harm.

The concept of "bottom" starts at the beginning of Alcoholics Anonymous in 1935, in the stories of its founders, Bill Wilson and Dr. Bob Smith. They famously met when Bill, newly dry and having just had a religious awakening, realized he needed to help other alcoholics in order to stay sober himself. On a business trip, he was introduced to Bob, who was known as a particularly tough case. Within six hours, however, Bill had convinced Bob to give sobriety a try. Their meeting forever changed the way America would see addiction.

Both of AA's founders had lost almost everything to alcohol before they created the program. Stockbroker Bill W., for example, had been fired from numerous Wall Street jobs because of his drinking. He had been forced to move in with his in-laws after going broke and had been hospitalized for detox numerous times. Dr. Bob, a proctologist, had a record of at least a dozen hospitalizations and more than 17 years of drunken arguments, broken friendships, and early-morning withdrawal symptoms before he finally quit. Most of AA's earliest members were what the program soon labeled "low bottom" drunks, some of whom had literally been on skid row and fallen as far as it is possible to go without dying.

One of the group's foundational texts, *12 Steps and 12 Traditions,* lays out why its creators believed that recovery could not begin unless an addicted person is first forced to feel completely defeated:

> Why all this insistence that every A.A. must hit bottom first? The answer is that few people will sincerely try to practice the A.A. program unless they have hit bottom. For practicing A.A.'s, the remaining eleven Steps means the adoption of attitudes and actions that almost no alcoholic who is still drinking can dream of taking. Who wishes to be rigorously honest and tolerant? Who wants to confess his faults to another and make restitution for harm done? Who cares anything for a Higher Power, let alone meditation and prayer? Who wants to sacrifice time and energy in trying to carry A.A.'s message to the next sufferer? No, the average alcoholic, self centered in the extreme, doesn't care for this prospect—unless he has to do these things in order to stay alive himself.
>
> Under the lash of alcoholism, we are driven to A.A. and there we discover the fatal nature of our situation.

AA didn't scientifically test these propositions; its founders were not educated in research methods. To Bill and Dr. Bob, basing AA practices on its members' experience was valid and sensible. If the two men felt that the "lash" was needed to bring people to the program, that's how it was. It made sense to those who had nearly died before gaining sobriety. And over time, the notion of the need to hit bottom became even more popular with treatment providers, who began basing their treatment on the 12 steps with the founding of Hazelden in Minnesota in 1949.

By 2000, 90% of all addiction treatment was 12 step based (the number has since contracted, but not much). As AA grew, celebrities began revealing their membership in it. And by the early '90s, rehab had become almost glamorous. The growth of 12-step programs and related rehabs and the publicity they received over the years made the idea that addicts need to hit bottom before they can recover into conventional wisdom. Not coincidentally, it also perfectly suits the punitive climate in which America deals with drugs.

Even AA itself, however, recognized early on that there were problems with "bottom." It noted in the first versions of its literature that many of the successful members who joined when the program grew beyond its initial hard-core founders were people who "still had their health, their families, their jobs, and even two cars in their garage" as well as "young people who were scarcely more than potential alcoholics." To salvage "bottom" as a requirement for recovery, AA and those who use its approach resorted to a rhetorical trick.

They called it "raising the bottom." Maybe in some cases, the reasoning goes, addicts and alcoholics do stop before the disease gets too bad. These people are said to have "high bottoms," which allow them to see the logic of quitting before they lose that two-car garage. For this lucky group, no additional consequences are needed to pressure them into abstinence because they can already see that their problems will not get better unless they stop drinking. For the rest, who have "low bottoms," though, recovery is still framed as impossible until so many bad things have happened that alcoholics and other addicts can no longer deny their "powerlessness," which is AA's first step ("Admitted we were powerless over alcohol—that our lives had become unmanageable").

From this perspective, the more punitively addicts are treated, the more likely they will be to recover; the lower they are made to fall, the more likely they will be to wake up and quit. Conversely, the more kindness and support they receive, the less likely they will be to stop drinking and using other drugs. It's easy to see how these ideas support both the current treatment system and the criminal justice approach: law and medicine aren't seen as conflicting, but are simply different means to the same end. For example, supporters of drug courts—which offer reduced sentences for complying with treatment ordered by a judge—often argue *against* attempts to reduce draconian sentences because they believe addicts won't seek help unless the "bottom" is as bad as possible.

Drug courts were initiated in 1989 as an attempt to mitigate the harshness of the drug war, but ironically, they can end up reinforcing it.

Consequently, drug court advocates have even fought policies to replace punishment with treatment, which seems illogical for people who see the condition as a medical disease. In 2000, for example, actor Martin Sheen wrote an op-ed opposing a California measure that required three tries at treatment before jail was an option. He wrote, "Without accountability and consequences, drug abusers have little incentive to change their behavior or take treatment seriously," implying that addiction is not a disease but a bad choice. The Betty Ford Center led the campaign *against* the initiative—despite the fact that this went against its own financial interest in moving funding from enforcement to treatment and even though it manages to treat alcoholics without criminal sanctions.

In fact, research finds that legally coerced patients typically do not do better in treatment than those who enter voluntarily, despite staying in treatment longer. There is also significant evidence that empathetic and empowering approaches that let patients set their own goals—basically the opposite of coercion—are far more effective than treatment that relies on confrontation and making people feel powerless. Nonetheless, widespread belief in "bottom" remains, driving people who claim to want to compassionately treat a "disease" to support harsh measures unlike those for any other condition. "Force is the best medicine," one drug court official told sociologist Rebecca Tiger, who has extensively studied these connections. The idea of "bottom" justifies both punitive coercion to force people into treatment and punitive forms of treatment themselves. But it is based on erroneous reasoning.

For one, the definition of "bottom" is extremely subjective. There's no way to tell who can recover with only a high bottom and who "needs" brutal punishment to achieve this state. Second, "hitting bottom" can only be defined retrospectively after a period of recovery. If the person has relapsed, by definition, they haven't really hit bottom since they are not in recovery anymore. Now, they will need to find a new low point before they can quit again. With at least 90% of addicts relapsing at least once, the only way you can really know for sure if someone's bottom is the true one is after they have died in recovery.

What this means is that every major relapse becomes a new bottom. And that makes a mockery of the idea that there is a clear lowest-ever turning point and that terrible losses are needed to get to it. Research shows that some people bounce back easily after a relapse; others get worse than they were before; others enter a static cycle of recovery and relapses, getting neither worse nor better.

And, of course, many die. Another huge problem with the notion of forcing someone to the bottom is that they may not survive it.

AA members and counselors sometimes use phrases like "bottoms with trapdoors" to try to rationalize this complexity. But the reality is that "bottom" is a narrative device in a story of sin and redemption, not a medical description of a key stage in recovery from addictions. Worse, the evidence on what makes for successful recovery actually contradicts the "bottom" story: people are actually more likely to recover when they still have jobs, family, and greater ties to mainstream society, not less. Indeed, the more "social capital" someone has—friends, education, employment, job contacts, and other knowledge that promotes links to the conventional world—the more likely recovery is. As soon as you think about it critically, it's easy to see why if you had to bet on whether a homeless, unemployed person or a successful physician is more likely to recover, your money would be safer on the doctor than on the guy on skid row.

Further, as you can see from the passages above, AA's program for dealing with addictions is explicitly moralistic. The steps start with a recognition of one's "powerlessness" over the addiction and move on to "surrendering" to a "Higher Power" to address the problem. They also include a full confession of all of the sins of addiction (in 12-step terms, a "moral inventory"), attempts to reform one's "defects of character," and making amends to people who have been harmed. While 12 steppers (including me at one time; about which, more later) vociferously claim that addiction is a disease, they don't treat it like one.

Imagine a psychiatrist telling a depressed person to surrender to God and take a moral inventory—or better yet, imagine this being proposed as a way to treat cancer or AIDS. Imagine group therapy for depression that involves putting the patient on a "hot seat" and having all of the rest of the patients try to break him down by hours of shouting, spitting, name calling, and repeatedly listing every flaw in his personality. Imagine therapy for leukemia that requires a patient to sit facing a wall, carry humiliating signs, or wear diapers. For no other medical condition is such treatment acceptable in mainstream medicine—and yet all of these tactics persist in addiction treatment today (though, thankfully, they are somewhat less common than in the past).

History again explains how this peculiar system came to be. American alcoholism and addiction treatment, like the rest of medicine, started with a succession of quack cures. Over time, however, as addiction proved resistant to most attempts to treat it, the area became a backwater. No one wanted to work with patients who were extremely stigmatized and largely unresponsive. By the early twentieth century, everything from psychosurgery to hydrotherapy, work cures, psychiatric hospitalization, many types of religion, and something called "Double Chloride of Gold" had been tried. The field was ultimately left to people who were willing to claim miracle cures based on little evidence—and to addicted people themselves, who discovered the importance of social support when founding and joining groups like AA.

As AA spread by word of mouth and positive media attention—including a 1941 article in the popular *Saturday Evening Post* magazine that quadrupled membership within a year—members and medical professionals who had seen people transformed began to import its ideas into treatment settings like hospitals. They also began to create residential centers for AA members who needed intensive support, starting with Hazelden.

This is where the real trouble with the "bottom" idea began. AA explicitly states that its steps are "suggested" and emphasizes that its program is voluntary. But when supporters founded treatment programs, that concept was abandoned. Hazelden's "Minnesota Model," named after the state where the program was located, became the template for the 28-day inpatient rehab and for most outpatient treatment available, even now. The Minnesota Model requires getting people to accept the ideas of AA, including hitting bottom. Although today these programs are typically gentle and try to persuade, rather than push, most of them went through a period where humiliation and emotional attacks were seen as acceptable ways to shove people more rapidly downward.

The use of humiliation and "attack therapy" reached its apogee at Synanon, a commune started by an AA member in 1958. Even now, any residential treatment center in the United States that lasts a minimum of three months (typically eighteen) and calls itself a "therapeutic community" owes its start to Synanon, and many counselors remain convinced that the tough approach to getting people to hit bottom is justified.

Synanon's founder, Chuck Dederich, had decided that AA was too soft on alcoholics and addicts. He began running therapy groups during which members played what became known as "the Game" for hours, even days, without breaks. The idea was to demolish the ego, using intimate secrets people revealed to find their weak spots and try to obliterate the "character defects" believed to be found in all people with addiction. At first, the idea was to subject residents to a year or two of such treatment, granting them greater privileges as they progressed toward graduation. In later years, people were meant to stay forever.

"Alcoholics Anonymous is based on love, we are based on hate, hate works better," Dederich explained. When a heroin addict proclaimed that the organization had solved his drug problem, Synanon began selling itself as a cure for what was then seen as an untreatable affliction. By the late '60s, it had won national acclaim and state officials across the country sent representatives to study it and create local programs. Only one state—New Jersey—bothered to evaluate Synanon before replicating it, discovering that the vast majority of participants dropped out within a few weeks and only about 15% stayed abstinent. But the data didn't matter: the media loved Synanon and soon even people without addictions began joining, believing that a life of confrontational, radical honesty among peers would bring enlightenment. Although Synanon didn't explicitly use the 12 steps, it took many ideas from them, most important, the idea of breaking people to achieve bottom. And in the '80s and '90s, therapeutic communities based on Synanon began reintegrating the steps themselves and recommending that graduates affiliate with 12-step programs.

While the 12-step approach is relatively benign when chosen, it can be devastatingly harmful when coercion enters the picture. Then, the need to hit bottom can be used to justify disrespectful and abusive tactics in order to prompt someone to feel powerless and desperate enough to try the steps. Since humility is seen as critical to this process, humiliating participants is acceptable. Since pride and confidence are antithetical to surrender, attempts are made to suppress or puncture them. Since social support can aid resistance, people are cut off from access to friends and family and even to kindness from each other during periods of punishment, which often involve being silent and totally shunned. Not surprisingly, this does not empower patients. It is explicitly intended to do the opposite.

And of course, as in the rest of health care, trying to force people to feel powerless, humiliated, and as though they are at the lowest possible point in their life is a recipe for harm, not help. Further, this is dangerous not only for the patients but for the providers who have to wield this power. In many cases, it goes to their heads. At least half a dozen treatment programs have literally turned into destructive cults when they applied these harsh methods, starting with Synanon itself.

In the 1970s, while states across the country like New York and California were creating their own copycat programs, Synanon's maniacal leader made members stockpile weapons, forced spouses to swap partners, and coerced men to get sterilized and women to have abortions. The only children he wanted in Synanon were those whom the juvenile justice system or parents were paying them to treat; others he saw as a drain on resources. Ultimately, Synanon's downfall began when Dederich ordered his henchmen to place a derattled rattlesnake in the mailbox of Paul Morantz, a courageous attorney who'd won cases against the group. The snake bit Morantz, who, fortunately, survived. (Dederich eventually did time for conspiracy to commit murder in the case; when arrested, the "ex-alcoholic" who had found a "cure" for addiction was drunk.)

Nonetheless, even today, virtually every publicly funded inpatient addiction program in the United States that calls itself a "therapeutic community" has its roots in Synanon. Phoenix House and Daytop, which were the two largest public providers of addiction treatment in the country in the '80s and '90s, were both copied directly from Synanon, and almost every other 3-to-18-month residential program in America was founded by Synanon members, founded by people trained by Synanon members, or founded by people trained by people who were trained at Synanon.

For decades, like their model, these programs used sleep deprivation, food deprivation, isolation, attack therapy, sexual humiliation like dressing people in drag or in diapers, and other abusive tactics in an attempt to get addicts to realize they'd "hit bottom" and must surrender. As of 1973 there was already evidence that Synanon-style encounter groups could produce lasting psychological damage in 9% of the normal college students who agreed to participate in a study, but nonetheless they continued to be imposed on vulnerable addicted adults and teenagers. Incredibly, many programs—particularly those that receive large numbers of patients from the criminal justice system—still use this approach. And all of this treatment rests on the mistaken idea that people with addictions must be broken and made to hit bottom before they can recover.

The "tough love" approach for parents and partners of addicted people is also deeply reliant on the idea of bottom. Tough love for families has part of its origins in Al-Anon, the program originally designed for spouses of alcoholics, which was founded by Bill Wilson's wife, Lois. Parents and spouses here are exhorted to stop "enabling" their loved ones' addictions in order to help them recognize their powerlessness. Although partners and parents are also, paradoxically, told that they didn't "create the addiction and cannot control it or cure it," Al-Anon simultaneously advises members that behaviors like covering up for job absences or paying debts or even arguing about drinking can prevent an alcoholic from reaching that all-important nadir.

Tough Love was also the title of a bestselling 1982 book and the name of a related support group, which had thousands of members during the '80s and '90s. Members urged fellow parents not to bail out their kids if they were arrested.

They were also advised to cut off all contact if their children did not become perfectly obedient, which typically required completing treatment, staying away from drugs, and being compliant with whatever other conditions parents set. Many Tough Love members became prominent within Al-Anon and treatment organizations. They advocated publicly for tougher laws and for harsh, Synanon-based treatment programs like Straight, Incorporated, which actually traumatized thousands of families. As with AA, however, Tough Love was never tested before it was widely accepted and implemented. While there are clear indications that it can sometimes do harm, no one knows how widespread the problem is because there is virtually no research on what happens to people whose parents or spouses decide to practice it.

One of the most distressing anecdotes, however, is the case of Terry McGovern, daughter of former presidential candidate and Minnesota senator George McGovern. A counselor told her parents to refuse contact with their depressed and alcoholic daughter if she kept drinking. Shortly thereafter, the McGoverns were devastated to learn that Terry had been found dead from exposure in a snowbank. While there certainly are times when parents and loved ones need to distance themselves from addicted children for their own sanity or to protect other children, there is no way to predict whether this will harm or help the addict.

A further outgrowth of the forcing "bottom" idea is the "intervention," which is the basis for the popular reality TV show of the same name. Again, the idea is to confront the addicted person—typically in a big meeting that can even include their boss—and threaten to cut off all emotional and financial support if he or she doesn't comply with recommended treatment. During the meeting, participants often harshly attack the addict, trying to "break through" to them and create a bottom. Again, this has a serious potential to backfire: there have been suicides immediately after interventions, most notoriously that of beloved '90s musical icon Kurt Cobain. (And since there are methods of intervening gently and respectfully that don't carry these risks, the approach is particularly misguided.)

In this context, however, it's not surprising that the criminal justice system is seen as an appropriate tool to fight addiction. There are few better ways to make people feel powerless than locking them up and controlling every aspect of their lives. There's no conflict between viewing addiction as a disease and as a crime if you believe punishment is the cure for the disease.

Of course, as we've seen, the data don't support this. In fact, a feeling of total powerlessness and helplessness is fundamental in determining whether a horrific experience will cause PTSD, which is not exactly a sign of what is likely to promote health. Indeed, PTSD actually increases relapse risk in addicts—and having PTSD doubles to quadruples the risk of becoming addicted in the first place. Moreover, a thorough review of the data on humiliation, punishment, and confrontation as treatment for addiction shows that it is not helpful, leading to worsening addictions and greater numbers of treatment dropouts. Over four decades of research, in fact, not a single study has supported the confrontational approach as superior to kinder and less potentially harmful treatments, according to a review of the data by William Miller and William White. Yet both our treatment system and our criminal justice system remain steeped in this ethos.

IN MY OWN life, I was extremely lucky that my parents hadn't been advised or hadn't chosen to practice tough love when I got busted. It's one thing to refuse to bail someone out when they are facing a few days or months in jail on possession charges. It's quite another if the consequences are a potential 15-to-life mandatory term. One of the many ways that racial and class inequities reproduce themselves in our criminal justice system, in fact, is that middle-class and rich people have the education to know how to fight back and the resources to do so.

Wealthy parents can afford to hire lawyers to protect their kids from sentencing extremes. When there are good treatment options available, people with money are more likely to find their way to them and be able to pay the fees required. Wealthy people also tend to believe that they *should* protect their kids from legal consequences for fear that having a record will damage college or employment prospects. Since punishment actually doesn't treat addiction, this leaves the poor and those who buy into tough love ideology incredibly vulnerable.

In 1986, though, I was in the grip of the system. That year, I read an article in the *New York Times Magazine* that terrified me. It was a positive profile of a New Jersey program called KIDS, an unapologetically harsh youth rehab. Treatment at KIDS consisted primarily of sitting rigidly in one particular position on hard chairs for 12-hour days and being subjected to constant emotional attacks. I would later write a book that exposed just how torturous and harmful that particular program was. I learned, for instance, that one thirteen-year-old girl, who didn't even have a real addiction, was held in it for thirteen years, eventually winning a $6.5 million settlement in a stunning civil case that exposed emotional, physical, and sexual abuse.

While the worst abuses weren't reported in the *Times Magazine* story, what it did describe still sounded like a nightmare to me. The article portrayed the program's use of bullying as therapy, with the reporter clearly buying into the idea that such tactics might be necessary to bring these bad kids back from the brink. Reading it then, I was horrified; I wonder now what would have happened if I'd seen an article that lionized an empathetic, kind, and supportive rehab program. I dashed off a letter to the *Times* comparing KIDS to a cult (it did later turn into one, as it happened) and argued that such an approach couldn't possibly help people with drug problems; the *Times* published it. Although I knew such "therapy" was definitely not what I needed, I had no idea what would work or how to find good treatment. What I knew of 12-step programs didn't attract me: most of the people I knew who went to meetings continued to use. It was another two years before I finally allowed myself to risk seeking help.

And I didn't completely manage to avoid tough love.

FIFTEEN

Antisocial Behavior

How can you tell when an addict is lying? His lips are moving.

—ANONYMOUS

I HAD GONE TO SEE MY counselor, Irene, with some misgivings. I already had the sense that she disliked me, because during group therapy sessions she would dismiss nearly everything I said as "intellectualizing" or "manipulating"—regardless, it seemed to me, of what I was attempting to express or of how deferential to her I tried to be. Still, I wanted to do everything I could to get better, so as I'd been directed, I was now sitting in her cramped office because I felt shaky and wanted extra support.

It was late August 1988, and I was at a 28-day inpatient rehab program called ARC Westchester (now closed). I'd been referred there after completing the seven-day hospital detox I'd entered nearly two years after my arrest. While the judge hadn't mandated further treatment, I knew that following the recommendations of treatment professionals that I get additional help could only aid my case. I was trying my best to do that. Since we were supposed to "open up" about what troubled us, I told Irene that I felt insecure about my looks and was worried that no man would ever love me.

"How can you even expect them to look at you?" she retorted, exclaiming, "You look like Jiminy Cricket." For a moment, I thought I'd misheard her, even though she articulated the words clearly. I am sure I wasn't able to disguise my look of utter incomprehension. I'd previously been unable to vocalize my fears, let alone share them with a therapist. I'd now tried to seek help when I felt low, attempting to follow the rehab's recommendations precisely. This bizarre statement was my reward. I didn't know how to react or even what to think. I started to worry that I would actually get in trouble for trying to do what I'd been told was the right thing.

The rehab I attended was actually among the best of its time. It was based on the 12-step-focused "Minnesota Model" originated at Hazelden and still widely used. However, what I experienced in treatment further illustrates the paradoxes of American addiction care, which moralizes while claiming to be medicine. As noted earlier, the overwhelming majority of American addiction treatment still uses the Minnesota Model. To improve rehab, a new perspective is needed.

Although the Minnesota Model emphasizes that addiction is a brain disease, what it really does at its best is teach ways to manage the aberrant learning at the heart of the problem. Unfortunately, by taking a one-size-fits-all approach and assuming that all people with addictions have the "defects of character" that the 12-step programs associate with "the disease," this model tends to undercut itself and undermine its ability to teach. By presuming that all addicts have antisocial personalities, the Minnesota Model can also bolster and promote ineffective ways of teaching and counseling.

My experience illustrates many of the issues here—and, sadly, it is still quite representative of what many people undergo in addiction treatment in the United States. What happens in many rehabs is still more like indoctrination than education. By telling patients to "take the cotton out of your ears and put it in your mouth" and seeing even sincere questioning as defiance, the programs diminish their own ability to educate and motivate change. Learners do not respond well to patronizing teaching that dismisses or punishes questions. And when every patient is seen first as a liar, a manipulator, a criminal, or worse, rehabs do not create an environment conducive to the curiosity and openness that inspires learning. Unfortunately, however, the constantly reinforced stereotypes about people with addiction as having a flawed and specific personality disorder mitigates against change and inclines counselors to use ineffective tactics.

Although I did recover, I'm not sure how much of my success came because of treatment—and how much was despite it. I'll try to sort that out here, based on what we know now from the data, by exploring the specific therapies I received. The next chapter will look more closely at 12-step ideology and what the research suggests about what we can do better by understanding the role of learning and individuality in addiction.

My counselor Irene's "expertise" seemed to come only from on-the-job training and, possibly, her own experience, although she didn't disclose it. Nationally, this lack of genuine qualifications is more the rule than the exception among addiction counselors, even now. There are no federal standards for accrediting counselors. Fourteen states have *no* educational or licensing requirements for those who want to sell addiction treatments. Only one state requires a master's degree—and in the rest, even a high school diploma is often not necessary. Worse, the on-the-job training counselors receive is often little more than instruction in stereotypes about addiction from people who were instructed by others just as ignorant as themselves. Although there are some wonderful people who do great work despite this, most of those charged with treating a complex disorder in which at least half the patients also have another mental illness are grossly underqualified.

To me, Irene looked to be in her 30s or 40s. She was slightly overweight with curly dark hair, already starting to gray. When she claimed that I looked like a sexless Disney character who represented a chirpy conscience, I was tempted to reply that she wasn't exactly a fashion plate herself. But I knew this was a bad idea: everything I'd seen so far in rehab suggested that compliance was always the best approach, even if you thought what was going on was completely nonsensical.

Her response to my attempt to seek her help had also thrown me. I couldn't believe that she really thought that my tortured relationship history and clear feelings of self-loathing were related to my looks. Why would reinforcing my fears about my appearance be therapeutic? I obviously knew that I didn't look my best at that moment, with my barely healing tracks and ratty hair. Was this some bizarre attempt at breaking down my personality?

Although I was actually more bemused than upset, I started to cry. In group, Irene tended to view my rapid tears as an attempt to emotionally manipulate everyone, but as had been the case in my childhood, they were an uncontrollable response. I wasn't just turning on the waterworks. My preexisting "emotional lability" had been further exacerbated by withdrawal. In rehab, I could cry over an AT&T commercial or a fuzzy kitten I saw on TV during the rare opportunities we were allowed to watch; given the amount of bullying I'd experienced because of my oversensitivity, crying easily was certainly not a trait I had cultivated. However, I didn't know how to begin to explain this. Anytime I tried, I was accused of making excuses or "rationalizing."

And, though I never actually thought that I looked anything like a cartoon cricket, I was still upset. Irene's response to me was so jarringly off from what you expect in a normal human interaction that it made me feel even more alone, confused, and unheard than I'd been beforehand. I decided that with this counselor, even though I wanted to make my usual attempts to be the A student, I'd better just do everything I could to demonstrate submission and forget attempting to learn anything useful.

I had been trying as well as I could to be open to the process of recovery. I was allowing myself to question my earlier ideas about addiction and paying attention to what I was being taught about "the disease." However, I still couldn't understand the point of these apparent attacks and the constant assumption of my ill intent. Indeed, the research shows that they are counterproductive. One study found that the more a counselor confronts alcoholic patients, the more they drink. Other research shows that it is not whether a counselor is herself in recovery that matters for success, but instead, how kind and empathetic she is—or isn't.

THE REHAB ITSELF was welcoming and located in a lovely setting. The building was wide open and airy—at least outside of the warren of group counseling rooms and offices. Situated on a golf course near a small heavily wooded mountain, it had a sparkling pool out back. The bedrooms were bright and large. It felt friendly, though I was apprehensive about what awful techniques they might use to brainwash me. It was a good thing that I didn't meet Irene until after I'd been there a few days.

Instead, I spent what seemed like all of the first day filling out forms. Apparently, the idea was that repeatedly writing out your story would help break your "denial" of being addicted—so I was told "no" when I asked if it was okay to simply Xerox the first one. When I finally finished, the admitting office handed me an insurance-related document. I had a moment of panic when I realized how much it cost, and that my insurance only covered 80%. The price was around $17,000 for 28 days—roughly the equivalent of a year at Columbia at the time. I was shocked. Before I signed, I told them that I couldn't afford to pay the other 20%. (These days, many programs like the Betty Ford Center—which still use mostly the same tactics—cost more than twice that.) The staffer at the desk assured me that they would take care of it and accept the 80% as full payment, although getting them to stop billing for the full amount later proved to be a challenge. Like it or not, I was now committed to stay for at least a month.

At first, even though I was malnourished, I couldn't eat—leading the staff to suspect I was anorexic. But within a week or so, all my appetites came roaring back. Cute guys suddenly seemed to be everywhere; I hadn't even realized that I had lost my sex drive along with my appetite for food. After a few days, I started to feel famished and began to eat ravenously and put on pounds. Embarrassingly, at the end of my addiction, I was so thin that I no longer needed to wear a bra. But after a week or so in rehab, the staff nurse told me that I needed to have my parents send one, immediately. Although I couldn't see the changes in my body, everyone else could—and that unsettled me because my perception was so clearly out of line with reality. Like tripping on LSD, this experience showed me how much my ideas and preconceptions could shape my world. In rehab, though, the malleability of both my mind and body now became obvious and frightening. All of what usually anchored my sense of my self felt like it was gone.

Our time was rigidly scheduled, which did provide some sense of structure and routine. We had to be at breakfast by 7 a.m. and finished by 8 for the morning community meeting. That was where we'd discuss chores—nothing major, just keeping the place neat—and any problems we might have with our fellow patients. Thankfully, this was nothing like the confrontational meetings or grueling make-work I knew were part of treatment elsewhere. A clear schedule and a sense of community are known to be helpful and that's what was provided.

Next, we had a lecture or film. Sometimes it would be on "time management" or "stress reduction"; at others it would focus on "prayer and meditation" or some "chalk talks" by a famous priest. The data suggest that films and lectures are nothing more than time fillers in rehab. A review that rated the most and least effective tactics ranked movies and talks at number 48 of 48, that is, essentially useless.

But one lecture did seem surprisingly helpful in my case. At ARC, the re-hab doctor gave the talks on pharmacology and how drugs affect the brain. At first, I thought these were dumb and oversimplified and that I already knew all the material. Then, in one class, a patient named Paula mentioned an area of research with which I was unfamiliar. She had been an advanced pharmacol-ogy student and she cited studies that suggested a link between the neurochem-istry of alcoholism and heroin addiction via a chemical called THIQ. The doctor agreed that the data were correct. Although I later learned that this re-search did not pan out, at the time I was impressed. I began to listen more closely and to take in more of what they were saying about the nature of addiction. This increased openness, even though it turned out to be sparked by a myth, would be crucial to my rehabilitation.

Next on the schedule was lunch—and after that was the time I really dreaded: two hours of group therapy led by Irene in a claustrophobic room with no windows. Worse, patients weren't allowed to leave for any reason, and there were no breaks. Inevitably, because the setup evoked my biggest anxieties, I would desperately feel like I had to urinate before the group was over. It was as though the situation had been designed to trigger my OCD: I was trapped in a crowded space with no access to the bathroom. This made the sessions horrific for me.

To try to minimize the problem, I would avoid drinking fluids. I would also run to the bathroom ten minutes before group started. I'd use the toilet—and then keep heading for the bathroom door only to stop myself and go back to be sure I'd fully emptied my bladder. I got stuck in a compulsive loop. Eventu-ally, I'd manage to drag myself out. Although I was never late, my OCD ritual made me more anxious, not less. Worse, I had no way of explaining it to anyone—let alone to someone who was cruel, as I'd seen Irene be. I was too ashamed to even describe it to a doctor; even just writing this now, I feel em-barrassment far out of proportion to the behavior.

And, as soon as the group started, I was flooded with fear. Like the person who is told to think of anything but a white bear, going to the bathroom would now be all I could contemplate. Soon, I felt physical pain in my bladder and then I really couldn't pay attention, my anxiety escalating and spiraling. I didn't care about people's stories, I didn't care about their feelings, I didn't care about whether they thought I was a snob—all I wanted was to go back to the bath-room. The lack of control terrorized me.

Irene, of course, interpreted this reaction as "resistance" and "manipulation." According to her, I wanted to flee my feelings—or show that I was better than everyone else. Because I'd told her I dealt drugs, she'd called me a "sociopath"—a term typically used to describe serial killers, conscienceless con artists, and sadists. She was determined to tear away what she assumed was my thick skin and break down what she saw as my intellectual defenses.

Although I certainly did try to defend myself with my intelligence, however, I was hardly a hardened criminal or antisocial personality. It is true that people with antisocial personality disorder (ASPD) are overrepresented among people with addictions. Indeed, the negative qualities associated with this diagnosis—like selfishness, lying, and coldly manipulative behavior—are often generalized to create the stereotype of the addict. Such people do exist and there are more of them in the addicted population than among people in general.

However, at least 82% of people with serious drug use disorders *don't* qualify for this diagnosis, let alone meet the extreme criteria that delineate a psychopath or sociopath (the terms have become synonymous), which is the far end of the antisocial personality spectrum. I certainly didn't, although my Aspie focus on intellectual issues and poor social skills may have somewhat masked this.

It's important to recognize how different ASPD really is from addictions, not only because the conflation of the two produces unduly negative stereotypes about addicts, but also because this shows how varied addicted people are and, again, why learning matters. Confusing ASPD and addiction has greatly increased the stigma attached to addiction—and failing to differentiate between these disorders has also helped support the criminalization of drugs. To give a sense of the problems that result, I'll explore the distinctions between the conditions below. I only wish I had known about them earlier.

ADDICTION WAS INITIALLY characterized as a type of antisocial or psychopathic personality disorder. It is easy to get this misconception when antisocial behavior is defined largely by lawbreaking, and drugs are illegal. The early psychological and psychiatric literature often discusses addicts as psychopathic—and perhaps this reflects the fact that the first research on addiction was typically done inside of prisons. But obviously, this is likely to produce a skewed picture of those with drug problems.

Today, the consensus based on research is very different. Now, as we've seen, addiction is defined solely by obsession, compulsive behavior, and consequences related to drugs or to addictive experiences like gambling. It is neither necessary nor sufficient to be cruel, selfish, or dishonest to meet addiction criteria. In contrast, however, ASPD actually *is* characterized by a fundamental indifference to other human beings and their rights: if you don't have a reckless disregard for other people, you don't qualify for the diagnosis. And so, while the typical person with ASPD is defined by a lack of concern for others, the careless and self-centered behavior often seen in addiction that is not linked with ASPD is primarily the result of the perceived need for escape through drugs, not the cause of it.

Further, it's also critical to distinguish between the "normal" antisocial behavior seen in adolescence, the social problems associated with the autism spectrum, and the antisocial behavior of ASPD. None of these is synonymous and yet all of them can be confused, especially in a climate where certain types of drug use are criminalized and disabled people are highly stigmatized. One of the most important implications of understanding addiction as a learning disorder is recognizing how different temperamental predispositions can lead people to develop it and distinguishing between them in order to treat people compassionately and effectively.

First, adolescence. As is well known, during the teenage years, many previously well-behaved children become almost unrecognizable, engaging in defiant behaviors like lying, stealing, vandalism, fighting, drinking, drug use, and many other types of mainly petty crimes and rebellious behavior. The prevalence of antisocial behavior among children increases by a factor of 10 as they hit their teens; what is "normal" during adolescence is not normal at other life stages. Fortunately, most kids genuinely do grow out of it. Indeed, most people who do not age out of teen misbehavior didn't really "age in" to it in the first place—they had problems that started long before they hit puberty, which simply continued or worsened.

In the most severe cases, these children qualify for a diagnosis of conduct disorder (CD). CD is the childhood form of ASPD. For these kids, trouble typically starts as early as nursery school. In order to get diagnosed with ASPD as an adult, in fact, people have to have met the criteria for CD during early childhood, even if they weren't formally diagnosed at the time. Qualifying symptoms include but aren't limited to bullying, animal cruelty, property destruction, fire setting, deliberately injuring people, repeated dishonesty, covert manipulative behavior, and a callous lack of concern for others, with such behavior occurring far outside of the normal range of childhood mischief. Basically, children with conduct disorder are a parent's worst nightmare. One review of the literature summed it up this way: CD is "a repetitive and persistent pattern of behavior in which the basic rights of others and major age-appropriate societal norms or rules are violated."

Conduct disorder is often found in conjunction with ADHD, but the two are not synonymous. In my case, I did not qualify for that diagnosis because my childhood behavior, while odd, wasn't deliberately destructive or malicious. Ironically, the guilt and self-hatred I felt about my childhood "selfishness" were actually evidence that I wasn't as bad as I thought I was—since another key symptom of CD and ASPD is lacking a conscience or having only a weak one. The combination of CD and ADHD, however, increases addiction risk around twice as much as ADHD alone.

Further, while children with ADHD are frequently inconsiderate, they don't tend to be deliberately cruel, or at least no more so than typical children. Youth with CD, however, are often coldly spiteful. Worse, they are generally immune to punishment—a quality they share with addicts, but that has a very different cause in these antisocial syndromes. Children with CD tend to lack fear. Because they aren't motivated by pleasing other people, they also don't respond with the dismay that ordinary children have about social rejection when they disappoint parents or teachers. This means that both social and physical punishments have little effect. If you don't care what your parents think, don't want adult approval, and don't fear pain, lectures, time-outs, and spankings won't deter you.

The same is true for those who don't grow out of conduct disorder, which is between 40% and 75% of those diagnosed. These people go on to develop ASPD in adulthood. At the very far end of the ASPD spectrum is what is known as psychopathy or sociopathy—a complete lack of conscience and concern for anyone other than oneself. This group includes serial killers, but more commonly, nonviolent con artists and white-collar criminals. (The fact that a significant percentage of conduct-disordered youth do grow out of it also suggests that ASPD and psychopathy need to be examined from a developmental perspective.)

So, although many counselors "joke" that addicts lie "whenever their lips are moving," this characterization is unfair. It may be more true of people with ASPD—but research on the addicted population overall (including people with both disorders) shows we are in general no more likely to lie about drug use or sexual behavior than anyone else, so long as our disclosures are kept confidential and not used for punishment. Addiction is not a personality disorder;

as noted earlier, there is no single set of personality traits seen in all people with addictions. And when you remove drug use and sales crimes from the crime aspect of the definition of ASPD, the rate among addicted people is significantly reduced.

ASPD is also extremely different from the social withdrawal, "empathy problems," and odd behavior seen in autism spectrum conditions. While some of my compulsive behavior seemed antisocial, in fact, it was related to sensory or emotional distress. It is true that autistic people do often have difficulty understanding or recognizing other people's feelings. However, when they do become aware of these emotions, they care about being kind and generally don't want to cause pain. Many are actually obsessed with justice and with behaving ethically. The problem is not being sure how to act. Rather than being uncaring or indifferent to the social world, autistic people have a hard time understanding the rules of interaction and emotional expression. In CD and ASPD, however, this concern for other people is missing.

Fortunately, only around 4% of the population has ASPD. Among addicted people, however, that figure is indeed much higher—with 18% of those who have some type of illegal drug use disorder also affected by ASPD and 9% of those with alcoholism affected. This proportion is higher in men (20%) than women (14%) and may be as high as 37% in IV opioid users, particularly men. The idea of the selfish, hedonistic, and callous addictive personality comes mainly from this group—but the vast majority of people with addictions do not fall into it. And the majority of people with ASPD—60%–70%—don't have alcohol or other drug addictions, either. Although there are also other personality disorders that can be linked with lying and manipulative behavior—notably borderline personality disorder—only 16% of people with alcohol or other drug addictions have this condition. Addiction is diverse—and people who come to it via different paths need different approaches.

FOR ME, WHILE I may frequently have been clueless about other people's feelings, I was not indifferent: if I did know that you were in pain or, especially, found out that I'd said or done something hurtful, I felt bad. If you were a friend or family member, in fact, I'd often be desperately anxious until I could find a way to make it right—even if that meant self-abnegation and taking blame I didn't necessarily deserve.

As for Machiavellianism, I was also unskilled. If I wanted something from you, you generally knew it because I was direct, often comically so. I didn't have the social skills or even the desire to scheme and toy with other people's emotions. It was hard enough managing my own. In fact, I actually found it hard to believe that other people really did this outside of soap operas. Sure, I might do something obvious like wait until someone was in a good mood to ask for a favor, but I thought things like pretending interest in one guy to make another jealous were wrong, and for a while, I even worried about the morality of flirting.

With no language to describe these issues, however, I couldn't figure out how to get Irene to understand me. To her, I was the bad person I always feared I was. Asperger's syndrome hadn't even yet become an official diagnosis. All I knew was that I was socially weird and oversensitive. I was doing my best to be a good patient, but this didn't remove my various compulsions or my need for ways to deal with my sensory problems. It didn't stop my claustrophobia, agoraphobia, or OCD.

It was obvious, however, that I had to comply with her demands and make it through her daily group in order to complete rehab. At the same time, I remained tortured by my anxiety and compulsive behavior. Frequently, I just couldn't stop myself from running out of the group to go to the bathroom when it got really bad. When Irene threatened to have me expelled from the program if I did it again, I panicked.

Thankfully, at this point, someone on the staff decided that I needed psychiatric evaluation. That would ultimately solve the problem, although not by actually determining what was wrong. The psychiatrist I saw was apparently a consultant to the rehab; unlike the general practitioner who was there at least a few times a week, he seemed to pop in only occasionally to deal with particularly complicated cases.

I doubt that he could have picked me out of a lineup after our visit—nor could I have selected him. The only description I can give is that he was an older male. I was in his consultation office for a maximum of five minutes. Within three, he had diagnosed me with bipolar disorder and started to write a prescription for lithium. I don't know if he was called in to put a professional stamp on an amateur diagnosis made by Irene or whether he immediately jumped to this conclusion himself. All I know is that it was the fastest diagnosis I ever received.

In fairness, I will say that I may have seemed manic depressive: I spoke rapidly with what might have appeared to be a symptom known as "pressured speech," and it was obvious from my frequent crying and occasional elation that my moods were unstable. But all of that can also occur during withdrawal—and I wasn't given enough time or questioned thoughtfully enough to reveal my compulsive behavior. All I got was a prescription.

Fortunately, for what I needed most at that time, it proved highly effective. A common side effect of lithium is increased urination: the perfect way around my issue with group. Since having to go frequently was now officially "physical" and not "psychological," I was given permission to leave as needed. That ended my conflict with Irene and allowed me to stay in treatment—though not exactly the way anyone probably intended. I took the lithium for a few months and then let the prescription lapse, with no apparent ill effects.

THE EVENING OF the day I had my first family group, I was awakened by a flashlight being shined in my face. I was furious: it had been emotionally exhausting, and I desperately needed sleep. Outside of the rehab's Family Week, I recall few other times in my life when I have been so wrung out by pure emotion. It was my third week in rehab. With the exception of this rude awakening, Family Week would be conducted with empathy and compassion—here, the staff didn't try to break me or make me see myself as sociopathic. Instead, they used kindness and support.

There was actually a good reason that the staff had to be particularly sure that everyone was in the appropriate bed that night: I learned the next morning that Judy, a patient who was in my group, had fled. Before that, the slight young brunette who often wore her hair in two braids had seemed to be doing well in fighting her heroin addiction.

Earlier the day she left, we'd all listened as her father described an incident where he'd found her, naked and near death from an overdose, in a filthy apartment. He was an anxious-looking man, his face furrowed with distress. Both his love and fear for her were clear as he spoke. And as I listened, I thought about my own father and glanced over at him. He, too, was paying attention, frowning as he concentrated. My mother, my nine-year-old brother, and my fourteen-year-old sister were also nodding their support. (My other sister, Kira, had been unable to attend due to her college schedule.) They were visiting my rehab daily, for a week of therapy.

Hearing Judy's father had a strong effect on me. During my addiction, I'd been so wrapped up in my own pain that I'd rarely even considered what my drug use was doing to those who loved me. I thought drugs were necessary. They were nonnegotiable and if others didn't like it, it was their problem. Why should it concern them, anyway? Now, I cried. I watched Judy sit stiffly, trying to deal with her dad's pain. It was excruciating. I began to see why my own father had been calling me every other day: he didn't want to run my life, but he was afraid he'd find I was dead. I also realized the reason why the rehab placed multiple families in the same group, beyond the obvious need for fewer therapists: it's easier to see your problems played out in someone else's family than it is to recognize them in your own. The next day, however, Judy was gone.

A COMMON—AND to my mind, misleading—theme in stories of addiction is the search for a single incident or memory, like Rosebud in *Citizen Kane,* that explains everything. Although there may be some cases where this is true—and research suggests that creating a coherent narrative out of your experience may help recovery from trauma, which is extremely common in addiction—much more often, addiction is overdetermined. As with many developmental disorders, there is rarely a single necessary and sufficient cause; these conditions unwind over time, with many influences, few of which alone can cause the problem. There are also many turning points in which the trajectory can change entirely, preventing a predisposition from expressing itself as a disorder at all.

In my case, my early temperament was an obvious factor; the bullying I experienced, in part as a result, was another. Genes that are now suspected to increase risk of multiple psychiatric disorders—like depression, autism, addictions, bipolar, and schizophrenia—may have been present, given my family's history of depression on both sides and gambling disorder on my father's side. My father's Holocaust trauma and my mother's early maternal loss are other influences, which could have affected my epigenetics, either through directly changing the reading instructions on relevant genes or by altering the way they parented and the emotional tenor of our household.

In addition, the cultural climate of the late '70s and early '80s, including the high prevalence of cocaine use, rising inequality, and the emphasis on individualism and careerism, probably also played a role. So, too, did my difficulties fitting in at Columbia and my interpretations of my social experience. All of these clearly shaped my development and how I learned addiction. But there is one aspect I haven't yet explored here, which was the focus of my third week of rehab: my relationship with my parents and siblings.

At the start of the family program, the counselors proclaimed that addiction is a "family disease," which often responds best to family therapy. Unlike with many of their other claims, there is good evidence that family therapy is among the most effective approaches. For adolescent drug and behavior problems, in fact, it is probably the single most well-supported approach—at least when the therapists use tactics like Multisystemic Therapy, Multidimensional Family Therapy, and Functional Family Therapy, which have been extensively studied. Given how hard it is to determine when adolescence ends and young adulthood begins, there is good reason to think it may be superior for young adults as well. But as with all talk therapies, the devil is in the details: the research on psychotherapy is quite clear that the most important "active ingredient" is a good rapport and a high level of empathy in the therapists.

This is where the rehab I attended shone. While the tactics used in Family Week could have been deployed in a shaming way, in my case, what occurred was not harsh or unnecessarily confrontational, although it was emotionally intense. Patients were asked to sit in a group with other families and each family member in turn had a chance to discuss the harm done by the addicted member. Later, the patient from that family could respond and discuss incidents where he or she felt harmed as well. The idea was not to blame and shame the addicted person but to clear up lingering problems in the family. No interruptions were allowed, nor were personal attacks. We used the now-clichéd "when you did X, I felt Y" formulation.

The therapist who led the group was a woman with short black hair and a friendly, open face. Unlike Irene, she seemed to respect her patients and had a calm, nonjudgmental manner. The first family to go through the process was that of my pharmacologist friend, Paula. In her case, there really was something of a Rosebud moment, which encapsulated the family's pain. Paula's brother had died in a car accident and she had always felt that he was her father's favorite. The whole family was haunted by an incident where, while drunk, her father said that he'd wished Paula had been killed instead. No one denied this. Watching Paula and her parents discuss it for the first time in a place where they could feel safe from each other's judgment was astonishingly moving. Everyone in the room was crying. Both the healing and the relief as well as the exquisite pain were palpable. It was hard to imagine that this opportunity to air the incident and forgive each other wouldn't help them to move forward, even though catharsis alone is rarely enough to change behavior.

The next family up had been Judy's—and the day she disappeared was even more wrenching. Both of her parents showed up for the group and participated even after they learned she had left. We all tried to give them extra support as her father cried and cried. Watching this scene unfold, I knew that I didn't want to put my parents through any more pain. I began to feel that if I did use again, the guilt would be unbearable. There was no way to deny the effect addiction had on the family after this.

My own family was the last to undergo the process. It was hard for me to sit still while they told me about the impact my addiction had had; it felt over-whelming. I wanted to explain myself, to say, "But I didn't mean"—but it wasn't permitted. I sat, fidgeting, a river of tears running down my cheeks. My mother spoke about how awful it had been when she saw me in withdrawal and could do nothing to help and how ashamed she'd been when I showed up strung out and skinny without even an appropriate dress to wear to my youngest sister's bat mitzvah. My father talked about how my distance from him had hurt him and how worried my court case made him. My sister spoke about the pain she'd experienced during a summer when my friend Amy had stayed with us and how she'd felt excluded because we were so wrapped up in getting high. I was moved and horrified by it all. But it was my younger brother who really broke my heart.

He was just nine years old, and he simply said that his problem with my drug use was that "I never got to know you." He succinctly expressed the truth about my addiction: it kept everyone, including myself, from knowing me. It crystallized ways of seeing myself that were distorted and wrong. Then, it anes-thetized me so I didn't have to examine them, and it kept me from developing better means of coping.

When my turn to speak came, I talked about how I'd felt caught in the middle of my parents' separation and divorce, which occurred in my last years of high school and early years in college. The fact that these were the same years that my addiction took off was probably not unrelated. I don't believe their mar-ital problems caused my addiction, nor do I buy the "family systems" interpre-tation that was presented to us during Family Week. This was the idea that I unconsciously created my addiction as an attempt to bring my parents back to-gether. Given all of the other influences that led me toward it, it would be facile to simply pin the Rosebud here.

However, I do know that I used drugs to hide both my emotions around the divorce and my difficulty separating from my mother. Indeed, as my par-ents' marriage fell apart, my mother sometimes behaved more like a friend than a parent, turning to me for support and making me feel trapped between her and my father, always at risk of betraying someone. At the time, I thought our closeness was better than the contentious relationships I saw in my friends' families during their teen years. But it had also been a trap. In group, I spoke about how I had difficulty finding boundaries. We discussed the idea of the "parentified child" who tries to take care of her parents and soothes their feelings, rather than herself. I talked about how hurt I felt when, at the end of my addiction, my mother wouldn't come to court and refused to talk about anything other than getting into rehab. I felt as though we'd gone from being too close to being completely distant.

When I addressed my father, the issues were different. I discussed how his criticism and seeming inability to find anything I did "good enough" had hurt me as a child; it was only later I would realize this was linked with his depression, not my inadequacy. I also spoke with my sister about how I'd been jealous of her because she'd been able to be both popular and smart, unlike me. I talked about how I saw myself as a failed experiment, with her as the new, improved version, and how this wasn't fair to her. I tried to apologize for the way that it had interfered with our relationship.

And finally, I faced my little brother. In his case, all I could do was thank him for being brave enough to say what he did and for being there for me through the whirlwind of emotions of the week. Although I can never know how much this contributed to my recovery, I can say that I found it valuable. At the same time, I was beginning to wrestle with the conundrum of the 12 steps.

SIXTEEN

The 12-Step Conundrum

Take what you like and leave the rest.

—ANONYMOUS

THE AA SPEAKER WHO FIRST GAVE me hope that its program might work was a doctor. He had performed abortions illegally, before *Roe v. Wade* was decided in 1973—and that grabbed my attention because it was unusual. He mentioned it early on in his "qualification," or AA talk—so-called because it is intended to reveal why the speaker "qualifies" as an alcoholic. I often found these speakers boring and sometimes depressing, but this guy seemed like he might be diverting, at least.

John was around twice my age. His looks weren't memorable and he didn't seem like the best candidate to spark identification from a 23-year-old female heroin and cocaine addict. But he told a lively tale, which included not just drinking but drug dealing and a clear relish of the high life. His tone and warm demeanor soon had me listening intently. He said he had many years in recovery and was visiting the rehab as a way of doing "12-step work," or "carrying the message." It must have been about a week or two into my stay.

Twelve-step meetings and their rhetoric were not just an extra recommendation here. They were the core of treatment, as they still are in most American drug programs. If I wasn't at one of the two daily hour-long AA or Narcotics Anonymous meetings led by a fellow patient or an outside speaker, I was in group counseling that aimed to teach us to "work the program." Or, I was at a lecture by a doctor or watching a video with a similar message about addiction as a disease and 12 steps as the only treatment.

John did two things that prior speakers had neglected. For one, he didn't pretend that all of his drug use was misery. Previous speakers had been all too mindful of the way that positive accounts of drug use can spur cravings. Staff called this "euphoric recall." Consequently, most meetings focused relentlessly on the negative. However, rather than keeping me from wanting drugs, which I thought about constantly anyway, this made me feel that the speakers had had an entirely different experience of getting high than I did. After all, if my drug use had *always* been so hellish, I'd never have continued long enough to get hooked. I felt as though such speakers were either disingenuous or had nothing in common with me. John, however, made clear that he had loved getting high.

The second way in which his story seemed different was that he talked mainly about what he'd gotten from recovery—not what he'd had to give up. He discussed falling in love and getting married. The joy he took in his wife and children shone in his eyes. He also described enjoying his career and activism, which made him feel like he was making a difference. His qualification was fascinating and quirky—quite the opposite of the catchphrase salad I associated with meetings. Perhaps most importantly, I could see that he was actually happy without drinking or taking drugs, not just trying to make the best of a bad lot.

Other speakers had always seemed to emphasize deprivation. They didn't have such clear contentment in their eyes—sometimes, quite the opposite. Following their dreary litany of depressing drug experiences, they'd emphasize things like taking it "one day at a time" and avoiding "people, places, and things" associated with drugs. This did not make me want to sign up. Instead, it made it seem like the main idea was dropping all my friends, giving up the "druggy" music I loved, and avoiding pretty much anything else potentially fun.

I was told, for example, that I should dump my boyfriend since he used and sold drugs. Even at my worst, I could accept that this made sense. But as I started recovery, people in the program told me that I wasn't supposed to try to find a new partner, either, at least not within the first year, which felt like an eon in my 20s. So sex and love were off-limits, too. Concerts, parties, and nightclubs, for that matter, were simply "triggers" to be avoided, not potential alternate sources of pleasure. The same was true for people we'd gotten high with—basically all of my friends. Even sugar and caffeine were frowned upon; the rehab banned them and a secret high point of my stay was the time my roommate and I snuck into some guys' room to eat forbidden candies and listen to a hidden radio. (It did no harm, unsurprisingly.)

Overall, the 12-step speakers I heard seemed to advocate denying yourself anything that might feel good—not just drinking and drugs, but sex, comfort food, and rock and roll. The only vice you seemed to be allowed was cigarette smoking—and I wasn't about to take that up now. It had the appearance of a Puritan life of drab flavorless vegetables and forced calisthenics. The talks that highlighted this sort of asceticism were the kind of teaching that every student hates—didactic and focused on what the teacher thinks is important, not on what the students really want to know and can use.

I was already having a hard time imagining life without *any* psychoactive substances ever, which is what the program prescribed. In the Minnesota Model, anyone with a drug problem is assumed to also have an alcohol problem. That's why we all had to attend both AA and NA meetings, regardless of our preferred substances. The idea that I'd have to drop my boyfriend, my friends, my music, drinking (which hadn't been a problem), and even dessert made me feel hopeless.

John took a different tack. He focused on gains, not losses. He said, in AA terms, that if I wanted what he had, all I had to do was not drink or take drugs and go to meetings. It sounded too simple. But since the counselors and speakers constantly stressed that the only alternative was "jails, institutions, or death," I began to feel like perhaps I could accommodate to it. Maybe, his example suggested, it wouldn't even be that bad.

It was here that my complex relationship with the 12-step programs began. I initially embraced them. But ultimately I discovered troubling difficulties with this approach, such as the use of harmful tactics to force people to hit bottom and the way it mixes morality with medicine. Without understanding the system at the center of American rehab, however, we can't improve it. The 12-step programs claim to be a way to address a disease. But in practice, their most effective strategies are simply excellent ways to address a specific learning disorder.

IN 1935, WHEN Bill Wilson and Robert "Dr. Bob" Smith created AA, both were members of a then-popular Christian revival movement known as the Oxford Group. The steps were based on the Oxford Group's principles of surrender to God, confession, restitution for harms done, prayer, and service to others through proselytizing. But AA soon split off from the Oxford Group, which had been founded by a controversial preacher named Frank Buchman. (Just before the split, Buchman did not help his reputation when he praised Hitler and sought to convert the Nazi leadership into becoming his followers.) AA wanted to be a more ecumenical organization and the Oxford Group did not want its primary business to be converting drunks, so the parting was mutual.

Narcotics Anonymous, Cocaine Anonymous, and all the other 12-step fellowships basically rewrite AA's steps and literature to substitute their problem drugs or activities for alcohol, making the program itself essentially the same in all of them. However, the culture of particular groups can vary widely, even within fellowships aimed at the same drug of choice.

So how did a Jewish girl from Manhattan embrace and later come to question what I now see was designed as a Christian form of recovery? And why don't I advocate 12 steps for all since they apparently worked for me during the critical first five years of my recovery? Well—like every other issue in addiction—it's complicated.

For me, it started in rehab, when I first heard John. I was immersed in a milieu where every authority seemed to endorse the steps as my only hope. Because medical professionals—from the hospital detox doctors and nurses to the rehab physician and counselors to the OB/GYN who'd told his AA story—accepted what seemed like religion as actual medicine, I thought that there had to be strong science behind it. Since these apparent experts told me both that addiction was a medical disease and that the treatment was meetings, confession, and prayer, I didn't notice the contradiction that is now obvious to me when stating these facts.

Instead, it was a relief to see addiction as a disease and to know that although I was responsible for my recovery, the fact that I'd gotten hooked didn't mean I was a bad person. In meetings, I'd hear others talk about how they had hated and blamed themselves—and how the steps were the key to making past wrongs right. I knew that I needed to find some way to get over my sense of failure about the disastrous wrong turn my life had taken. Since I'd left Columbia, I'd found my future unbearable without drugs.

In the program, I learned that if I could just stop hating myself for hating myself, I didn't feel quite so bad. While the steps didn't formally address the issue, in meetings I'd hear people talk about swinging from grandiosity to self-loathing, often described in AA's earthy language as "feeling like the piece of shit in the center of the universe" or being "an egomaniac with an inferiority complex." I knew exactly what they meant and felt understood. Another slogan that helped was "don't compare your insides to someone else's outsides," meaning that most people try to present the best parts of themselves, but you can't hide your negative aspects from yourself. This concept is especially useful in the age of social media.

Simply hating myself was bearable if unpleasant; hating myself for doing so on top of that was not. And hearing friends, whom I knew to be good and kind, describe their own self-loathing helped me see how incorrect self-perceptions could be. This allowed me to ease up on myself, which reduced my need for escape. I began to learn the critical recovery skill of self-compassion.

Part of what the program gave me was hope—what AA calls the "power of example." Seeing people similar to me get better made a real difference—and I still think that this is often a crucial element in recovery. Although research shows that whether a *counselor* has his or her own addiction history does not affect outcomes, *some contact* with people who have been there and recovered often matters.

In fact, research suggests that the supportive community that 12-step programs provide is the main active ingredient in their success, when they work. The data are clear that social support aids both mental and physical health—and that people with more of it are much more likely to recover. Social support is the single most important factor in mitigating severe stress and trauma, which often contribute to addictions. Love doesn't always cure all—but without it, healing from psychological and learning disorders is almost impossible. We all tend to learn best when we feel safe and curious and want to connect and win our teachers' respect.

I also found some specific elements of the 12-step structure and literature helpful, beyond its large, welcoming community. For one, when you go to meetings daily, you hear thousands of stories and comments. Few other opportunities exist to hear people discuss their challenges and the mundane thoughts that disturb or distress them. With friends, partners, and families, we often want to hide this part of ourselves, so as not to upset or bore them—or, sometimes, because we don't want to display weaknesses or give those who might hurt us additional weapons.

But in a meeting full of strangers or semistrangers, the need for positive self-display is less pressing. For me, listening to a well-known model discuss how ugly she thought she was or hearing a renowned journalist discuss how he saw himself as a failure came with the hopeful suggestion that there was some possibility that my own negative ideas might be similarly deluded. And just hearing others admit their anxieties, fears, and hopes inevitably helps ground you.

Similarly, the same clichés that originally put me off sometimes held important truths. The rhyme "put gratitude in your attitude" still sets my teeth on edge. However, I recognized that I did genuinely have a tremendous amount to be thankful for and that focusing on this, in part simply by taking up mental energy, could push out negative thoughts and make me feel better. In the same way, the cheesy sign that said HALT, which stands for "Don't get too Hungry, Angry, Lonely, or Tired," gave me clues about what physical and emotional needs to take care of when I seemed to want drugs out of the blue. Sure enough, I'd often discover that a "drug craving" was actually hunger, irritation, or a need for sleep or social contact, which I could manage without resorting to heroin.

I later realized that the 12-step slogans are basically the groups' collective wisdom about how to deal with stress, anxiety, and other issues that could lead to relapse. In fact, many of the same ideas found in these often corny and haphazardly introduced sayings are the backbone of cognitive behavioral therapy (CBT) for addictions. This type of CBT is focused on the learned aspects of addiction and recovery and is one of the most effective treatments currently available, though sadly, it can be hard to find rehabs and therapists who actually utilize it as designed. CBT is far more systematic than randomly hearing slogans in meetings and doesn't complicate matters by moralizing—but nonetheless, if you go to enough meetings and get a sponsor, you are likely to learn much of what CBT recommends eventually.

Ironically, however, I also found the explicitly moral and spiritual parts of the program helpful—and yes, these are the same ones I still strenuously object to as medical prescriptions. But especially in my first months and years of recovery, I found hope in the idea of a Higher Power.

When I entered rehab, I was in that large, diverse group of Americans who call themselves "spiritual, but not religious." Culturally, I identified as Jewish:

I'd had at least five years of Hebrew school, was bat mitzvahed at 13, and was proud of the emphasis on education, intellectual achievement, social justice, and debate that I see as key to our traditions. I also desperately wanted to believe in an afterlife, since I remained terrified of death. In conjunction with my psychedelic experiences and reading on Buddhism, this produced a set of spiritual ideas that can be summed up like this: We're all part of some greater consciousness; we're here to learn and help each other; and in death, we somehow reunite with this life force.

Consequently, I wasn't averse to the idea of a Higher Power. Rather, I thought it odd that it should be involved with medical care. I was given no other choice, however, so I tried to go with it. The first step—"Admitted we were powerless over [alcohol/drugs]—that our lives had become unmanageable"—for me, became obvious. I could not control my drug use; it was destroying my life. If I continued to try to "manage" it rather than quitting, I would not get better. Clearly, I needed abstinence and was not going to be able to "just use cocaine and heroin on weekends" as I might have preferred. Though I didn't buy the idea of total powerlessness that some 12-steppers promote, I could see that any further attempts to use heroin or cocaine moderately would not be wise.

The second step, which is "Came to believe that a Power greater than ourselves could restore us to sanity," was harder to swallow. There are obviously plenty of mentally ill people who, despite sincere prayer, are left uncured. Still, I had always wanted to believe there was some God or force for good that cared— so I provisionally accepted it. After that, the third step of "Made a decision to turn our will and our lives over to the care of God as we understood Him" was no more difficult. I knew I needed help and support, so I decided to try praying for it.

And soon, I began to see what I interpreted as evidence that God was working in my life. For one, I was transformed physically: by the end of rehab, I no longer looked as though I was wasting away. That seemed like a blessing. As the months went by and I tried to pray and meditate, my cravings declined; this, too, seemed like divine intervention. My prayers had also seemed like they'd literally been answered when my ex-boyfriend Matt did enter treatment, days after I'd said the St. Jude Prayer for him with my Catholic roommate during rehab.

So, when the subway arrived on time, I began to see it as evidence of God's love; when it was late, that was a lesson in patience. When people were kind, I could feel God's presence; when they were not, it was a signal to move on. The program encouraged us to see the world as intimately shaped by a God who wants us to recover—and since I so urgently wanted to get better, I accepted it. I avoided thinking about the solipsism that can mar such a view. I brushed aside questions of theodicy, which was always an obsession for me as the child of a Holocaust survivor.

And frankly, going from shooting up 40 times a day and thinking I couldn't live without drugs to a completely sober life often felt miraculous. It was like flying from black-and-white Kansas to Technicolor Oz, from a dull, dank, gray hell to a riotous heaven, from being a frog to being a princess. In the context of what I was told by hundreds of people in daily meetings, and the support I felt from their care and acceptance, it was hard to see it any other way.

Moreover, I knew that my drug use was linked with my worldview. When I thought that the universe was friendly and that pain could lead to insight or at least sometimes had some meaning, bearing it made sense. If, however, I believed that "life sucks and then you die," I needed anesthesia and felt there was nothing wrong with seeking it. As Nietzsche put it, "He who has a why to live for can bear almost any how." I needed to have a reason to suffer; otherwise, why not get high? In heroin, I'd found something to protect me from life's ups and downs—to quit, I'd need a good reason to endure them. The program seemed to offer what I had been looking for all along: a way of seeing that would give me a community, an identity, and a way to make sense of the distress in the world.

AA's ubiquity in popular culture at the time also helped paper over its contradictions for me. In 1995, *New York Magazine* ran a cover story suggesting that it was a good place to find dates; the 1992 hit film *The Player* showed a nonalcoholic Hollywood producer attending meetings because that's where the deals were being made. In the late '80s and early '90s, the "recovery movement" was genuinely fashionable. Celebrities like Carrie Fisher, Robin Williams, Eric Clapton, and dozens of other A-listers were vocally public about their recovery via "support groups," which everyone knew was code for AA. A major network television show that focused on a comedian in AA, John Laroquette, ran from 1993 to 1996.

And, it was rare to see an addiction expert in the media support any other approach. Indeed, if anyone dared, the backlash was often vicious—as I'd soon find out for myself when I started writing about the conflicting data. But before I read the research, I had no idea how far away the positive media portrayal of 12-step programs was from the data collected by scientists.

Once I'd done the first three steps, however, I found the rest helpful. The fourth step, to continue in sequence, requires taking a "searching and fearless moral inventory." In practice, what you do is make a list of people you think you have significantly harmed, including people you resent or feel angered by. The fifth step is just disclosing this list to another person.

For my confessor, I chose a woman I admired. Geraldine was five years sober and was calm, wise, happy, and kind. She had short, gorgeous auburn hair and hazel eyes. She lived in an elegantly appointed, sunny duplex in the East Village, with a spiral staircase. When I arrived, she gave me a cup of herbal tea.

With my characteristic scrupulosity, I had typed out a 22-page single-spaced list. It included short descriptions of each incident, person, and offense; my family, friends, exes, and drug customers were on it. Without separating the pages of the dot matrix printout, I put it on her marble table like a lengthy scroll and began reading. She showed no signs of distress about what was clearly going to be a long session. Instead, she listened, with what felt to me like endless empathy and patience, without being overwhelmed, shocked, or bored. In fact, she admitted having done many of the same things that I'd felt were uniquely shameful.

I had gone into Geraldine's apartment heavy with self-loathing and moral disgust. But when I left, I felt accepted and acceptable, airy and light. In sharing my sins and foibles, I realized that I'd always felt subhuman because the standards I had for myself were superhuman. The self-hatred this evoked and the depression it probably helped create were key drivers of my addiction. To make matters worse, I had exacerbated the problem by being hard on myself for being hard on myself, creating a recursive loop of metadistress. Now, I was just human—not extremely bad, nor extremely good—and that was fine. I didn't have to be exceptional to be okay. The fourth and fifth steps had taught me an important lesson.

My "moral inventory" was not an exercise in self-debasement, as it can be for some people and as I'd feared it might be. It was not a humiliation or a trauma—the way it often is in rehab when it is forced. Nor was it something that could be used against me, the way counselors often do when people are open about their flaws in confrontational treatment groups. Rather than feeling shamed, I felt liberated. Essentially, the ritual helped me to start to put my past behind me—and was followed up later by making amends for whatever wrongs could still be righted, through the rest of the steps.

This allowed me to feel a sense of progress, offering both a way out of my recriminations and a way to see my behavior and character as changeable, not predetermined; malleable, not fixed. If you want self-esteem, I was told, do estimable actions. By changing my behavior, I learned that I could change myself: I could see that in how much my life had improved since I'd stopped the drugs. And, it turns out, it's really hard to see yourself as a terrible person when you are spending much of your free time attempting to help others. By engaging in the 12-step process, I learned a new and more adaptive way of seeing myself and the world.

And although listening to someone's fourth step might seem like it would be a drag—especially sitting through 22 pages of it—when I later did it for other people, it felt like a privilege. I found myself enjoying sponsoring newcomers, chairing meetings, and other forms of "service." What seemed tedious turned out to be joyous.

SO HOW CAN we reconcile the moral and medical? Is addiction, like light, inherently dual—not either a wave or a particle, but alternately a sin or a disorder, depending on context? Is it, unlike other psychiatric, neurological, or medical conditions, a special case, which does inherently require a moral and spiritual solution? Or is the way we treat it now a historical relic that will one day be seen as useless, primitive, and potentially harmful as we now see ice baths, insulin coma, and frontal lobotomies?

From the problems I've seen in 12-step-based treatment, I think spirituality and medicine need to be kept separate, and that the current addiction treatment system needs to be completely overhauled. But I also believe that 12-step programs don't get everything wrong. They have simply been misused and asked to take a role that they should never have played in professional care.

Although addicted people are neither uniquely spiritually needy nor especially prone to sin, social support, meaning, and purpose do matter for recovery. And, we can reconcile the conflict between morality and medicine without making addiction an officially stigmatized "special case" and the only condition for which mainstream medical care involves prayer and confession. The trick is disconnecting self-help from professional treatment and recognizing the strengths and weaknesses of both. Our health system isn't well designed to cope with learning disorders like addictions or conditions like type 2 diabetes, where new types of behavior need to be taught and supported. Self-help can be part of the solution in these cases, but it's important not to sell this as mediciné, nor to rely on amateurs as experts or promote any one group or philoso-phy as the best or only way to recover.

It's impossible for me personally to know whether 12-step groups were es-sential for my recovery—or whether another method would have been just as effective. In fact, when someone says to me that the 12 steps or any other treat-ment were the "only thing that could have helped," I sometimes ask whether the person has an addicted identical twin who failed all other approaches to pro-vide a counterfactual control. The lack of this type of evidence—obviously, with far larger numbers, not just one case—is the problem.

Indeed, the reality is that anecdotes alone simply cannot provide proof of efficacy. Data-driven approaches, like randomized controlled trials, are what give modern medicine its edge over priests and shamans. The reason that we can now prevent or manage killers like polio or AIDS is that the scientific method allows researchers to determine what causes them and then test and rapidly dis-card ineffective approaches. Anecdotes, in contrast, can make quack treat-ments seem effective and thereby deter progress for years, even centuries. A critical part of why addiction treatment fares so poorly in contrast to general medicine is that it has not been held to a standard that requires scientific proof. It isn't even held to the Hippocratic oath of ensuring first and foremost that no harm is done.

When such standards *are* applied to 12-step-based treatment, it is clear that what I was told about it in rehab and in meetings is not true. AA is not the only way, nor is the only alternative "jails, institutions, or death," as the slogan has it. AA and treatments like Twelve Step Facilitation are *not* the only effective approaches. They aren't even superior to other treatments when compared directly. A 2006 Cochrane Review—the highest level of medical evidence, often used by countries with national health plans to determine what treatments should be covered—summarized the data plainly: "No experimental studies un-equivocally demonstrated the effectiveness of AA or [Twelve Step Facilita-tion]," the authors concluded.

Moreover, the research that does show AA to be effective is overwhelmingly flawed by what is known as "selection bias." Basically, this means that it looks better than it is because the people who stick with it are different from those who drop out. In fact, when you force people into AA, they do no better—and sometimes do worse—than when given alternatives or no treatment, accord-ing to a review of the research. One study, for example, included over 200 work-ers mandated by their employers to get help who were randomized to AA, hospital-based treatment, or a choice of treatment. The AA-mandated group fared worst, with 63% later requiring additional treatment, compared to 38% for the choice group and 23% for the hospital group.

That's not to say that AA is useless, however. When you look at people who voluntarily continue to attend, like I did, they *do* have significantly superior outcomes. The problem is that as a result, it's hard to tell if these folks stay sober because they join AA—or if they join AA and stick with it because they are already highly motivated and more likely to succeed anyway. Selection bias is a fancy term for what happens when you compare fundamentally different groups to each other in treatment—and incorrectly conclude that the treatment, not the preexisting differences, is what mattered.

For example, one study followed 628 alcoholics—half women, mainly in their 30s, mainly unemployed and white—for 16 years. Before the study, they hadn't been in AA or treatment. They enrolled in the research either in detox or at a center where they'd sought a referral for treatment. To get a sense of the level of the severity of the alcoholism in this sample, consider that within 16 years, 19% of these relatively young people were dead.

At the end of the study, however, 67% of those who'd attended AA for 27 weeks or more in their first year were abstinent. Those who attended for over 52 weeks during years four through eight were nearly twice as likely to stay sober than those who went less often—with 71% of the frequent attenders achieving abstinence, compared to 39% for the others. That seems like quite a respectable success rate—until you start counting the dead and considering that only 28% of the group attended AA for 27 weeks or more initially. And there are numerous other studies that report roughly the same disappointing results. On average, 70% of those referred to AA drop out within six months, according to Scott Tonigan, a researcher at the University of New Mexico who has studied 12-step programs for decades.

However, the minority who do find these programs amenable are more likely to recover. Greater participation in recommended activities like helping others and having a sponsor is also connected with better outcomes. That means that it's still hard to tell whether these particular people would do equally well without it. But given that 12-step programs do provide social support and boost other psychological factors that are known to be connected with recovery, they probably do have positive results for some people. Moreover, the fact that these programs are free and available pretty much worldwide 24/7 makes them a resource for an underfunded, stigmatized area of health care that will remain important for the foreseeable future.

All that said, it is also clearly true that great harm has been inflicted by trying to force 12-step morality into a medical and criminal justice system. Some of its concepts, like the idea of hitting bottom discussed in chapter 14, are used to uphold harsh criminalization of drug use and punitive methods of treatment—both of which are ineffective and often harmful. Others—like the idea of powerlessness and the claim that the steps are the only way—can actively interfere with recovery if they are taken to heart. The mixture of institutionalized 12-step treatment and unchecked power over patients' lives can and has led to the formation of destructive cults posing as treatment providers at least a dozen times in the history of American treatment. This has produced large organizations that have inflicted traumatic stress on tens of thousands of people, often teenagers, such as Synanon, Straight, Incorporated, and the World Wide Association of Specialty Programs and Schools (WWASP).

Without doubt, the uniquely moral nature of the way we treat addicts as both sick and criminal also reinforces stigma. This is one reason why survey research finds, for example, that two thirds of Americans support employment discrimination against people with addictions, compared to just 25% who think discriminating against people with other mental illnesses is okay. It's also part of why there is so much resistance to accepting that addiction is a health problem. Arguing that addiction is a disease and the treatment is incarceration, prayer, and confession does not make the case that it's not a sin. It does the opposite. When the American Society of Addiction Medicine defines addiction in its literature as a disease with "bio-psycho-socio-spiritual manifestations," it undermines its case that it is a medical disorder better than any opponent of the disease concept could. No other mainstream medical society ever uses such terminology.

Further, the 12-step idea of powerlessness is disempowering. Research shows that the more someone believes in the idea that addiction is a disease over which he is powerless, the worse and more frequent a person's relapses tend to be. While some people can interpret "being powerless" in a benign way, others decide that it means they have no control not just over drugs but also over multiple spheres in their lives, including political action. This can be particularly damaging to women and members of minority groups, who are already faced with far too much powerlessness and do not have anything of the overweening sense of entitlement it was intended to address in the white men who designed it.

Teaching women that their only hope is 12-step groups can also leave them vulnerable to sexual predators. Unfortunately, predators are overrepresented in the program both because of the elevated prevalence of antisocial personality disorder among the addicted and because sex offenders are often court mandated to attend. One study found that 50% of women in AA had experienced behavior often euphemized as the "13th step," in which men try to seduce or coerce women who are new to the program. I personally narrowly avoided being raped by a longtime AA member; I had trusted him in part because the program teaches implicitly that people who have many years in recovery are admirable. I know at least two women who were not so lucky.

Youth who are forced to attend may also be targeted and victimized. Young marijuana smokers and drinkers can also be exposed to users of other drugs, who not only tell fascinating stories about cocaine, heroin, and methamphetamine but also know where to get these substances. It is not true in addiction care that "doing something" is always better than doing nothing: bad or inappropriate treatment can demonstrably make things worse. Because of these risks, 12-step groups should never be mandated for teens—and if attendance is suggested as an adjunct to treatment, it should be accompanied by clear warnings and only done in cases that are already severe.

Fortunately, though, it is possible to reduce the risks of 12-step programs while preserving potential benefits. This can be done by understanding addiction in the context of other disorders and in light of normal human behavior. First, addiction is far from unique in affecting the way the brain determines the motivational value of specific experiences like food, relationships, drugs, and romance. Immoral behavior is not isolated to addiction. All people have the capacity to behave badly—and many medical and psychiatric disorders can exacerbate it. Anyone who has ever "cheated" on a diet knows how hard it is to change habitual coping behavior—and addiction affects the brain's valuation systems in much the same way that overwhelming thirst or starvation does. This makes resisting temptation much, much harder.

Neurodevelopmental disorders like depression can also change how the brain sets its values, stunting lives by draining the pleasure from them and making nothing seem worthwhile. Schizophrenia, too, often includes harmful anhedonia symptoms that can warp motivation. Autism may make social experiences less attractive or pleasant; fear and anxiety can also drive harmful behavior. Antisocial personality disorder, unsurprisingly, profoundly affects values and, in its extremes, is linked with some of the most brutal crimes in history.

Indeed, pretty much any medical, neurological, or psychiatric condition that affects behavior can affect the brain's valuation systems and thereby morality: obesity obviously prioritizes eating, for example, and hoarding disorder can make piles of worthless possessions seem more important than people. Even a simple case of the flu can, at least temporarily, alter drive and ambition, making going back to bed pretty much the only alternative as immune cytokines produce exhaustion and lethargy. Our brains are embodied—much of the problem with the debate over addiction and psychiatry more generally is a refusal to accept this and our ongoing need to see "physical," "neurological," and "psychological" as completely distinct.

Despite all of this, however, we do not address the moral outcomes of any other medical, psychiatric, or learning disorder by making patients engage in confession and restitution. No psychiatric hospital would dream of forcing depressed patients to write "moral inventories"—and yet many rehab programs require it. No mainstream treatment for bipolar disorder or OCD demands submission to a Higher Power, nor is this recommended in schizophrenia—let alone in care for flu, stroke, or dementia. In conjunction with symptoms of many medical and psychiatric conditions, people often become irritable, inconsiderate, selfish, dishonest, or even violent, but treatment does not include confession, prayer, or amends. In fact, it would be considered outright offensive and abusive—if not malpractice—for a medical professional to demand this.

The 12-step programs and the medical system are a terrible mix. These groups are not and never were intended to be medical treatment. They were designed as mutual help organizations, not professional therapy. Like faith-based groups for people struggling with divorce, like religious congregation members who band together to visit the sick, like spiritual support groups for cancer patients, 12-step programs are meant to be lay groups where afflicted people help each other. Just as no one would trust a brain surgeon whose only training was having had a brain tumor removed, we shouldn't think that simply having recovered from addiction makes someone an expert.

Consequently, indoctrination into 12-step ideology should not be the core component of rehab—it shouldn't be part of professional care at all. It should not be mandated by courts or used coercively in any context, particularly for women, minorities, and youth. This is not to say that social support doesn't matter—but many options need to be made available. Such support also needs to be freely chosen—or rejected—by individuals, if it is to help.

Further, regardless of the fact that the idea of a Higher Power can be made acceptable to some atheists, every U.S. court decision made on this question has found that mandatory AA violates the separation of church and state. Yes, it's possible to see a 12-step group or a doorknob as your Higher Power—and Jews, Muslims, and Buddhists can make it work for them—but that doesn't mean that the program doesn't have an underlying Christian theology. Even though AA can be taken as "spiritual not religious," it doesn't make sense to require people to find some form of belief within its framework and then undertake moral education as part of their medicine. The risks of allowing providers to use punishment, humiliation, and emotional attacks to attempt to create a sense of powerlessness or force a "bottom" are too great. Abuse in institutions is common and difficult to rein in even when punitive measures are banned. It's inevitable when staff are instructed to see patients as liars and con artists and told to deliberately try to make them feel helpless.

To top it all off, AA's own guidelines say that members should never be paid to "carry the message" of the program to others. This clearly violates Tradition Eight, which says that AA should be "forever nonprofessional" and that people should not be compensated for "our usual 12th-step work." Proselytizing is intended as voluntary service and cannot serve its spiritual purpose if money changes hands. But because state standards are weak and no national ones exist, thousands of counselors whose only expertise is in the 12-step model are paid for such work every day.

In this context, it's especially insane—given the extremely limited resources available—that government and insurers pay incredibly high prices for what can be had for free in most church basements. Rehabs often charge over $1,000 a day and tend to suggest a minimum of 30 days—and many hours of each of those costly days are spent in 12-step meetings or lectures. Other therapies that have been shown to work as well and sometimes better—like motivational enhancement therapy, medications, and cognitive behavioral therapy—require trained therapists and doctors. Why not use all of the treatment time to offer these treatments, which aren't available for free? Teaching practices like surrendering to God and taking moral inventory in rehab not only wastes time and money, but it also reinforces stigma and undercuts the idea that addiction is a medical problem—specifically, a developmental disorder.

The relationship between 12-step programs and treatment should be this: these programs may be recommended, perhaps even available on-site during treatment, but should never be forced or made part of the official rehab curriculum. Secular alternatives like SMART Recovery, LifeRing Secular Recovery,

Women for Sobriety, and, for those who seek to moderate their drinking, Moderation Management should be given equal weight. If any preference is given to suggesting that people try 12-step groups it should be made on the explicit basis that these organizations are still, by far, the most readily available worldwide and are free. While people in recovery from addictions should be welcomed and given educational opportunities and priority in hiring in the field, this can only work if their qualifications go beyond their own experience and they understand that different people need different therapies.

Further, if any 12-step recommendations are given by medical or legal authorities, the programs' pros and cons should be explicitly discussed. Women and youth must be warned about the possibility that not everyone in the program has good intentions; no matter how long someone has been in the program, that person may still be a predator. In fact, if the program represents the general population of addicted people, around one in five members will have a personality disorder that is marked by manipulative and dishonest behavior. The problems with "hitting bottom" and "powerlessness" should also be explained.

For me, AA's most important slogan was "Take what you like and leave the rest"—and this message should be strongly imparted to anyone who is referred. I did personally find the moral and spiritual aspects of 12-step programs helpful—but I also know that the opposite is true for many people, and I can't say that what worked for me was specific to addiction.

Moreover, addicted people are far from the only folks whose misbehavior can damage the social fabric. Learning to repair relationships and treat yourself and others humanely, coming to understand the restorative power of volunteering and service, and engaging in practices like meditation and prayer that focus the mind beyond the self can be helpful to many human beings, regardless of whether or not they have addictions. The 12 steps are one system for doing so—not the one true way.

The mistake our treatment system makes is in pretending that these teachings are therapy and the only way to manage the morally stigmatized disorder of addiction.

SEVENTEEN

Harm Reduction

Harm reduction values life, choice, respect, and compassion over

judgment, stigma, discrimination, and punishment.

—HARM REDUCTION POSTER

MY CALENDAR NUMBER WAS CALLED, AND I walked steadily to the front of Judge Snyder's cramped courtroom. The justice flashed me a brief smile, with what seemed like unusual warmth. Then she briskly asked my lawyer, Don, and the assistant district attorney assigned to my case to approach the bench. I tried to listen, but I couldn't make out what they were saying. I heard a burst of genuine laughter at one point, however, which was highly unusual in a room usually filled with tension and tears.

My appearance on September 9, 1988, was the polar opposite of the previous one, just four months earlier. No longer weak from withdrawal, I wasn't sickly. No longer painfully thin, I felt fat; I'd gained some 40 pounds, which actually put me at a healthy weight. My skin wasn't pale, but tan. After 30 days of rehab, I'd spent three months in a halfway house in Arizona on the outskirts of Phoenix, getting plenty of sun on my several-mile walks to and from my job as a maid at a Marriott Courtyard hotel. On some weekends, I had swum miles in Saguaro Lake, surrounded by canyons and cactuses.

Now, I no longer had the distorted pupils of addiction or withdrawal. My hair was still a terribly pale bleach blond, but at least it no longer stuck out in stubby clumps of frizz. Both my parents were with me, tentatively proud of how far I'd come. I didn't realize it, but my looks testified to my recovery at least as clearly as words could. The judge saw as near complete a physical transformation as is possible without surgery. But while I was no longer shaking from withdrawal, I still feared a possible 15-year mandatory prison term.

After Judge Snyder announced my next appearance date, Don guided me out into the hallway. "She's going to give you a chance," he said in his Brooklynese. "If you stay off drugs, she's going to try to work it out so that you stay out of prison." He cautioned that this might not ultimately be legally possible but said that it was still great news. Then he told me why they'd been laughing.

Judge Snyder was stunned by how much better I'd looked. But she did have one complaint. "That hair color," she said to Don. "Tell her to do something about that hair." I laughed, but it took me at least another year before I realized she was right and moved to a more natural color.

IT WOULD BE another four years before my case was finally resolved, years during which I completed college, wrote freelance articles for *The Village Voice* and *Spin* magazine, and began work as an associate producer for PBS's *Charlie Rose* talk show. (Bizarrely enough, I got the *Charlie Rose* job after answering an ad in the *Village Voice*; I used the article in which I'd disclosed I'd been an IV drug user as one of my writing samples.) During that time, I attended daily 12-step meetings and appeared in court at least once every month or two, updating the judge on my life.

Facing a felony charge is a major life stress, which has been compared to losing a close friend or facing a life-threatening illness. I can certainly attest to that. Often, it felt as though I was at the mercy of a relentless and unthinkingly brutal machine, like an ant in front of a tractor. Talk about powerlessness: in court, it seemed as though I were in a foreign country and barely able to speak the language, helpless to do anything but watch as my entire future was debated in ways I couldn't understand or affect.

But actually, I had it easy. My family was able to pay for a good attorney. We had health insurance. I was white. It's impossible to know how much the privileges of race and class were what enabled me to get decent addiction treat-ment, return to college, and help my lawyer demonstrate to the court that I could soon be a respectable taxpayer. They certainly mattered, however.

I had chosen not to reenroll at Columbia, both because of the expense and because I was afraid that the environment might be risky for my recovery. In-stead, I attended Brooklyn College, where I became an A student again. To sup-port myself, I got a job as a receptionist. This made it simple for me to show the court that I was making progress, something that would have been far harder for someone who hadn't previously had my advantages. In fact, when I needed letters of support to be sent to the court, my wonderful statistics and psychol-ogy professor, the late Bart Myers, concluded his by joking that if the judge had anyone else like me on her rolls, she should send them to his classes.

My race alone made me stand out in court. At most of the more than three dozen appearances I made, I was the only white defendant. The evidence was stark: the judge and most of the lawyers were white, while the accused were pri-marily people of color. Although studies show that whites use drugs at the same rates as African Americans and actually sell more—and I certainly saw high rates of use and sales by whites in the Ivy League in the '80s—people like me aren't typically the ones being arrested and prosecuted.

My multiple court appearances were mainly what are known as "calendar dates," where I had to show up to prove I was still drug-free and to allow the judge and the lawyers to discuss the next steps for the case. On these days, judges also accept pleas and impose sentences. They are basically stultifying bureau-cratic routines interspersed with an occasional moment of ritualized drama. The lawyers, mostly white, would address the judge. One by one, their black and Hispanic clients were led away in chains. I would watch those who'd been sentenced kiss their spouses, partners, or children good-bye, would observe the mothers try to quiet the screaming babies, and silently weep, both in sympa-thy for them and in fear that I was looking at my own future.

Even now, these disparities remain a national disgrace. These days, the life-time odds of going to prison for a black man in America are 1 in 3—a figure that has doubled since Richard Nixon declared war on drugs in the 1970s. This rate is more than five times higher than for white men, who have a still shock-ingly high six percent chance. The racial inequities are clearly driven by drug enforcement. Between 1980 and 2011, the annual number of drug arrests of black men rose 164%, while arrests for property crime and violent crime fell: homi-cide arrests fell by 46%, robbery arrests by 27%, and burglary by 42%.

Moreover, racial disparities occur at almost every point in the criminal jus-tice system. A 2003 analysis found that black people are ten times more likely to get arrested for drug crimes, compared to whites. African Americans are also ten times more likely to be sent to prison for these offenses, according to a 2009 study. Conviction rates, too, are higher. In federal prison, on average blacks serve nearly as long for drug offenses as whites do on average for violent ones, 4.89 and 5.14 years, respectively. And 82% of people sentenced under the harsh fed-eral mandatory minimums for crack cocaine are black.

This shouldn't be surprising, given the racist origins of our drug laws, all of which, as we've seen, were passed by stirring fear and using blatantly racist im-ages and rhetoric. And it has been enabled by a concept of addiction that remains more moral than medical and was born in the same racism as drug prohibition.

During my time in court, in the late '80s and early '90s, it was the peak era of New York's harsh Rockefeller laws. This legislation had been passed in 1973 so that then Governor Nelson Rockefeller could look tough on crime—and im-plicitly, black people—in a planned run for president. The laws mandated terms of 15 to life for possession of four or more or sales of two or more ounces of narcotics, even on a first offense. That made the crime I was charged with, possession with intent to distribute more than four ounces, an A-1 felony.

It was equivalent to the sentence for second-degree murder and more seri-ous than rape—and because those violent crimes did not have mandatory sen-tencing requirements, the result was that, in New York and many other states that followed its lead, first-time nonviolent drug offenders often served more time than rapists and killers. One judge, upon sentencing a defendant whom he believed did not deserve a lengthy sentence, said, "[T]he mere possession of four ounces of this controlled substance is considered by the State of New York to be just as serious as the taking of a human life. That to me is an absolute atroc-ity. . . . However, I am obliged to enforce the law however stupid and irrational and barbarous it be." The percentage of New Yorkers in state prison for drug crimes more than tripled between 1973 and 1994 as a consequence of the Rocke-feller laws. More than 80% of those sentenced—who had to serve many years with no possibility of parole—had no history of violence.

Tellingly, the same harsh rules initially applied to marijuana—not just co-caine and heroin. Starting in 1973, New Yorkers unlucky enough to be caught selling two or holding four ounces of weed were looking at a decade and a half in prison. But given how likely such laws were to enmesh white kids, marijuana wasn't included for long. Once white middle-class college students started getting

15 years for pot, political pressure was placed on the governor and state legislature. By 1979, cannabis no longer triggered draconian sentences, illustrating once again the different way we treat drug crimes depending on who we see as the "typical drug addict" or dealer. By 2008, *more than 90%* of New York State's drug prisoners were minorities. (It took until 2009 for significant reform to be enacted—at that time legislators removed most of the mandatory minimum sentences, except for the A-1 felonies, which were reduced.)

However, even as far back as the '80s, it was patently obvious that the drug war was failing and the courts were staggering under the sheer weight of the numbers being put through them. Between 1985 and 1992 alone, the number of New York City residents sent to state prison annually for drug crimes quintupled, going from 2,000 to 10,000. The figures were similar nationally: the American incarceration rate rose over 400%, with the number of state and federal prisoners going from 200,000 in 1973 to 1.574 million in 2013. If you include the population of state and county jails, our incarcerated population is 2.7 million; and if you add in people on probation and parole, the total correctional population is close to 7 million, or three percent of the entire U.S. population. We are now the world's largest jailer, with 5% of the world's population, but 25% of its prisoners.

The courts, judges, prisons, jails, police, probation officers, and every other aspect of the system have been overwhelmed by this tsunami, with overcrowding, staff shortages, paperwork issues, and other problems leading to frequent delays and errors. But for decades, politics also meant that this truth was unspeakable by those who could do anything to change it. Opposing any part of this crackdown was "soft on drugs" and, therefore, political poison for both Democrats and Republicans, who had to be sure they outdid each other in proving their fealty to absolute war on drugs.

Indeed, one of the most extreme ironies of the drug war was that by the time I got busted in 1986, New York City had become the nation's capital of crack. When that drug had initially begun to spread, much of the trade, especially on the East Coast, moved through this city. I was in the cocaine business at the time; I watched as many freebasers went from having to buy powder to make smokeable cocaine for themselves to being able to purchase ready-made crack on dozens of corners in Manhattan alone.

This market grew exponentially during the worst part of my addiction, in a state where, when crack hit, we'd already had nearly a decade of experience with the Rockefeller laws. Even far short of the A-1 felony level that I faced, these laws were severe. If harsh, mandatory sentencing did deter selling and use, New York should have been the last place to develop a crack problem. By pioneering tough sentences, we should have been spared a crack epidemic, or, at least, should have been less hard hit.

The fact that the opposite occurred—and that Congress went on to pass what were essentially federal versions of these laws focused on crack in 1986 and 1988 once this was already clear—was a bleak illustration that drug policy was not really about preventing or managing drug problems. For all the media hype about the threat of crack gutting the middle class—there were over 1,500 stories about the drug published in the *Washington Post* alone in the 12 months between October 1988 and October 1989, for example, and in just one month in 1986, 37 stories on the evening news about it—the people being taken away in handcuffs were overwhelmingly not suburban whites. In fact, crack never did hit the suburbs the way it devastated poor communities.

However, when white kids *do* start being known as the ones who are being arrested and locked away, the solutions for drug problems become very different—as occurred with marijuana under the Rockefeller laws and as we are seeing now in the public response to the rise in heroin and prescription opioid addiction. Viewing heroin as a white problem, politicians call for the expansion of harm reduction programs—like needle exchange and access to the overdose antidote naloxone—not crackdowns or locking up more users and low-level dealers for longer. Activists advocate more treatment, not more punishment.

But although the press claims that heroin use is now a white thing for the first time, in fact, the majority of heroin users have been white since the 1970s. And from 2010 onwards, blacks have actually been underrepresented. Moreover, if you look at the data rather than the headlines, the group that is now hardest hit is not the upper middle class—instead, it's lower middle class, poor, and working-class whites who have fallen prey to stagnating incomes, debt, job loss, foreclosure, and financial insecurity. Rates of heroin addiction in people making less than $20,000 per year are triple those for people who earn $50,000 or more. As whites increasingly face the same kinds of economic hardships that preceded the '80s crack epidemic in minority communities, it is not surprising that they are becoming more vulnerable to hard drugs. Addiction disproportionately kicks people who are already down.

*　*　*

AFTER THE JUDGE told me she would try to avoid incarcerating me, I became less frightened about my future. However, my legal fate was still unclear. Both the judge and the prosecutors could see the racial disparities in front of them. They did not want to make them worse. They certainly didn't want to let the few white people who had gotten indicted off the hook.

Consequently, the prosecution refused to compromise: they wanted hard time. The judge, however, wanted to give me probation. Since I was employed, in recovery, and helping others to kick, she saw no reason to send me to prison. The assistant district attorney argued that I was being given preferential treatment because I was white; Judge Snyder noted that she had given the same breaks to some black and Latino defendants who had also put drugs behind them. As this process continued through multiple delays, years went by.

The official end to my encounter with the Rockefeller laws was so bureaucratic and anticlimactic, thankfully, that I can barely remember it. According to records that are now sealed, it occurred on July 17, 1992, nearly six years after my arrest. That day, Judge Snyder granted a motion by my lawyer to dismiss the case "in the interest of justice."

I had four years off drugs, a cum laude degree, and a budding career in public television and freelance journalism. For Judge Snyder, the stalemate had gone on long enough. Since the prosecution would not allow me to plead guilty to a crime that would permit a noncustodial sentence, she dismissed my case outright. This not only spared me prison time, but means that I have no criminal record. In her decision, she concluded, "It is clear that the defendant has made extraordinary strides toward her rehabilitation. Most importantly, she has become a 'role model' for people who have drug problems by virtue of her success after drug addiction, and she continues to be an example for people who suffer from narcotic and alcohol addiction. These achievements would be destroyed by the imposition of a jail sentence at this time and society would no longer benefit from her work."

I was extraordinarily, exceptionally, almost extravagantly lucky. As a result, I felt and continue to feel obligated to do all I can to make sure that others are able to be treated with similar mercy. If we can move away from the idea that addiction itself is a sin and get past a medical model that is still rooted in morality and racism, we will do much better for far more people and for far less money.

* * *

AND AS MY court case was wending its way through the system, an alternative to the war on drugs was starting to take shape. It's called harm reduction. Since the 1980s, this philosophy of drug policy has gone from a tiny, radical movement to a mainstream approach that threatens the foundation of world drug prohibition. Unlike current policies, harm reduction recognizes the crucial role of learning in addiction and the failure of punishment to solve drug problems. It is also based on the idea that even during active addiction, people can learn and change. That insight probably saved my life—and it can save many more if we adopt its principles more widely.

Two months after I'd first shared needles in 1986, a harm reduction worker visiting from San Francisco taught me how to protect myself from HIV. At the time, half of New York's IV drug users were already HIV positive, but little effort was being made to reach us. We were thought to be hopeless unless we kicked drugs, unable to learn and change if we weren't in recovery. Consequently, I didn't know AIDS was even a risk for addicts, despite reading two newspapers a day. The little coverage the disease got back then was almost entirely focused on gay men, with addicts mentioned as an afterthought, if at all.

I was about to shoot up in the East Village kitchen of my friend Dave's apartment when Elise—a friend of his who came from the Bay Area and worked for a group fighting AIDS in IV drug users—stopped me and told me of the danger. I watched intently as she showed me how to clean a needle. Elise emphasized that it was best never to share works. But if there was no alternative, drawing bleach all the way into the needle and expelling it at least twice and then rinsing the whole syringe thoroughly with water could reduce the odds of infection. After her demonstration, I cleaned my works compulsively, just as I used them. In fact, I never used without precautions again. I later learned that Dave was HIV positive and that several other people with whom I could easily have shared syringes were also infected. I was outraged that I had unwittingly been at such great risk and that no one cared to save the lives of people I knew and loved.

Once I'd been educated myself, I taught everyone I knew who shot drugs about the virus and about safer practices. I carried small vials of Clorox wherever I went. I graffitied instructions on cleaning needles on bathroom walls in diners and restaurants in the East Village frequented by IV drug users. I tried to conduct a survey of other drug users to illustrate that I wasn't unique and that given the opportunity, injectors would change needle-sharing behavior, even if they wouldn't or couldn't give up drugs. After I completed rehab, I began covering AIDS and addiction issues, trying to ensure that no one else would have to die of ignorance.

The harm reduction movement, which had protected me from HIV, had its start in Liverpool, England, and many of its pioneers were current or former drug users themselves. It owes its name to a 1987 article published in a British newsletter by Russell Newcombe, which was headlined "High Time for Harm Reduction." In his article, Newcombe argued that policy makers must accept that drug use is ineradicable and focus instead on minimizing damage associated with it. In essence, harm reduction is the idea that the goal of drug policy should be to minimize negative effects of drug use, not use itself. And by 1987, it was becoming clear that the biggest danger associated with illegal drugs was HIV.

To fight AIDS, harm reductionists argued for two critical weapons: needle exchange programs and the expansion of methadone maintenance. In 1984, a group of IV drug users in the Netherlands known as the Junkiebond had fought the spread of hepatitis B by working with the local health department to distribute clean syringes and teach safer use—so there was precedent for this idea. As for methadone, even as far back as the 1980s, it was the only treatment known to reduce injection drug use and related mortality significantly in the long run. Unfortunately, at the time, both the United States and the United Kingdom had policies that facilitated the spread of AIDS: some laws made clean needles hard to get, and methadone programs had restrictions that prevented people from getting or staying in treatment. In America, abstinence-only and 12-step rhetoric about the need to hit bottom before recovery could occur—and the claim that addicts either couldn't learn to change needle-related behavior or wouldn't bother—dominated media and political dialogue on the issue.

By 1988, however, British harm reductionists had managed to convince the conservative Thatcher government of their logic. That year, the British government's Advisory Council on the Misuse of Drugs released a report endorsing easier access to methadone and clean needles, saying, "the spread of HIV is a greater danger to individual and public health than drug misuse. Accordingly, services which aim to minimise HIV risk behavior by all available means should take precedence." This was harm reduction's essential insight: drug use and even addiction aren't necessarily the worst that can happen.

Harm reduction's fundamental principles are these: stop trying to fight drug use, most of which does not cause harm. Don't focus on whether getting high is morally or socially acceptable; recognize that people always have and probably always will take drugs and this doesn't make them irrational or subhuman.

Instead, work to find practical methods that reduce risk and minimize damage—and understand that everyone can learn, just not all in the same way.

For instance, shooting drugs doesn't inherently spread HIV and other viruses, but sharing syringes can—so provide clean needles. Opioid overdose risk is not static; it's linked with fluctuating tolerance—so provide maintenance drugs and connect those who want more intensive services to them. Most use isn't linked to addiction or significant health consequences—so don't worry about it. Instead, find out what plagues those whose drug use is harmful, risky, or chaotic, and teach and empower them to make changes that improve their lives.

From the harm reduction perspective, drug use and even addiction isn't necessarily irrational; it's a response to the environment and people can learn to make better choices about it. However, these choices will depend on their own values, which may differ from those that governments or health care workers share. Harm reduction inherently respects the dignity of drug users, recognizing that whether or not they want to stop using drugs—and whether or not their choices around drug use are totally free—they can and often do make decisions that can improve their lives and health if given the opportunity. It puts users at the center of policy, embodying the disability rights' movement's mantra of "nothing about us without us." Harm reduction explicitly embraces the idea that all drug users, including those who are actively addicted, can learn if taught and provided appropriate support. It emphasizes access to such education and care as a key response to drug problems.

In alcohol policy, harm reduction (though not labeled as such) has long been both successful and relatively uncontroversial. The "designated driver" is a harm reduction approach: it accepts that people will drink to excess and works to cut drunk driving, not drinking. Media campaigns against driving while intoxicated also fall under this rubric. Just as needle exchange is neutral on needle use per se, campaigns that criminalize or stigmatize taking the wheel while drunk do not argue that drinking itself is wrong. Their target is impaired drivers.

Together, media campaigns against drunk driving, stricter laws, and the designated driver produced one of the greatest successes in public behavioral health. Since public concern began to grow in the early 1980s, the percentage of deaths associated with drunk driving has fallen from 53% in 1982 to 31% in 2013. The total number of drunk driving deaths in the United States has been cut in half, from over 21,000 in the early 1980s to just over 10,000 in 2013 (with help, as well, from improved auto safety, mainly in the form of seat belt laws).

But unlike the United Kingdom, the United States fought tooth and nail against harm reduction for drugs other than alcohol. As late as 2008, Alan Leshner, the former head of the government's National Institute on Drug Abuse, published a paper in a scientific journal calling for researchers to avoid the term because it was controversial and linked to legalization, which he seemed to see as unthinkable. For years, representatives of the U.S. government even lobbied at international meetings to get other countries to abandon harm reduction policies, or failing that, at least avoid using the phrase. When AIDS was a rapid death sentence, abstinence-based and 12-step treatment providers typically opposed harm reduction on principle, ignoring data and the fact that the people they were supposed to be serving were dying. Religious leaders and politicians—both left and right—spoke out against it. Activists like ex-addict Jon Parker and members of the AIDS activist group ACT UP had to get themselves arrested in order to legalize needle exchange in New York, which happened in 1991, only after a judge ruled that the health threat of AIDS was a more important public health concern than the laws against needles. Even now, a ban, sponsored by conservative senator Jesse Helms, on federal funding for needle exchange programs remains in place, despite the fact that the data are unequivocal.

To take just one example, compare the spread of HIV in the United States and the United Kingdom. Great Britain acted early with harm reduction to fight the virus. That prevented an epidemic in IV drug users. Indeed, with expanded access to methadone and clean needles, AIDS infection rates never rose above 1% in needle users in England. Consequently, because there was no pool of infected drug users to spread the disease, there was also no heterosexual epidemic and no epidemic among babies infected during pregnancy and birth.

In contrast, in New York, which acted late, in some neighborhoods the prevalence of HIV in people with addictions went over 60% and the prevalence of the virus in the *general population* was 1%. In New York alone in the 1990s, a thousand babies a year tested positive at birth for the virus and two thirds of these babies were born to addicted parents. But after New York's harm reduction programs took off and syringe access was legalized, these rates plunged. In 1992, IV drug use accounted for 52% of all AIDS cases. By 2012, the figure was 3%. The state now calls syringe exchange "the *one* intervention which could be described as the gold standard of HIV prevention" (emphasis in original). Because of both needle exchange and antiviral drugs, New York no longer has any "AIDS babies."

The power of harm reduction, however, goes far beyond needle distribution. The human interactions that take place at harm reduction programs show its deeper, but hidden, benefits and how it helps teach change. Because harm reduction recognizes that people learn best when they feel welcomed, respected, and safe, it reaches people who were thought to be beyond help and hope. Despite years of writing about it, I still find the experience of harm reduction in action to be deeply moving.

For instance, when I went out on the streets with the early needle exchangers before it was legal, it was like being with the Pied Piper. From nowhere, the activists would soon rapidly be surrounded by IV drug users of all races, gender variants, sexual orientations, and social positions (the only significantly underrepresented group was young people, probably because word about the exchange spread first among more experienced users). Many looked the part: they had visible tracks or the sickly mien of severe addiction. Quite a few, however, did not: there were men in suits, old folks with canes and in wheelchairs, women with strollers, and couples you might expect to find snuggling at a movie theater, not holding hands waiting for needles.

What was most amazing, however, was the emotional climate. The crowds felt benign, not threatening. Their interactions didn't have the tension of an illegal drug transaction—nor did they have the general air of dreariness, depression, and despair I had experienced at my methadone program. Instead, participants were grateful and quietly protective. Many were obviously moved that anyone cared at all—let alone enough to risk arrest to try to help them. It was as though they couldn't believe their eyes and for at least a moment, their masks of distress and pain would lift.

Most of the first needle exchange clients were people hardened by years of addiction and stigma. They were primarily folks that "respectable" people walked on the other side of the street to avoid. The experience of having someone give anything to them freely, without strings, petty bureaucratic rules, or moralizing, was almost completely alien to them. Virtually every other encounter they had with mainstream culture—doctors, methadone and other drug counselors, charities, government agencies, even stores—came with a big dose of shame like a lecture, a religious pitch, a urine test, or some other potent reminder of their low status. Nearly every social contact was itself an admission of defeat, helplessness, or need.

Not so with needle exchange. Although it is called an exchange, free needles are also given to those who do not have used ones to return. Although return is strongly encouraged to allow safe destruction of potentially infectious waste, typically, nothing is officially required. People must simply be what they already are: IV drug users. And, it turns out, the unexpected kindness of being helped without expectations is itself a potent psychological intervention. The idea that someone sees you as both being able to behave responsibly about your health and worthy of a chance to live is even more powerful. This opens doors to hope in surprising ways.

For one, every time I covered needle exchange, I found stories like Erin's, which I told *Spin* magazine in 1991. When I first met her, she reminded me of how I had once looked: painfully thin, wired on coke, pale, and unkempt. She had stayed up all night to meet the exchangers because, "They were the only decent thing in my life. They listened to me and they had the best works." But the next time I saw her, I literally didn't recognize her. She had 74 days free of drugs; a woman at the needle exchange had helped get her into recovery. During that time, she'd put on 25 pounds. Now, her eyes were lively. She was smiling. "Eventually," she said, "I want to do needle exchange."

But even when the positive outcomes weren't as obvious, they were visible. Dan Bigg, a harm reduction pioneer who was the first to distribute the overdose-reversing drug naloxone to people at risk through his Chicago Recovery Alliance, defines recovery as "any positive change." In that light, its glimmer is visible during most harm reduction encounters. You could see it clearly in the interaction between workers and drug users, which were respectful and supportive. You could see it in the eyes of the participants, which showed life when they were treated with dignity. And you could see it in the way drug users tried to take care of each other, bringing clean needles to those who could not attend the exchange themselves.

Critics who say that such programs "send the wrong message" could not be more wrong. Needle exchange and harm reduction don't say: Go on and kill yourself with drugs, no one cares. They tell people—both drug users and non-users—that everyone deserves life and dignity and that being addicted shouldn't be a sentence of death or exile from humanity. To some of the most downtrodden and powerless people, these programs say: I believe in your ability to protect yourself and others. I believe you can do something that matters. You don't need to be forced to make good choices. This teaches an extraordinarily valuable lesson.

When people start to be valued by others, they start to value themselves. And even when their drug use remains unchanged, harm reduction nearly always increases the amount of warm, social contact that the most traumatized and marginalized people have. Because this is essential to coping with trauma, it provides a foundation for further growth. Harm reduction is the opposite of tough love—it is unconditional kindness and imbues what looks to outsiders like irredeemable ugliness with startling moments of transcendent beauty.

MY PERSONAL BREAK with 12-step programs began when I learned about harm reduction and what the science really shows about addiction. But I didn't act on it until I developed serious depression in my seventh year of recovery. I had never been a dogmatic stepper who believed that being on psych meds wasn't "really recovery"—an attitude that was dominant in meetings in the '80s and '90s. In fact, I encouraged many of my friends who were depressed and in the program to see psychiatrists and follow their recommendations, despite the widespread opposition and derision. I thought ignoring medical advice was dangerous—and the official position of AA was that members should not "play doctor." Still, I resisted trying meds myself. To some extent, I had bought into the program idea that "pain [is] the touchstone of all spiritual progress" and that "escaping" any emotional pain via drugs could only be dangerous and coun-terproductive.

I had yet to realize that too much suffering can be even more life-constricting than addiction: being trapped alone in bed in emotional or physical pain can be as bad or worse than being trapped alone in a room shooting up. Nor did I recognize my own bias, which made me secretly feel morally superior to people who had to rely on maintenance treatments like methadone or psych meds like antidepressants.

In 1995, however, I hit a wall. I began to dread everything and everyone. Nothing made me feel good or even just okay. For most of my recovery, even during my worst times, I felt motivated in my work—but now that, too, seemed bleak and pointless. When I couldn't stop crying at work, I dragged myself to a psychiatrist who prescribed one of the serotonin-enhancing antidepressants.

To my surprise, the first day I took Zoloft, I was reminded of my earliest recreational drug experiences—with psychedelics. Not long after I'd swallowed my first pill, I felt a sensation I remembered from using LSD. It was a sense in the pit of my stomach that things were about to get weird. (This is probably due to the fact that both drugs affect serotonin receptors, most of which are—oddly enough—in the gut, not the brain.) Indeed, soon I was also having mild hal-lucinations. Whenever I closed my eyes, I'd see complicated, brightly colored geometric patterns, just like the vivid structures I'd seen while tripping. Sometimes, the images would linger even after I opened my eyes, especially if I was looking at a blank wall. I called my psychiatrist, who told me to halve the dose, saying the hallucinations would pass. I was skeptical, but I took her advice.

And two days later, the psychedelic patterns were indeed gone. The depression and pleasurelessness, however, were as strong as ever. I even found myself missing the trippy sensations. At that point, I became aware of yet another reason why I'd continued using cocaine long after it had ceased to be at all enjoyable. Although I certainly had what felt like a neurochemically driven compulsion to shoot coke, the distraction of experiencing something, anything at all, was also preferable to blank anhedonia, if only barely.

Ten days in to my antidepressant odyssey, I felt the first therapeutic effects. I noticed that I wanted to write. To top that off, I actually felt better while writing and after finishing an op-ed. That tiny boost in pleasure and drive gave me some optimism, enough to reduce the dread that kept me from wanting to do anything or see anyone. This led to a startling conclusion: I now saw that a crucial part of my depression was a complete inability to feel good in any way. Without that, I was just as far from being able to change as I'd been when I was completely numb and overwhelmed by my addiction. This pain hadn't led to growth—but to stagnation.

The fact that depression is joyless isn't itself especially shocking, of course. But what I did find remarkable was that I hadn't even noticed the loss of so many different types of pleasure and reassurance. I recognized their reappearance with great relief and with wonder that I'd not even been aware of their absence until they came back. Because I'd been incapable of experiencing pleasure, it seemed I was also incapable of recalling it or imagining its possibility in the future. That, of course, probably helped produce my overwhelming apprehension. If you don't even know how to remember good experiences, you're not exactly going to see the future as friendly.

When the medication fully kicked in, I felt transformed—similarly, in fact, to the way I'd felt when I first found heroin. But I wasn't euphoric or blissed out. Instead the Zoloft provided an overall sense of comfort and safety, a buffer that reduced my oversensitivity. That, far more than the euphoria, had really been what had hooked me on the heroin. I wasn't greedy for constant ecstasy; what I wanted was not to feel so overloaded and overwhelmed all the time. Historically, in fact, opiates were among the first antidepressants—and there is now research suggesting that the opioid maintenance drug Suboxone can help with depression that is resistant to other treatments.

For me, on antidepressants, I wasn't "better than well"; I was the way I am when I'm okay. My constant mental atmosphere of objectless dread was now replaced with a neutral or slightly positive emotional valence. In other words, I stopped fearing encounters with friends and dreading the phone. I took pleasure in simple accomplishments. If something awful happened, I felt appropriately upset; the difference was that now I no longer cried uncontrollably while watching families reconnect in TV commercials. I felt competent and far less needy. The reassurance I'd sought from 12-step meetings and from frequent phone calls to my sponsor and other members for repeated affirmation didn't seem so necessary anymore. I felt okay, just as I was. I could hate myself less because my selfish needs and intrusive worries were genuinely less pressing.

Paradoxically, what the drug gave me was greater control over my own thoughts and behavior and more self-sufficiency. I could still choose to act impulsively when irritated, but I could also more easily choose not to do so. I could reach out to friends when I felt lonely, but I didn't feel like my loneliness was the result of a secret defect that meant either that no one could really love me or that if they did, I could not feel it or believe in it. This reframed my perspective on recovery, forcing me again to recognize how physiological psychology can be. It reduced my emotional intensity—and allowed me to be more empathetic. Another paradox: Zoloft did this by making me less sensitive, which meant I could tamp down my distress at other people's pain and actually try to help them, rather than running away or withdrawing because it was so upsetting.

While some studies suggest that all antidepressant action is actually a placebo response, this "numbing" effect of SSRI's is consistently replicated. In fact, even some of the psychiatry critics who think that these drugs are mainly placebos see this reduction in emotional sensitivity as a true pharmacological effect. Somewhat illogically, however, many of the same critics denounce it as a horrifying side effect, ignoring the possibility that these "placebos" may work for some people through precisely this pharmacological mechanism.

As with similar claims that methadone and other types of maintenance automatically turn people into emotionless zombies, however, this defies more than logic. Not everyone has the same wiring and not everyone needs to be more sensitive. Sometimes, less is more. Too much emotion can be just as likely to cause social problems as too little. For people whose addictions are driven by sensory and emotional overload, such relief can be critical. Indeed, people on the right dose of a maintenance drug—whether an antidepressant or an opioid—aren't "numbed out"; they are at the right level of sensitivity to allow optimal emotional functioning for them. Since people start out tuned differently, where they start will affect what they need to get to the right place. If your baseline is too far above normal, you will need a different remedy than if your baseline is far below it.

Of course, for a person who is undersensitive, reducing emotionality could truly be awful and perhaps even produce suicidal thoughts or pathological ennui. For a sociopath, it might reduce empathy even further, which could be catastrophic. In other people, it might be neutral. However, for me, it was positive. The action of a drug depends not just on its pharmacology but on the person's baseline. If you are starting out way too high on a dimension—even a seemingly good one like sensitivity—going lower may be helpful. The wide range of responses to antidepressants probably has to do with the huge variety of natural human wiring, which can make the same drug into a poison, a panacea, or a placebo, depending on the dose, the timing, and the patient.

Consequently, the contention that there is one true way to recover and that people who need antidepressants or opioid maintenance drugs should be seen as second-class citizens or "not truly recovering" is as absurd as arguing that radiation is the only treatment we should offer for cancer and claiming that those who do well with chemo "aren't really" cancer survivors. Different pathways lead to different types of addiction; different wiring and different needs require different treatments. That some of these may have opposite effects in one person compared to another is not surprising, given the variety of reasons that people take drugs and the variety of risk factors for addictions. Discriminating against people merely because they have a particular chemical in their bloodstream—regardless of how it actually affects them—is just as irrational as doing so on the basis of other physical characteristics, like race or gender. What matters is what works best for each individual. And what works best will depend on a complex interaction between a person's biology and what they have learned.

WHAT ALL OF this adds up to is that addiction is driven by learning—not just exposure to a particular substance, having certain genetic traits, or being traumatized. Dependence on a substance, person, or experience to function isn't its essence—a learned compulsion that continues despite punishment is what matters. To effectively address addiction, then, we need to recognize the enormous variety that is created by the fact that it is learned as people mature, which results in all kinds of different outcomes and needs. One single treatment cannot possibly fit all. However, because learning matters, there are commonalities that all successful approaches to addiction will share.

For one, to help people overcome learning problems, they must be treated with compassion and respect. All types of students do better in schools where they feel welcomed and safe—in fact, schools that generate a real sense of community have been found to have students with fewer drug problems, compared to those that treat kids as though they are suspects and must be supervised by dehumanizing measures like drug testing and metal detectors. People learn best in environments where they feel connected to others—not places dominated by a sense of threat and fear.

Harm reduction recognizes these social and learned components of addiction. It "meets people where they're at," and it teaches them how to improve their lives, whether or not they want to become abstinent. As harm reduction pioneer Alan Marlatt put it, "Harm reduction does not try to remove a person's primary coping mechanisms until others are in place." Instead, it allows people to learn new skills and then move away from drugs, rather than attempting to force them to do the reverse. In many cases—particularly among the most traumatized and disenfranchised people—this may be the only path to recovery. The poor prognosis seen with current abstinence-only treatment models may, at least in some cases, be a result of failing to understand this.

Either way, when we recognize addiction as a learning disorder, the critical role of compassion in preventing and treating it becomes clear—as does the inappropriateness of a coercive or criminal justice approach. The same factors that ease learning of complex material in any setting will help addicted people, too. Language, for example, which is one of the hardest intellectual challenges a human must master, is learned almost entirely without punishment. Babies learn to talk not because we hit them if they don't, nor because we lock them in closets when they use bad grammar. Instead, children start speaking because it's fun and rewarding and connecting.

And just as both children and adults learn better when they feel secure and wanted, so, too, do addicted people respond better to treatment that is calming rather than threatening and sees them not as bad people, but as those who are lacking information and skills that teachers can share. Being a student is a more empowering role than being a patient, as rehabs obliquely acknowledge by calling completion "graduation." But truly recognizing the role of learning in addiction requires far greater transformation of our systems for dealing with it than this. The rise of harm reduction is only the start.

EIGHTEEN

The Kiwi Approach

Make one drug illegal and another, more dangerous one will take its place.

—MIKE POWER, *DRUGS 2.0: THE WEB REVOLUTION THAT'S CHANGING HOW THE WORLD GETS HIGH*

ONLY A FEW WEEKS EARLIER, COSMIC Corner had been well stocked with legal, recreational drugs. Under a stripy, candy-colored marquee located in the center of the city, with a distinct '70s vibe, the store, one of a chain of such retailers, sold psychoactive products with names like Mind Trip, Kronic Skunk, WTF, Stargate, and (my personal favorite name) Everest Tibetan Toot. Most of these products promised an experience similar to that of smoking marijuana; some claimed to be comparable to cocaine; others mimicked drugs like ecstasy. All contained newly created, legal, synthetic drugs. Thousands of doses in small, often condom-sized packets were being sold.

In most countries, newly synthesized psychoactive substances like these reside in a gray zone. If their contents haven't specifically been banned and their packages are labeled "not for human consumption," sales are, at least technically, not against the law. (Worldwide, it is generally illegal, of course, to sell new and untested chemicals as foods or drugs.) These so-called legal highs are developed rapidly, often within weeks after governments have officially prohibited their predecessors. Unsurprisingly, given the fact that these new chemicals often haven't been tested on animals—let alone humans—some have been linked with psychotic reactions, seizures, and deaths. Nonetheless, in the United States, they are commonly sold at convenience stores, gas stations, and head shops that sell other drug paraphernalia, as well as online.

In late May of 2014, however, I was outside a Cosmic Corner in Auckland, New Zealand. From November 3, 2013, to May 8, 2014, over-the-counter sales of several dozen new psychoactive substances had the official governmental seal of approval and could be sold for human consumption. These were truly legal "legal highs." Cosmic Corner had the license to prove it. Experts from Europe, the United States, and the United Nations were watching encouragingly as New Zealand took on an unprecedented drug policy task. The Kiwis were trying to enact, for the first time in world history, a rational system for testing, approving, and regulating recreational drugs. In order to create drug policy that genuinely reduces harm and minimizes the odds that addiction will be learned, this is the kind of innovative thinking that is needed.

In 2013, the UN agency in charge of drugs admitted in its *World Drug Report* that "the international drug control system is floundering, for the first time, under the speed and creativity of the phenomenon known as new psychoactive substances." By 2012, more than 1 in 10 American high school seniors reported having taken such a substance—typically one marketed as a substitute for cannabis—in the past year. That made legal highs second only to marijuana itself in popularity, and it was almost certainly the fastest rise of a new recreational drug category since such statistics have been kept.

I had traveled to this small island nation at the bottom of the globe to learn more about its radical alternative approach to these drugs. The world's first agency to set safety and efficacy standards for medicines was what is now known as the American Food and Drug Administration (FDA), which was founded in 1906. Today, nearly all countries have medical regulators that perform similar functions.

Until 2013, however, no nation had ever even tried to create such a system for nonmedical drugs that alter consciousness, despite their widespread use. Our current way of regulating—or simply banning—nonmedical drugs was not arrived at by any scientific or rational process, as we've seen. No sane policy maker, for example, could justify allowing commercial sales of a drug like to-bacco, which cuts life expectancy by ten years on average, while using incar-ceration to punish those who take marijuana, which is not associated with increased mortality at all. Our policies are not based on rationality, risk assess-ment, or logic—just history.

To create strategies that actually can reduce addiction and drug-related harm, we need to evaluate them scientifically. One of the key insights of harm reduction is that drug risks can't be considered in isolation. Pharmacological dangers associated with particular substances need to be weighed against harms related to black markets, like disease, crime, and corruption—as well as harms linked to enforcing drug laws, like incarceration, and harms linked to alterna-tive substances that users might take instead. This is where understanding the role of learning in addiction is especially important. Pharmacology, dose, use patterns, use locations, drug prices, and cultural beliefs all influence the development of addiction. None of these can be controlled through outright prohibition. But they can all be shaped by regulated markets.

New Zealand has begun to try to do just that with its legal high regula-tions.

NEW ZEALAND BECAME the first country to attempt to create an FDA for rec-reational drugs for several important reasons, two of them unique to its his-tory and geography. The first is its isolation from the trade routes for traditional illegal drugs like cocaine and heroin. "The boat doesn't stop here with cocaine," says Matt Bowden, a steampunk guitarist and legal high manufacturer who is himself probably the second major reason his country became the first to take the plunge.

Known locally as the "godfather of legal highs," Bowden, 43, is the second of seven children born to an IBM computer systems engineer and a piano teacher in Auckland. With hazel eyes and tousled brown hair, he smiles as he shows me through his gleaming white and chrome lab, the crown jewel of his company, Stargate International. Here, he hopes to create new drugs that can meet the safety standards of the Psychoactive Substances Regulatory Authority— which his own lobbying helped create.

As a fan of psychedelic drugs and the rave scene, Bowden became a mar-keter of "herbal highs" when he wasn't able to make a living in music. (These are products like Red Bull or the now-banned "natural" stimulant ephedra, which was sold in the United States as a health supplement.) Informally, he began studying pharmacology in the '90s, looking for substances that were not illegal and might be sold as safer highs. His timing was perfect: at the turn of the twenty-first century, the Internet and on-demand manufacturing had just begun to allow drug connoisseurs to order new substances of interest from Chinese or Indian labs, no questions asked. Soon, many were doing so and comparing notes in online forums.

Not since the nineteenth century—when an earlier wave of globalization made cocaine, opium, and marijuana as easy to buy as aspirin—had there been such a burgeoning and unregulated pharmacopeia. And by all indications, the future promises only more acceleration. In 2014, researchers at Stanford demonstrated that it's possible to produce opioid drugs like morphine using a genetically modified form of baker's yeast; in 2015, researchers demonstrated the success of a process that could allow mass production of these substances. Some of the scientists involved are already calling for strict controls to prevent the yeast from leaking to traffickers—but if self-reproducing organisms that can produce an unlimited supply of drugs get into the hands of individual users, the entire market could change forever. Soon, even the production of traditional illegal substances or illicit versions of pharmaceuticals could become a highly decentralized cottage industry, posing the same kind of regulatory challenge that the specter of 3-D-printed firearms holds for the project of gun control.

In the '90s, when Bowden began his journey, New Zealand's club scene was rapidly becoming infatuated with methamphetamine, devolving into a quagmire of lost jobs, paranoia, and poor health as users moved from snorting to injecting. This wave swept up both Bowden himself and his soon-to-be wife Kristi, a Penthouse Pet. When they met, she was working as a stripper and already injecting. And, when he gave her an ultimatum that said either you stop or I'll start, she handed him the needle. "He very gallantly dove off the deep end with me," Kristi says. As their addictions worsened, a meth lab explosion killed one of his close friends. Another friend crashed through a glass door at a party, and, believing himself invincible, fatally impaled himself with a samurai sword. Bowden knew there had to be a safer alternative drug somewhere.

In the pharmaceutical literature, he came across intriguing data on a substance called benzylpiperazine, or BZP for short. Proposed as an antidepressant in the 1970s, it was never approved because testing for abuse liability showed that stimulant users liked it. Otherwise, BZP didn't seem to have major safety issues—and as it turns out, it's actually not highly addictive. "It doesn't reward binge behavior," Bowden says. "The next day you feel like you don't want to have any more for a while." Best of all, BZP was not on New Zealand's official controlled substances list, nor was it banned by international drug laws.

Consequently, Bowden thought BZP could be a safer substitute for meth. Clubbers like him wanted something to provide energy for all-night dancing. Rather than ignore this apparently perennial demand, why not offer a different substance that served the same function but didn't provoke binges or overdose? Bowden says he always saw his work as harm reduction, though he also admits he stood to profit from it.

In 2000, he began to have small quantities of BZP made in India. Soon, Bowden himself ended up turning to the drug in much the same way methadone is used in heroin addiction, a kind of DIY replacement therapy. "I addressed my addiction issues at a biological level," Bowden says, noting that he saw many others use BZP to reduce or eliminate their methamphetamine habits. "It was working," he says. "People were able to get their lives together, hold down a normal job, and end their involvement with organized crime. We thought it was pretty important."

Then, Bowden began to market BZP to the masses. He sold it in head shops and convenience stores, competing directly, he hoped, with meth dealers. Before long, he was selling millions of doses. Within several years, a study found that 20% of all adult New Zealanders had tried "legal party pills"—as products made from BZP became known. And, nearly half of the respondents who reported having used both legal party pills and illegal party drugs said they used BZP to replace the illegal drugs.

Both Bowden and New Zealand were extremely lucky in the drug he chose to sell. Many substances tested far more extensively by the pharmaceutical industry turn out to have disabling or even deadly side effects that aren't discovered until the drug is widely marketed. The heart problems caused by the now-discontinued painkiller Vioxx are just one recent example. Even large clinical trials will often miss rare but dangerous effects.

Fortunately, however, BZP doesn't seem to cause lasting damage. "Twenty-four million [BZP] pills were consumed by 400,000 consumers on 9½ million occasions over 8½ years with no deaths and no lasting injuries," Bowden says, citing a 2007 analysis he commissioned from an independent researcher. Research conducted for the New Zealand government basically confirms his position. It found that there were no deaths associated with BZP alone, but one or two have been reported when it is taken with other drugs. (In these cases it is not known if the other drugs could have accounted for the deaths by themselves, so it's not clear how much or whether BZP contributed.)

As party pills spread, the media was as skeptical as you might imagine it would be toward new recreational drugs available to kids 24/7 at corner stores. Soon, sensational coverage of BZP replaced stories about the meth epidemic. "Have New Zealand authorities been caught napping on the latest chemical craze?" asked one fairly typical article. But Bowden was undaunted and also had some media savvy. He managed to show up in virtually every story, ready with a quote about how BZP was "a safe alternative" to illegal drugs. Early on, he had his eye on the long game, arguing that new psychoactive substances ought to be regulated, not banned. Personally, he preferred to be a legal drug lord.

By 2004, however, in the face of mounting public concern, Bowden and BZP were facing a government investigation. Here, too, Bowden met the scrutiny head-on. He went directly to the Ministry of Health—often dressed in rocker regalia, sporting waist-length blond cornrow braids—and began meeting with groups like the Expert Advisory Committee on Drugs, which advises Parliament. Stargate's response to the country's meth epidemic, Bowden argued, was consistent with New Zealand's national drug policy of harm reduction, which its government had officially embraced years earlier. Bowden argued that the best way to cut harm was to come up with safer drugs that could be credibly marketed as such.

New Zealand is a small country of 4.5 million people with a small government that is occasionally capable of surprising levelheadedness. When the Ministry of Health responded to the uproar by ordering an expert committee to study BZP, one member of the team was Dr. Doug Sellman, a former director of New Zealand's National Addiction Centre at the University of Otago in Dunedin. Sellman looks the part of the distinguished researcher, with short white hair, glasses, and an air of genial authority. Given that it was his job to investigate BZP, Sellman decided to try it himself, as an added data point for his research.

"I went down to Cosmic Corner," he says. "They gave me this little packet. I went home on a Friday night and took it and sat in front of the television, waiting." Nothing happened at first. "But the extraordinary thing was that I was still sitting in front of the television at four o'clock in the morning and I just didn't need to sleep," he says. The next day he had what he describes as the world's worst hangover. "I went back to the [committee] and said, 'You don't have to worry, really, about this drug,'" he says, citing the lack of euphoria and the aftereffects.

In April 2004, Sellman's committee released its report, concluding that BZP didn't fit the definition of a dangerous drug and therefore shouldn't be outlawed. However, it didn't fit into other legal categories like food or medicines, either. The report suggested that the government consider creating "new categories of classification that can incorporate some levels of control and regulation." Quietly, it began doing so.

In the meantime, the party pill market exploded. Before long, it was taking in $15 million annually. Bowden now had numerous competitors, many of whom cared nothing about harm reduction. Some sold BZP in insanely high doses. Users also began taking it to extend alcohol binges. By 2006, 40% of New Zealand men between the ages of 18 and 24 reported having used party pills at least once in the previous year. Moreover, a distressing 1 in 100 users said they'd visited an emergency room as a result, with symptoms like seizures, shaking, and confusion. By 2007, BZP was seen as a national threat. Although it wasn't causing deaths or lasting harm, the next year, the political pendulum swung and BZP was prohibited.

AS MANY OTHER countries have since discovered, outlawing one legal high doesn't solve much. The Kiwi BZP ban set off an early version of a pattern now familiar globally: the legal highs arms race. One substance would be banned, only to be replaced by another, sometimes more harmful than the last. "I've banned 33 separate substances, 51 or 52 different products, and they keep being reformulated and reappearing," New Zealand Associate Minister of Health Peter Dunne told a local paper, looking back at the worst period, in 2013. By then over 4,000 New Zealand stores were selling legal highs.

Once again, political pressure mounted for something to be done. But what, exactly? Banning a whole possible universe of drugs is, it turns out, a very different matter from banning an individual drug—which is precisely the problem that the rise of fast supply chains, decentralized manufacturing, and the Internet have forced on the world. Countries cannot outlaw everything that gets you high; that would rule out alcohol, coffee, tobacco, and maybe spices, sex, and rock and roll. They also can't easily prohibit drug analogs, preemptively banning anything that's chemically similar to known drugs like pot or coke.

America has tried, but such legislation is hard to enforce because some molecular look-alikes act in differing and even opposite ways—and some drugs that appear quite dissimilar chemically can have virtually identical effects. For example, a substance with a chemical structure similar to that of the active in-gredient in marijuana might turn out to actually block marijuana highs; mean-while, a drug with a completely different structure might offer the perfect stoner experience. Which of these two is the "marijuana analog"?

A chemist might argue for the first one, while a pharmacologist could make the case for the second. The inability of experts to agree on what an analog really is makes prosecuting these cases complicated—and the U.S. Supreme Court re-cently ruled that prosecutors must prove that sellers know the drug is an analog if they are to be convicted. Our loose regulation of "health supplements," which allows many "natural" substances to be sold without testing makes matters even murkier. Testing of these health store products frequently finds new psy-choactive substances that are sometimes also sold as legal highs.

And, to make it even more complex, a blanket ban on any chemical that can substitute for an illegal drug runs the risk of accidentally outlawing, say, the cure for Alzheimer's. Once a drug becomes a controlled substance, research on it is restricted and often requires expensive licensing fees. Many companies just abandon such compounds, even when they are promising.

The more you look at these problems, the less possible it seems to ban one's way out of them. And as New Zealand came to grips with its legal highs crisis, the solution that eventually rose to the fore was what Bowden had championed all along: allow drugs to be sold, but only the safest ones.

OVER THE YEARS, Bowden became as much a professional lobbyist as a drug maker, though he has sold many products, ranging from an ecstasy substitute to synthetic cannabinoids. In late 2004, he founded a trade group called the So-cial Tonics Association of New Zealand. He chose the word "tonic" specifically because he wanted to avoid the moral stigma associated with "drugs"; "tonic" also harkens back to the nineteenth-century world where medical and recre-ational drug use weren't really separate.

And as a lobbyist, he's been surprisingly successful. As it became clear that the new legal highs market was dwarfing the old market for BZP, the coalition in favor of regulation grew. Bowden hired Chen Palmer—a law firm headed by a former prime minister—to further legitimize and solidify his efforts. Together, they helped draft the language of what would become the Psychoactive Sub-stances Act 2013.

As public demands for action heated up, the lobbyists made the case that regulation was the only way to keep the drugs out of the hands of kids and protect consumers. Importantly, proponents presented their case not as a call for legalization, but as a kind of crackdown—a measure that would make the industry responsible for proving that its products were safe. This framing may prove critical in getting such regulations passed elsewhere, if other countries ultimately want to follow New Zealand's lead. The idea is not to make more drugs available but to try to get safer ones to push out the more dangerous and recognize that regulation offers more control than prohibition. "We are reversing the onus of proof. If they cannot prove that a product is safe, then it is not going anywhere near the marketplace," said Dunne, the associate minster of health, in a press release issued during the debate. "The new law means the game of 'catch up' with the legal highs industry will be over once and for all."

In the summer of 2013, the Psychoactive Substances Act passed by a margin of 119 to 1. It's not clear whether all the politicians who voted for the law fully realized that they were creating the world's first regulatory system for recreational drugs, but that's what they did. The law mandated that legal psychoactive substances could only be purchased by an adult at a licensed outlet. To determine which substances ought to be legal, it required manufacturers to conduct clinical trials and sell only "low-risk" products that passed those tests. To preside over this new market, the law created the Psychoactive Substances Regulatory Authority—an office whose unenviable first task was to sort through the million gory details of what all of this might actually mean in practice.

For instance: How do you define a low-risk recreational drug? The question is more mind-bending than it might first appear. Accepting the level of risk associated with, say, cigarettes would legalize virtually anything short of absolute poison; after all, half of heavy smokers die from their habit. If you set the level of risk associated with alcohol as your standard, numerous deaths and injuries would still be acceptable: booze causes six percent of deaths worldwide.

However, if the bar is set at the risk level associated with marijuana, which does not increase death risk even in heavy users and is less addictive than alcohol or tobacco, few new drugs of any kind would be legalized. Nor, for that matter, would many other forms of recreation. The odds of death from swimming? Approximately 1 in 57,000. Climbing Mount Everest? Nearly 1.5 in 100, roughly the same as those for someone who shoots heroin for a year. While marijuana may occasionally lead people to take stupid risks, there has never been a single reported overdose death. The challenge of regulating drugs rationally requires thinking about risk in new ways.

Other unprecedented tasks loomed over New Zealand's regulators as well. They had to determine how to regulate manufacturing, imports, and sales. They had to devise clinical trial protocols for drugs with benefits that aren't medical. For example, how do you scientifically measure and weigh qualities like transcendence and euphoria against risks to health? The higher the value placed on these intangible qualities, the higher the level of acceptable risk will be when balanced against them. Indeed, the U.S. FDA itself, which was only recently given regulatory control over cigarettes, ran into trouble with a similar task in 2014. The agency had conducted a controversial risk/benefit analysis of e-cigarette regulations, which was required under a new law aimed at making regulations less onerous to businesses. The analysis included a measure of "lost pleasure" to smokers who quit. When this factor was weighed against the health benefits of the regulations, it made the regulation seem more costly and less useful.

The questions that measuring pleasure during addiction raises rapidly become philosophical. For instance: How much pleasure is an addicted smoker really getting from cigarettes—and how much of this is simply relieving withdrawal caused by the habit itself? Surely some smokers still enjoy at least some of their cigarettes—but most report that this is not the case most of the time. How do you accurately quantify pleasure and loss of it here? In the smoking case, after a public outcry, the FDA was forced to change tactics and is no longer permitted to weigh loss of pleasure in smoking as a strike against smoking regulations.

But these issues are where values shape science and where the rubber hits the road in recreational drug regulation. Typically, pleasure and transcendent experience have not even been seen as worth consideration in discussions of recreational drug laws; only harms are counted, with benefits seen as being products of user delusions or simply ignored. More generally, in the West, unearned pleasure has been labeled as sinful—the opposite of valued. Whose values should determine how much risk is acceptable from a drug, particularly one that can be addictive and therefore entangle some proportion of its users in a state that is antithetical to enjoyment and that typically reduces productivity? When is it okay to value pleasure over health? Why is potentially fatal skydiving okay but not LSD—a drug that is not linked with addiction, mental illness, or overdose death? Moreover, if a drug can be shown to improve productivity or creativity, how should this be weighed against risk? In a multicultural society where values vary widely, these questions are truly vexing.

New Zealand is the first country to try to face them, and it hasn't yet begun to find even provisional answers suitable for their country. Indeed, before regulators could start wrestling with clinical trials, their most pressing challenge was to figure out how to manage the transition to a regulated market—a task that nearly derailed the whole process. Needless to say, none of the nearly 300 products that were sold before the law was passed had gone through testing; very little was known about them at all. However, banning everything right out of the gate would have sent consumers en masse into the arms of the black market. And so, in a calculated risk, the law allowed drugs that had already been sold without reported problems for at least three months to stay legal.

To detect problems, the Psychoactive Substances Regulatory Authority set up an online tracking system that allowed doctors and hospitals to report adverse events connected to any given product. With data from these reports, the agency started informally rating every product on the market. To do so, they used a modified version of a numerical scale devised by the Poisons Information Center in Freiburg, Germany.

Minor risks to any bodily system rated a 1; moderate risks a 2; and severe, permanent, life-threatening risks a 3. Because the Freiburg scale was designed to measure acute poisoning, not addiction risks, the New Zealand regulators added measures for withdrawal. Minor effects were weighted differently depending on how common they were, with greater frequency meaning higher scores. Any substance that scored 2 or more on the scale was banned. Of the nearly 300 products on the market, only 41 survived this first test. These were granted interim approval. Legal sales began on July 18, 2013.

NEW ZEALAND'S INTERIM regulations hit rough water almost immediately. The trouble stemmed in part from an unforeseen problem, one that initially appeared to be not a curse but a blessing.

On the eve of the regulated market's debut, the number of stores selling legal highs in New Zealand fell by 95%. It's not entirely clear why this happened, but it appears that some store owners were reluctant to go through the hassle of acquiring licenses. Conflicts between local and national authorities over the approval of specific sales outlets may have also helped winnow the numbers.

On its face, the sharp drop-off in retail outlets seemed like a big win. Police, for one, said the market was much easier to keep tabs on. Certainly, fewer outlets reduced the odds of sales to children. But what political operatives call the "optics" of the situation were terrible. Users concentrated near the rare stores that remained, and so did TV cameras. Suddenly legal highs were more visible than ever. Images of young, unhealthy-looking people hanging out near legal highs shops "created the public impression of it being this huge problem," Dunne says.

These media portrayals also showed that race is not absent from New Zealand's policy debates: virtually all of those pictured to illustrate legal high stories were Maori people, the nation's original inhabitants. Chris Fowlie, who owns The Hemp Store, which has sold legal highs in Auckland since 1999, expected media calls about the new law since he frequently got them about marijuana issues. But none came. "It's because our customers don't fit their narrative," he says. "They weren't the homeless winos, they weren't young funny-looking teenagers with facial piercings, or the kinds of things that they want to take a picture of and put on the news." Journalists didn't see unharmed users as being a story—so they didn't look for them when they covered the law, even though this meant that their stories weren't representative.

That 2014 was also a national election year in New Zealand was another bad break for the Psychoactive Substances Act. Journalists and politicians called the policy a disaster. Incumbent politicos claimed that they hadn't been aware of what they were voting for. Not expecting any substances to pass legal muster, they seemed to have thought that regulation would be de facto prohibition. In May, less than a year into the experiment, Parliament passed a revision to the Psychoactive Substances Act. It banned all of the substances that were then on sale. It also prohibited animal testing in New Zealand to determine legal high safety, which basically makes doing so impossible.

Many in the New Zealand media called the amendment the practical end of the Psychoactive Substances Act. But Dunne insists otherwise. "It's not that we suddenly had a U-turn," he says. The only real change, he says, is that the original batch of untested products that crept in under the wire are now illegal. Moreover, nothing prevents a manufacturer from doing preclinical testing outside New Zealand, where it remains legal to study drugs in animals, and then moving to human trials with safe substances, as long as the animal results aren't cited in the approval application.

"I think that's possible," Dunne says. Donald Hannah, who served as the first head of the Psychoactive Substances Regulatory Authority, confirms that a manufacturer who is relying on human tests that show a low risk of harm "should be able to make a valid application." In 2015 as I write, local planning for the siting of stores to sell currently nonexistent products continues, as required by the law. Regulators are licensing manufacturers and universities to do research. They are continuing to wrestle with what "low risk" means and will accept applications from manufacturers that wish to submit products for approval.

Bowden hopes to have one on the market in three years, although the total ban on his products—through the law he pushed to pass—has left him in financial trouble, forcing him to sell his lab and his other properties.

A STRATEGY LIKE New Zealand's to test and regulate low-risk recreational drugs can make use of the new understanding of addiction as a learning disorder and inform future efforts. By selecting drugs with low risk of addiction for approval, such policies can encourage people to replace more harmful drugs with less harmful ones. By setting pricing through taxation or other measures and regulating the location and times of sale, they can affect the context of drug use, ideally fine-tuning this as new data come in to further reduce addiction risk. By limiting use to adults, this system can also minimize youth access at the time when they are at highest developmental risk for addiction (though enforcement here is tricky and ideally does not involve any criminalization of young users, just education and support). Also, by restricting advertising, these kinds of policies can limit the risks associated with industry pressure for sales—although a complete advertising ban and even limiting sales of psychoactive substances to nonprofit entities like governments would probably be better than allowing full-fledged commercialization.

New Zealand's attempt to regulate recreational drugs came at a good time in international drug politics. If such a system had even been considered in the 1990s, the United States would have immediately squelched it, in our unofficial role as the world's narcotics enforcer. While many Americans are unaware of this aspect of foreign policy, until recently, the United States was the primary supporter and enforcer of the global drug war—in part because it was one of the few things Cold War enemies could agree on. Consequently, with the backing of Russia and China, America has largely determined what is acceptable under international drug law, which was first codified in the United Nations' 1961 Single Convention on Narcotic Drugs.

Over the last half century, using threats of cuts in aid or trade sanctions and other heavy-handed tactics, the United States has stopped many countries from liberalizing drug laws. Contrary to popular belief, marijuana isn't legal in the Netherlands (the laws just aren't enforced)—and American pressure is a big reason why. According to David Bewley-Taylor, a scholar of drug control at Swansea University, the United States pressured Jamaica to back away from marijuana decriminalization in 2001 and deterred Mexico from reducing drug penalties in 2004. We have threatened Canada over its harm reduction policy of providing safe injection rooms in Vancouver and also tried to push Europe, the United Kingdom, New Zealand, and Australia away from harm reduction in the '90s and '00s.

But since President Obama allowed Colorado and Washington to legalize recreational use and sales of marijuana following ballot initiatives in 2012, the United States itself is probably now violating international law. (Because we have traditionally been the ones who interpret and enforce these laws, it's hard to know exactly; of course, we say we are not.) And with even federal drug control officials slowly embracing harm reduction officially, we have remained silent on New Zealand's law.

Meanwhile, other international bodies are being surprisingly open-minded. Drug policy officials from Britain, the Organization of American States, and the European Union have cited the New Zealand program as a possible model to emulate. And when Kiwi health official Dunne attended the UN Commission on Narcotic Drugs meeting in Vienna in 2014, he said there was "genuine interest" from around the world. In October of that year, the premier scholarly journal in addiction research, *Addiction,* devoted an entire issue to New Zealand's experiment.

Of course, any sensible system for regulating the recreational drugs would have started by trying to legally control marijuana, not new psychoactive substances. Cannabis has been used by millions of humans for thousands of years and a U.S. government agency has spent decades—and over $100 million between 2008 and 2014 alone—trying to prove it a high-risk drug, with little luck. Legal highs, in contrast, are brand-new substances with no history of human use, and some are already linked with fatalities.

But at the time that New Zealand began its regulatory journey, considering marijuana as a possible legal high was politically impossible. After the backlash forced the ban on existing products, however, the local debate over marijuana legalization was revitalized. After all, the vast majority of legal highs are substitutes for weed, which users actually prefer. Those who choose synthetic marijuana tend to do so to avoid drug testing or because they want to take something that is legal. As is often the case under our irrational system, users are driven toward more harmful and concentrated drugs by our attempt to prevent use of less harmful ones.

Now that four states and D.C. are experimenting with marijuana legalization, however, we will be able to learn more about what types of controls are most successful. These states, as well as New Zealand and Uruguay, which has recently legalized marijuana on a national level, could all provide examples of what works—and what doesn't. Each one has created a somewhat different system, which should allow at least some comparisons to be made.

To further refine such regulation it will be critical to view addiction as a learning disorder. We know that the dose and pattern of use matters in determining whether someone gets hooked or not—so how can we control availability in creative ways to minimize the development of addictive use patterns? We also know that developmental timing matters. Are there better ways of educating kids and delaying the start of use? Where a drug is consumed is important, as are people's expectations of it. How can regulation and education best manage these factors? Temperament, people's interpretations of their own experience, trauma, and access to healthy ways of managing psychological distress all matter profoundly as well in whether or not addiction is learned—so what can we do to intervene early, without doing the harm that potentially stigmatizing diagnoses and some treatments can cause? Moreover, knowing that punishment is not an effective way to teach people with addiction to recover, how can we find the best alternatives?

A few pioneering organizations right here in the United States are beginning to try to answer these questions.

NINETEEN

Teaching Recovery

Pleasure is the correlate of learning and development. We take pleasure in things as we come to know the world better—as we come to be engaged in and at home in the world.

—ALVA NOE

HOWARD JOSEPHER, NOW 76, DIDN'T KNOW he was practicing harm reduction when, in 1988, he was asked to help design a program to educate ex-offenders with histories of needle use about AIDS. He didn't know that the program he would help create would have life-transforming "side effects" that went far beyond HIV prevention. Nor did he know that it would someday have over 10,000 graduates and be part of a multi-million-dollar agency that he would found and run. But Josepher did know a great deal about injection drug use and drug treatment—not just professionally, but from long personal experience. He was about to become a pioneer in compassionate, education-based treatment.

Like many people with addictions, Josepher says that from his earliest memories, he always felt that something in him was missing. Raised by immigrant Jewish parents in the Bronx, he was an anxious child, who'd become terrified of the dark and even, for a time, afraid to leave his house when he was around seven. That same year, his mother had fallen into a depression so severe that she was given ECT. His father was an alcoholic. High school football provided some joy and respite as he grew up, but poor grades kept him out of a college where he could play.

Searching led Josepher to the Beat scene, where he attended poetry readings by Allen Ginsberg and Gregory Corso in Greenwich Village. It brought him to LSD and avant-garde acting classes—and, ultimately, to heroin. By 1963, he was addicted. The rest of his story is a tour through the highs and lows of the last five decades of addiction treatment. And ARRIVE, the program he founded and still runs, is one form that treatment can take when addiction is understood primarily as a learning disorder.

When AIDS began decimating IV drug users in the 1980s, Josepher was a social worker in private practice who had put his addiction problem far behind him. He was asked to consult on a research project that had received a grant from the National Institute on Drug Abuse. The creators of what became ARRIVE knew that many people leaving prison were at high risk of relapse to IV drug use—and that the only way to prevent the spread of AIDS in this group was to reach those who had no desire to quit.

That meant that any program they designed would have to be, literally, user-friendly, and it could not require abstinence as a condition of participation. To attract newly released drug users, ARRIVE was designed as an 8-week certification course aimed at teaching users how to avoid HIV or cope with it if they were already infected. It offered valuable job training, certifying graduates to work as peer counselors for others in the same situation. Such jobs were among the few that this group was uniquely qualified for—and ARRIVE grads eventually came to staff many of New York's AIDS and harm reduction programs, with some even founding or leading their own agencies and many serving on executive boards.

The ARRIVE program was built on a fundamental insight about addiction and learning. Presented in 24 class sessions, it was deliberately designed to be the equivalent of a college course worth four to five credits. "In 'treatment,' the focus is on individual pathology. You're looking at what's wrong with the person and how you go about fixing them," Josepher says. "[But] to engage someone in a teaching dynamic, they're a student and their only job is to be open, to be receptive. You're not telling the person there's something wrong with them." Previously, when he'd worked in traditional drug treatment, Josepher had been taught that active drug users were unreachable and unteachable. But when he began his private practice, he found that he could genuinely help actively addicted people and that, as other harm reductionists were also discovering, many of them could make other healthy life changes long before they were ready to give up drugs.

ARRIVE's developers also realized that in order for people to learn to make healthier choices, they needed to feel capable of achievement and have attainable goals. Josepher starts each semester with a motivational speech that takes a very different perspective on power than 12-step programs do. Instead of focusing on the weaknesses of those with addiction, ARRIVE highlights their strengths. "How many times did you go to bed with no drugs and no money and get up the next day and get high?" he asks—and participants always smile or laugh in accord. "That's power," he says, explaining that what his program will teach is ways of focusing this drive and strength in a new direction.

The classes teach specific material about HIV and addiction, in order to make the certification useful both personally and professionally. Although not designed to be so dauntingly hard as to be intimidating, it's also not made to be easy. This is a key insight from research on learning: there's a sweet spot between impossibly hard and boringly easy where students thrive and learn best. Josepher also believes that part of the value of the program is learning that you can overcome challenges. Only by doing so, he says, can people gain self-esteem. For some participants, graduating from ARRIVE is the first time they have successfully completed any educational program.

But ARRIVE isn't just groundbreaking in terms of AIDS. It is one of the few organizations where people in all stages of addictions and with varying types of recovery are encouraged to mix. Some participants are on methadone; others are abstinent and are enthusiastic participants in 12-step programs. Some have given up cocaine but still smoke pot and drink alcohol; others continue to inject heroin. As long as people are not disruptive and do not engage in illegal activities on-site, the program has no interest in what substances may or may not be flowing through your veins and brains. In ARRIVE, you define recovery for yourself—and everyone's path, whether it includes methadone or 12-step programs or church or simply managing your use better on your own, is accepted. All that is asked is that you show up ready to learn and treat other participants with respect.

Traditionally, treatment providers have found the idea of mixing active and recovering people horrifying—and many still do. Their main concern is that current drug users can lead recovering folks to relapse, especially when they interact in the context of providing information about safer injection practices. Moreover, many 12-steppers have been taught to see other forms of recovery as inferior. Consequently, bringing into the same program people on maintenance, people who are still using, abstinent folks, and those somewhere in between seemed like a recipe for conflict, not healing.

Over the last 20 years, however, I've visited ARRIVE many times—and what I've seen is almost exactly the opposite. (I should disclose that over time, Josepher has become a friend.) As the community initially faced the triple threat of AIDS, IV drug use, and crack cocaine, old antipathies broke down. As the virus spread, philosophical differences became less important than stopping the deadly disease. While ARRIVE developed and grew, it became clear that people who were dying, people who were traumatized, people who were so broken that their only pleasure came in crack or heroin didn't need any more shame or pain. They were already about as humiliated and rejected as possible.

What they needed was hope, not more fear—and love, rather than more loss. The obvious need for empathy during New York's AIDS epidemic made many who originally opposed harm reduction rethink and then embrace it. In a published evaluation, ARRIVE was associated with reductions in both injection drug use and other HIV risk behavior. These data also show that it does not do harm when it mixes people in various types of recovery with those who continue to use drugs.

Because no one was mandated into ARRIVE, it had to make itself as appealing as possible, initially paying participants $10 to attend each class and participate in the research. But now, people clamor to attend, largely because of the good reputation it has among treatment providers, probation officers, and other social service agencies. At its offices, which recently relocated to downtown Manhattan, near Battery Park, a warm atmosphere of support and respect prevails. This is evident from the way you are welcomed at the reception desk to the frequent hugs exchanged in the hallways. One graduate, Elsa Gonzalez, described the program as being like "a husband to me." She added, "ARRIVE stepped in, it held me up and said, 'Go ahead, you can do it.'" At the time she joined, she had been off drugs for two years, but was homeless—and terrified. ARRIVE helped her stabilize her life and leave welfare for employment and a new apartment.

For many, especially in the early days, ARRIVE's classrooms were the only place where they could talk about their HIV status and their drug use. As with needle exchange, participants rapidly came to feel protective of the program, and no one wanted to be the one who made others feel unsafe. To promote a peaceful atmosphere, mindfulness is taught throughout. But this isn't just sitting: classes open with five minutes of dancing, recognizing that many addicted people—particularly those with ADHD—find that suddenly trying to keep still is the opposite of calming. The program is tailored to the way people with addictions learn best. In addition, political activism around addiction and HIV issues is encouraged, making ARRIVE participants and graduates a familiar sight at New York rallies and demonstrations, which Josepher often helps organize. This can help participants find a new sense of purpose and connection to the community.

Josepher himself is an inspiring and charismatic speaker. Tall, thin, with gray hair and warm eyes, he has the energy and look of a much younger man. In addition to ARRIVE, his overall organization, Exponents, now includes a more traditional outpatient drug treatment program, as well as services like trauma groups, nutritional aid, GED classes, and HIV testing. It also offers AR-RIVE "postgraduate" classes, which teach people how to understand and manage the complex medical information around HIV and hepatitis C.

Exponents has 48 employees—nearly three quarters of whom are ARRIVE graduates. Although it isn't formally based on defining addiction as a learning disorder, its design is especially congruent with it. Because it offers a safe space to discuss an issue of great concern to current and former drug users—HIV and now hepatitis C—this takes the spotlight off of drugs and addiction, allowing participants to engage from a position of strength. They are seen not primarily as sick people who must be helped to get well but as students who are learning new skills. By focusing on other goals first, and by accepting people as they are, it allows people to address addiction issues at their own pace. Once they begin to feel competent and worthy, they become ready for additional challenges.

Sometimes, the ARRIVE model suggests, it is better to teach implicitly, by example, than it is to try to direct people to change their behavior. This, of course, is not only true for addiction. Canadian Mary Gordon founded a school-based program for children called Roots of Empathy that has been shown to reduce bullying and increase compassion. She often says that "empathy can't be taught but it can be caught." The same may be true of recovery.

ARRIVE exemplifies this aspect of how, in learning, social contagion can be more powerful than explicit instruction. The focus on other health issues moves participants away from an identity based on drug use and minimizes opportunities for negative types of peer pressure that can result from repeated discussions of past and present highs. Learning by example is often much more powerful than learning through instruction. The ARRIVE program offers a model of one way to help addicted people learn—and catch—recovery.

AND IT'S NOT just social service agencies that are picking up on this new, more compassionate and learning-based view of addiction. Even police and prosecutors are starting to get on board. In Washington State, a program called LEAD, for Law Enforcement Assisted Diversion, is working with some of the hardest-to-reach addicted people, who typically view cops with great suspicion. Most LEAD participants are drug users and low-level dealers. Some are prostitutes. Nearly all have mental illnesses and addictions—often both. Many have been arrested dozens of times. Frequently homeless, they inhabit the most forsaken areas of the city.

But in downtown Seattle, LEAD officers are no longer the enemy. Rather than constantly arresting and rearresting the same people for petty drug possession, prostitution, and nonviolent drug-linked crimes like shoplifting, the city is trying a new approach. Before entering LEAD, Misti Barrickman, who was homeless and addicted to heroin, lived in a tent in a wooded area near Seattle's Belltown region. She hadn't had an easy life. Raped by a relative when she was just two years old, she was later molested repeatedly by a female babysitter. Barrickman was raised by her father, but she did not get along with her stepmother, who beat her frequently and even threatened her with a gun. She also suffered from painful scoliosis. Introduced to Oxycontin by coworkers at a Subway when she was in her teens, Barrickman ultimately moved on to heroin. By age 16, she had dropped out of school. Over the course of seven years on the street, she estimates that she was arrested "at least 50 times," typically for drug offenses.

For much of this time, Barrickman actually wanted to enter a methadone program. However, the waitlist was a year long—and every time she got arrested, she was dropped back to the end of the line. Her last arrest was just five days before she was finally scheduled to start treatment—and that's when, fortunately, she was enrolled in LEAD. Taken to jail, Barrickman asked to see the sergeant in charge, pleading with him just to give her a chance to get into treatment. She was sitting in her cell, starting to suffer withdrawal, when the arresting officer returned.

She recounts the story with tears in her eyes in a recent documentary: "He comes back and he tells me, 'For your own good, I hope that you make it to that methadone appointment.'" Although she didn't quite fit LEAD's requirements at the time, he had gotten her an exception and was able to enroll her. Set free and helped to start medication, she is now in college and has more than two years in recovery.

Not everyone who participates in LEAD, however, wants to stop taking drugs—and that's not a disqualification. Tim Candela, a social worker involved in the program, described the approach to the *Huffington Post* like this: "Everyone has goals. . . . I guarantee you once you work towards those goals and a couple of things come to fruition, you'll come back and say, 'I want to deal with my addiction.' And it's a lot more meaningful than if I say, 'You need assistance.'"

Like ARRIVE and other harm reduction programs, LEAD "meets people where they're at" and uses the kind of compassion and encouragement that help all kinds of learning to take place, whether in elementary school or during addiction. As with needle exchange programs, participants tend to be moved by the simple fact of being valued unconditionally precisely when they expect to be rejected or shamed, and by being seen as capable of changing for themselves.

LEAD is the rare law enforcement program based on harm reduction. The key idea is not to force people into abstinence and punish them if they fail, but rather, to connect them up with services that can help them meet their own goals—like finding a house or a job. By doing so, LEAD cuts arrest and incarceration rates and reduces revolving-door criminal justice costs that typically have had no positive impact on drug use or crime. In Washington State, an average arrest for such low-level crimes leads to a jail term of around three weeks—and the whole arrest, incarceration, and prosecution process typically costs around $3,100 per incident. In contrast, case management for a month by LEAD costs $240.

Recently, a controlled trial of the program was conducted by researchers at the University of Washington. It found that participants had 58% lower odds of rearrest, compared to controls who were simply arrested, locked up, and then freed as usual. After diversion to LEAD, the proportion of the group who got rearrested fell by 30%—but for those who were not diverted, this rate increased 4%. Official evaluations of LEAD's cost-effectiveness, its influence on getting homeless people housed, and its success in improving quality of life for both participants and the community are under way. So far these results also look promising—so much so that cities like Santa Fe and Albany are already starting similar programs and dozens of other cities, including New York, have expressed interest.

LEAD also has significant advantages over drug courts, which are far more expensive and which often continue the dysfunctional relationship between America's legal and drug treatment systems. Although drug courts have been praised by everyone from Newt Gingrich to the Obama administration, they only reach 120,000 defendants a year, a tiny proportion of those with drug problems. In drug courts, moreover, judges dictate what treatments people can receive, using sanctions like increased court appearances to punish small infractions—or short jail terms for major violations like relapse. In contrast, LEAD is not punitive. It recognizes that addiction is defined by resistance to punishment and that the timetable for recovery may not be the one the authorities would prefer.

Indeed, the whole idea of LEAD is to keep nondangerous participants entirely out of the justice system, where judicial oversight itself may not only interfere with good treatment practices but also actually *increase* incarceration rates. This can happen in several ways. For one, the court system doesn't tend to do high-quality clinical evaluations, meaning that defendants may be forced into treatment that is inappropriate or even harmful. With most of these evaluations conducted by people employed by prosecutors, few of whom have doctorate or even master's-level degrees, psychiatric issues are often missed, which leads to treatment failure, prompting incarceration.

Second, drug courts are founded on the misconception that punishment must be part of treatment. This leads them to favor programs that are tough. Not only does this bias them toward ineffective treatments and reduce success rates, it also props up abusive rehabs. Over one third of patients treated for substance addictions—including alcohol—have been referred to treatment by the criminal justice system, and in some programs, the proportion is closer to 80% or 90%, according to Kerwin Kaye, a sociologist who has studied drug courts and treatment in New York. This means that many programs come to view the legal system, not their patients, as the customer they serve. These providers have little incentive to improve care, fire disrespectful staff, provide decent living conditions, and avoid brutal disciplinary tactics, because their patients' only alternative is incarceration—and because punitive treatment is seen as the most helpful. Further, if the patients relapse, it's not the poorly trained counselors or ineffective treatment program administrators who wind up behind bars.

Obviously this approach is contrary to good practice—particularly for opioid addiction, where the standard of care is maintenance medication. Drug court judges often take a dim view of maintenance, in part because it is not punitive. According to a 2012 study, only about a third of all drug courts permit participants to start maintenance as the treatment component of their program, and 50% ban it entirely. Such policies have resulted in at least four recent deaths, where either denial of maintenance access or forced cessation have been immediately followed by overdose. There may be many more—but no one knows because drug courts aren't required to track mortality when they report outcomes.

In essence, the drug court system allows judges to practice medicine without a license—deciding what medications and treatments are acceptable in a way that they never do for other illnesses. No judge, faced with a mentally ill defendant, says that he can take only Haldol but not Risperdal, or that talk therapy must be psychoanalytic, not cognitive. However, drug court judges frequently require that patients be denied or tapered from medications that are the standard of care, and often, they only allow them to attend certain programs.

In LEAD, in contrast, participants decide whether or not to get treatment and can select programs that are conducive to their own goals. Since LEAD is based on harm reduction, it encourages maintenance because that saves lives. It does not have a punitive ideology that interferes with treatment choice. Further, by avoiding arrest in the first place, the program can't help but lower incarceration rates and reduce the loss of employment opportunities that comes with conviction. It does not further entangle participants by making relapses into crimes.

LEAD also has another key advantage over drug courts: at least in Seattle, where it was founded, it has brought together groups and systems that often see each other as adversaries and work at cross-purposes. It was founded through a unique collaboration between Lisa Daugaard, a public defender who had sued the city's police department for racial discrimination, and Steve Brown, the commander of that same department's narcotics unit. The program was designed with input from law enforcement, prosecutors, city government, harm reduction experts, and community groups. When these organizations work together, not only are individuals helped, but counterproductive policies like dropping people from methadone waiting lists for being arrested can also be discovered and changed.

Ultimately, says Daugaard, the idea of LEAD is to make the police department more like the fire department. When a fire breaks out, the sight of that red truck is comforting—firefighters are seen as helpers to welcome, not en-

forcers or occupiers to be feared or to try to evade. When police officers participate in LEAD and understand harm reduction, their relationships with people
on the street also start to change. They learn about the trauma that so often
underlies addiction and begin to understand the condition not as defiance but
as self-medication that has become dysfunctional. They also begin to see people
as they change for the better, rather than interacting only when something goes
wrong.

One officer, Victor Maes, told the *Huffington Post* that he first began to believe
that LEAD might work when he noticed that two chronic offenders weren't on
the streets anymore. "These two guys were out constantly every day, smoking
cocaine, wheeling, dealing, facilitating deals, assisting others to get dope, shoplifting," he said. But after they joined LEAD and went to rehab, he didn't see
them. One even came into the precinct to show him that he had money and
wasn't buying crack—and to share pictures of his family.

In many ways, LEAD can be seen as a tentative move toward decriminalization. In Gloucester, Massachusetts, which has been hard hit by increased prescription opioid and heroin use, the local police department is also taking a
step in this direction. Users who voluntarily turn in their drugs are not arrested
but helped to access treatment. Other police departments are considering this
approach as well. These efforts are a good start—but much more systematic
change is needed. Police enforce laws—unless these laws are changed to make
drug possession for personal use legal, people with addiction will face prosecution for what they do every day in order to cope.

The existing system that most closely approaches this strategy is that of Portugal, which removed criminal penalties for all personal level drug possession
in 2001. As in LEAD, the idea is to avoid arresting people whenever possible. Indeed, the Portuguese police no longer seek to make drug possession arrests at
all, for those with ten days or less of a supply of recreational drugs including
marijuana, heroin, crack, and MDMA. However, if drug users do come to police
attention, they are referred to a "dissuasion committee," which is deliberately
made to feel far more like a quiet conversation than a court appearance. Each
committee has three members—typically a psychologist, a social worker, and
an attorney—and the goal is to determine if the person is addicted. If so, treatment referrals are made and light sanctions—like fines or being restricted from
certain neighborhoods—may be imposed. Treatment is not coerced, however, only recommended. If the person is not found to be addicted, the matter
is often just dropped.

Since the law was changed nearly 15 years ago, Portugal's results have been astonishing—at least to those who predicted that such a move would produce a chaotic, drug-crazed country. HIV infection rates in drug users have fallen, the number of people in treatment rose 41%, and drug use rates by teenagers declined. Most promisingly, the rate of injection drug use fell by nearly half. And while use by adults rose slightly, the same trend was seen in other European countries that did not change their drug policies, suggesting that this rise was not associated with Portugal's new laws. Based in part on these results and on the failure of global criminalization to reduce drug use, the World Health Organization endorsed the decriminalization of all drug possession for personal use in 2014.

In the United States and even in Seattle, however, there's still a long way to go to get even close to such a change. Many people continue to be arrested for drug possession and far too many remain incarcerated. Addiction treatment remains both difficult to access and often low quality, when people can get in. But LEAD offers one way to move toward a less punitive, more health-focused drug policy. And the more we understand that addiction is a learning disorder and that drug use is often a rational response to an unfair and painful world, the closer we'll get to good policy.

I'VE OFTEN WONDERED how different my life would have been if, as a child or teen, I'd known ways of defeating depressive thoughts before they became engrained in my brain. Could I have avoided the addiction risk that my temperament carried by having better ways of coping early on? If I'd had a way to be less hard on myself—and I hadn't labeled myself as irreparably "bad" or "unworthy"— would I have needed so much anesthesia? If I'd had social skills training and known that my sensory and emotional issues didn't mean that I was "selfish" or "bossy," could I have managed better?

A prevention program that has been tested in the U.K. and Canada and is currently being trialed in Australia is attempting to answer these kinds of questions. The data so far suggests that, at least in some cases, heavy alcohol and other drug use and the symptoms of mental illness that can drive it can actually be averted. This program takes an explicitly developmental view of all of these conditions, recognizing the critical roles of learning and self-labeling.

Developed by psychologist Patricia Conrod, who has joint appointments at King's College, London, and the University of Montreal, it's called "Adventure,"

"Preventure," or "Co-venture," depending on the study and population. The fundamentals are the same in all three "venture" programs. Working with adolescents aged 13–16, the idea is to target those with temperaments that increase risk for both addiction and mental illness and provide specific tools for coping with the unique problems that each temperament presents. "A mental health approach to alcohol and drug prevention looks like it's much more effective and promising than simple drug education or alcohol education," Conrod told me.

To avoid issues related to labeling children as "high risk" or as having personalities outside the norm, Conrod and her colleagues devised a clever, innovative strategy. Months before the program itself actually begins, students take a personality test, known as the Substance Use Risk Profile Scale (SURPS). However, the reason for the test is not disclosed to the kids—nor are the results. Depending on which profile is found, Conrod says, this 23-item scale can detect "between 80 and 90% of all future drug users," due to their high scores.

SURPS measures four traits that each increase both addiction and mental illness risk in different ways. Those who score outside the norm on any one trait are later invited to participate in two 90-minute seminars that are described to them as teaching skills for success. Rather than being labeled as being deviant, the participants are told that they were selected by lottery to be permitted to enroll—and in the studies, nearly all of the kids who are selected opt in. (Kids who test high on more than one trait are placed in the group for which they got the highest score.)

The first trait measured by SURPS is the most obvious: impulsiveness. This is quantified through responses to items like "I usually act without stopping to think" and "I often don't think things through before I speak." A frequently related trait called "sensation seeking" is measured by items that address a desire for novelty or an interest in taking up hobbies like skydiving or motorcycling. Together, impulsiveness and sensation seeking are often linked with ADHD, which increases risk for addiction both by making it more likely that people will try drugs and by making it less likely that they will be able to resist urges when they want to stop. These traits are also found in conduct disorder, which often co-occurs with ADHD, and is linked to the development of both addiction and antisocial personality disorder in adults.

The last two traits put kids at risk via caution and compulsiveness, rather than recklessness or boldness. The first of these is "hopelessness." It is measured by agreement with statements like "I feel that I am a failure" and is an obvious precursor to depression. The final trait, which researchers call "anxiety sensitivity," is measured in responses to items like "It frightens me when I feel my heartbeat change." This predisposition generally involves an overreaction to physical sensations associated with fear—and is linked to panic attacks and panic disorder.

The program's goal is not to change personalities, which would be impossible in two 90-minute classes. Instead, it aims to give adolescents skills that allow them to cope with their vulnerabilities better. Little emphasis is placed on alcohol or other drugs themselves. Instead, for example, the impulsive kids are educated about this trait and taught cognitive skills for managing it, while the sensation-seeking teens are taught healthy ways to meet these needs. The depressive adolescents, meanwhile, are instructed in tactics that allow them to recognize distorted thinking and counter it—for instance, to see a rejection by a peer as one event, not a sign that all you will ever experience is rejection and that you deserve it. "If you're walking into a room and feeling like everyone hates you and if you drink whenever you feel that way, you never learn to check out your assumptions and just numb the thought," Conrod says.

The program is taught by classroom teachers, who are given intensive training. This is based on evidence showing that prevention programs taught by familiar people, rather than, say, cops who stop in for a few weeks, are more effective. And for such a brief intervention, Conrod's program packs a powerful punch. A study of over 2,500 13- and 14-year-olds in the U.K., whose schools were randomly assigned either to teach the program or to do nothing, found that overall, drinking was reduced 29% in the Adventure schools. Reductions in drinking were not just seen among the high-risk kids who took the classes—the entire school had a lower drinking rate, presumably because the high-risk kids were less likely to overtly or covertly influence their peers' drinking.

Looking at the high-risk kids alone, binge drinking was reduced 43%, compared to high-risk kids who didn't go to a school offering the program. And symptoms of problem drinking like difficulties in school or dangerous drinking-related behavior were reduced by 42% in high-risk students and 24% in low-risk students in the experimental group. A second study found that the odds of conduct problem symptoms were reduced by 21% in the impulsiveness intervention group over two years. While the results for anxiety and depression were less clear, the effects on substance use suggest that even if the program isn't sufficient to reduce these symptoms long term, it may still reduce the use of drugs to cope with them.

The "venture" programs suggest that taking a developmental view of substance use disorders not only accurately describes these conditions but is useful in preventing them. They target high-risk youth while avoiding the dangers associated with labeling them as such. They also help prevent social contagion around teen drinking and drug use, while providing coping techniques that help those who do try substances to avoid heavy use. By mitigating the development of particular dysfunctional thinking patterns and thereby reducing the harmful use of substances to cope with them, they are far more likely to prevent addiction than lessons focused simply on drugs and their effects. And, they are far less likely to make kids curious about drugs than the substance-focused programs do.

BY VIEWING ADDICTION as a developmental disorder, we can devise far more effective ways of preventing and treating it. And, by understanding the role of learning in addiction, we can also create better policies to regulate substances to minimize the chances that it will be learned. ARRIVE, LEAD, and the "venture" programs all take advantage of this perspective. There are many others that do so as well that I could not include for space reasons: among these are contingency management, cognitive behavioral therapy, and the Good Behavior Game.

But to truly improve the way we handle addiction, we'll have to go much further, ending the criminalization of drug use and finding better ways of regulating psychoactive substances and potentially addictive behaviors like gambling. New Zealand's model suggests one approach—as do the developing marijuana regulations in Colorado, Washington, Oregon, Alaska, Washington, D.C., and Uruguay.

Fortunately, there are some general principles that can help.

TWENTY

Neurodiversity and the
Future of Addiction

If I got rid of my demons, I'd lose my angels.

—JONI MITCHELL

IN 2016, AS THIS BOOK IS released, the United Nations will hold a special General Assembly on drug policy. These high-level talks will be the first time in history that the world as a whole will formally consider a more health-oriented drug policy, rather than setting a timetable for law enforcement to impose a "drug-free world." (Prior UN plans had us achieving this exalted state by 2008.)

For anyone concerned with health, the bankruptcy of our current approach has been obvious for decades. By 2014, the World Health Organization had announced that it favors complete decriminalization of all drug use and personal-level possession. And now, no serious drug policy analyst can muster a case for mass arrest and incarceration of users. Even hard-core drug warriors think treatment is a better option for low-level offenders, although they wish to retain the threat of incarceration.

Most important, the United States is no longer an obstacle to significant reform, having recognized the virtues of harm reduction, complete with hiring a former alcoholic—who is "out" about his addiction—as "drug czar" and allowing him to headline the 2014 conference of the Harm Reduction Coalition. As drug policy expert John Walsh of the Washington Office on Latin America think tank told the *New York Times*, "There is near unanimity that the focus needs to be on health and public health. . . . That is very significant, considering most of the policy remains focused on enforcement and interdiction."

Seeing addiction as a learning disorder offers new insights to guide better strategies. Below is a brief summary of the key implications of this paradigm, based on the issues explored in the previous chapters.

Drug Exposure Alone Doesn't Cause Addiction

As we've seen, only 10%–20% of drug users become addicted to substances like marijuana, alcohol, cocaine, and heroin. This statistic in itself means that simply trying a drug cannot cause addiction; if it could, the problem would occur in every user. Even among people who report intense euphoria, addiction can't occur until a drug has been taken repeatedly and the brain has learned that the substance can be relied on to cope with emotional challenges. Healthy users, who make up 80%–90% of those who try drugs, often become scared when a drug is too euphoric, rather than purely delighted. They frequently say things like, "It was so good; I knew I couldn't risk taking it again" in response to the same experience that people who go on to develop addictions respond to by using more as soon as possible.

To learn addiction, a person must choose to continue using—and such learning through repetitive dosing is most likely if a person is already short of coping skills, extremely stressed and disconnected socially, suffering from childhood trauma, predisposed to mental illness, or otherwise genetically or environmentally vulnerable. The highest risk period is adolescence, when the brain is physiologically prepared for deep learning and when, psychologically, the skills necessary to cope with adult emotions and experience are still nascent.

Consequently, without learning specific, personal cues that drive addictive behavior, addiction cannot occur. Drug exposure isn't sufficient: a person needs to associate the substance and the trappings of the experience with pleasure, stress relief, or both over and over again in order to get hooked. From Rat Park to Central Park, where, when, with whom, and how you take drugs plays a key role.

Set and Setting Matter—and Can Be More Effectively Controlled Through Regulation, Not Prohibition

Since addiction is learned in specific cultural and physical environments, the "addictiveness" of particular drugs themselves can actually change as societies decide how to manage them. Legislation and social mores both influence the mental and physical environment in which drugs are taken. However, because there is always a demand for mood-altering substances and the most vulnerable people are the most likely to get addicted, absolute prohibition tends to deter those who aren't at risk and harm those who are, by criminalizing them.

The pattern of the experience, the dose, the situation in which addictive behavior occurs, a person's stage of brain development, and the expectations that people bring to their experience all matter. So, too, does an individual's background and even simply what is fashionable at the moment. All of these factors can influence not just whether someone takes a drug—but how addictive that drug is to that person at that time.

This means that the form that addictions take will vary depending on family, social, and cultural factors, again, precisely because addiction is learned. The way that a country manages access to drugs and potentially addictive experiences will shape what addictions are possible and, to a large extent, how harmful those addictions will be. A heroin addiction in a nation that allows heroin maintenance and has a strong social safety net, for example, will be quite a different experience from one in a country where use is harshly criminalized, no maintenance is allowed, and little effort is made to help those on the margins.

The difference can be as large as life and death—or a chaotic life lived on the streets versus one in an ordinary home with productive employment. Whether affected by drug laws and social services or by the climate of expectations and social rules that surround particular drugs, learning makes set and setting matter. An understanding of these facts should determine the future of drug policy.

An excellent example of how changing set and setting can reshape addiction can be seen in tobacco control. Since American states began restricting when and where people can smoke and raising cigarette taxes, the proportion of smokers who smoke heavily and daily compared to that of "social smokers" who use only occasionally—typically, when drinking—has changed dramatically. In 1980, for example, 26% of college students had smoked at least once in the past month—and around half of them smoked a half pack or more daily. By 2013, there were far fewer college smokers overall. In addition, a much greater proportion of them were "social" smokers. That year, 14% of college students had had at least one cigarette in the past month—but only 17% of the smokers smoked a half pack or more a day.

Cigarettes themselves haven't changed much—and if they have, it's in the direction of becoming more addictive. Nor is it likely that the number of people with predispositions to become addicted dropped. And tobacco has not been prohibited. What has changed is the cultural and physical environment where cigarette addiction is learned. By reducing the number of times cigarettes can easily be smoked and the number of situations where smoking is socially accepted, the opportunity to learn addiction has been reduced. If you don't learn to smoke at your desk at work or at school, you won't have environmental cues and memories of doing so to prompt smoking in those settings. Both lessened repetition of the habit itself and fewer cues for smoking make addiction less likely to be learned and engrained.

This has important implications for policies related to other drugs. To reduce the odds of addiction, keep the times and places where drug use is socially acceptable limited. Complete prohibition can't allow this level of control because once people violate it, the lives of those who are addicted become restricted to marginal spaces where they can take drugs as often as they can afford it. This is why the people who do smoke heavily today are much more likely to be poor and/or to have serious mental illness. In these cases, people have more need for escape and tend to live outside of mainstream culture. Sensible restrictions on places and times in which drug use is permitted can reduce harm, in part by reducing the odds that addiction will be rehearsed enough to be learned.

People Can Learn, Even While Addicted

Although the traditional disease model suggests that people in the grip of addiction have no free will, this has never fit the facts. As noted earlier, addicted people don't get high in front of the police; they plan specifically to ensure their supply and to avoid being detected. They frequently end addictions when circumstances change, such as when they fall in love, get a new job, or graduate college. On the other hand, addicts certainly also behave compulsively and apparently irrationally.

Seeing addiction as a learning disorder accounts for this paradox far better than either a disease model of complete slavery to drugs or a moral model of completely free choice. If what changes in the brain during addiction are the areas that set priorities, this will skew the ability to choose well, but it will not eliminate it. Any attempt to address addiction through public policy, whether it involves designing rehabs or treating people who are committing predatory crimes during their addictions, must recognize these facts.

For policy makers, this means that harm reduction programs like needle exchange, overdose prevention training, and programs like ARRIVE and LEAD are likely to be effective and should be prioritized. Learning does not require abstinence, and it may be necessary for people to develop alternative coping skills before recovery is possible.

Punishment Cannot Solve a Problem Defined by Its Resistance to Punishment

Although people can certainly learn during addiction, the condition itself is defined by persistence in the face of punishment. Consequently, it is unrealistic to expect that arrest, incarceration, and other coercive approaches will effectively treat it—and, as described earlier, that is what the data overwhelmingly show. Arresting drug users does not stop addiction, and incarceration actually seems to exacerbate it.

Since addiction doesn't respond well to punishment, there is little reason to involve the criminal justice system in cases of drug possession or addiction-related dealing that don't harm anyone other than those who are currently taking drugs. (For predatory and violent crimes committed in order to support an addiction, a system like drug courts could work—so long as the judge and prosecutors are "prescribing" a proportionate punishment that fits the crime and addiction treatment decisions are left to medical professionals and based on sound data.)

All People Learn Better When Treated with Respect

Because addiction is fundamentally a problematic type of learning, programs to prevent and treat it should rely on what is known about how people learn. Educational research shows overwhelmingly that fear and threat are not effective teaching tools—but respect and empathy can work. For instance, beating children is no longer seen as a best practice in schools (though some schools in the southern United States, unfortunately, continue to allow it)—but this is not only because it is inhumane. Corporal punishment and the use of shame and humiliation are also ineffective. People learn best in a challenging, but encouraging environment—not a climate of terror and fear. They learn best when they respect and feel respected by their teachers—and want to please them, rather than evade detection of errors. They learn best when driven by enthusiasm and passion—not mandates.

For example, studies find that the more autonomy and control people feel they have over their goals, the more likely they are to succeed in tasks requiring self-restraint—whether they involve dieting, sticking with medical treatment, or quitting smoking or drinking. A feeling of autonomy is exactly the opposite of what our current punitive treatment and criminal justice responses promote, however.

People with addiction are human; their learning disorder is limited to compulsive behavior around drugs that resists punishment. Treating them like other learners with this one difference in mind will be far more effective than assuming that what they need is to suffer more so that they can hit bottom. Acknowledging the humanity of addicted people should lead to treatment that is both more effective and more compassionate—because more compassionate treatment *is* more effective. An expansion of the harm reduction approach, which is based on these ideas, is now widely supported by research evidence.

Treatment Must Be Reformed to Be Respectful

Our current treatment system is based on the false assumption that the main reason people with addictions don't seek help is that they are enjoying their drugs too much, consequence-free, to want to stop. The reality is quite different: many addicted people are aware that they have a problem, but like anyone else, they do not want to be maltreated and they know that rehab is often not only humiliating but ineffective. Rather than a shaming, moralistic system that mainly attempts to compel people to accept the prayer, Higher Power, and confession of the 12 steps, we need an attractive, available, evidence-based system of treatment that "meets people where they're at" and teaches rather than preaches.

It's true that our current medical system isn't very good at managing conditions that have a social and psychological component: just telling an obese person to lose weight is about as effective as simply telling an addicted person to quit. But that doesn't mean that we should medicalize self-help groups that don't have the expertise to deal with all of the aspects of a complex and highly variable disorder. If we use self-help to support behavior change, participation must be voluntary and there need to be many options.

We need a system that actually is a system—one that starts by evaluating people's needs, rather than assuming that because someone takes drugs excessively, that's the core of the problem and their main need is to attend a support group. Instead, using a case management approach, addicted people could start at some type of central referral system—either via their insurance companies or a government hotline—and then be thoroughly evaluated and connected with the types of services they actually need. Today, people tend to get the kind of services that a particular program offers, whether or not it actually suits their condition, in part because agencies are looking for patients and have an incentive to accept anyone who shows up and can pay. If evaluation is separate from the financial imperatives of treatment providers, people are more likely to get appropriate care. Identifying co-occurring disorders early on can also help make sure these do not go unaddressed in one-size-fits-all rehab and lead to relapse.

For some, help will look like ordinary medical care—getting prescriptions from a doctor to treat depression or for maintenance with Suboxone or methadone. But most will also need a psychosocial and/or educational component, where people are taught about the nature of addiction and given the specific help they need to learn to cope with it in their daily lives. In some cases, this might be cognitive behavioral therapy. In others, it might involve new housing. Some people might be referred to classes like the ARRIVE program, others might do job training, higher education, mindfulness training, or psychotherapy. Residential treatment should only be used for the most severe cases—not only because it is expensive, but because it isn't more effective than intensive outpatient care. Moreover, what people learn in the artificial setting of a rehab is hard to transfer: since they are eventually going to have to deal with the real world, it's better to connect them with support immediately where they live whenever possible.

Understanding addiction as a learning disorder means creating individualized approaches to address it—on the biological, psychological, social, and cultural levels. It means using models that view participants as students who need to be empowered by education—not defective people who can't make good choices unless they are forced and who need to be humbled and then indoctrinated. It means trying to create a system for social care and psychoeducation that doesn't have to be entirely "medical" or aimed at "treating disease" but is evidence-based and can improve mental health care and services for people with all types of developmental disorders. Because social support and education are critical to recovery from many of these conditions and because the medical system is poor at providing these services, we need a better way to bridge treatment, education, and ways of getting socially connected.

While a reformed system might still refer some people to 12-step meetings for additional support, it would not include professional treatment based on it. There's no need to pay for what people can get for free—and no need to twist the voluntary steps into a coercive and often harmful approach that encourages abusive behavior by treatment staff in order to create a sense of powerlessness. A better treatment system would also never make referrals to mutual help groups without discussing both the potential risks and benefits—and would always offer alternatives for those who prefer them.

Primary Prevention Should Focus on Coping, Not Drugs

Prevention programs also need to be remade in light of the learning disorder model. Trying to eliminate drug use is futile. Fear-based prevention tends to deter those who wouldn't become addicted anyway, and it can actually make drugs more attractive to those who are looking for ways to be "cool" or have intense experiences. As in the programs described in the previous chapter, prevention efforts should focus more on teaching children to cope with their specific temperaments in healthier ways—and less on drugs. Such prevention should start early in life, helping kids understand their temperaments and their emotions and how to maximize self-control. Programs that aim simply at teaching self-regulation reduce drug problems without the risk of glamorizing drugs.

Some prevention programs should also be targeted to specific issues. However, these initiatives need to avoid labeling "high-risk" children as much as possible, to reduce the chances that children will take the label as a prophecy. Indeed, children should be taught explicitly about "fixed" and "growth" mindsets—so that they see good qualities like intelligence or kindness not as something you have to be born with but as characteristics that can be developed. Students should also be given information about labeling and how it can influence people's sense of themselves. Knowing how negative self-concepts can be internalized may help reduce the damage they can do.

In addition, prevention should teach kids about common developmental disorders and mental health diagnoses in an age-appropriate fashion aimed at reducing stigma. This could help those who might find a diagnostic label useful to discover that they are not alone in having particular constellations of symptoms. But kids should also be taught the limits of current diagnoses and the problems associated with seeing too many conditions as mere abnormalities and illnesses—so that those who might attach too much weight to a label can reject it or at least avoid perceiving it as an absolute truth that limits their potential.

Harm Reduction Is the Most Important Goal of Drug Policy

Since people always have wanted and probably always will want ways to alter their moods, drug use will likely always be with us. We can either accept this and try to reduce associated damage—or aim for the unrealistic ideal of a drug-free world, regardless of the damage done while seeking it. Opponents of harm reduction have long feared that the harm reduction philosophy is incompatible with absolute prohibition—and, in this case, they aren't wrong. Once the focus of policy is not simply eliminating drug use, harms related to strategies for fighting drug use must be taken into consideration. If harm related to enforcement, incarceration, corruption, black-market-associated violence, the spread of disease, and the impure drug supplies that result from prohibition become part of the calculus, it shifts drastically.

This is especially true once addiction is recognized as a learning disorder dependent on context. It's cheaper and easier to regulate a market and control factors like price, places where use is permitted, times when it is acceptable, and purity than it is to try to legislate and enforce it out of existence. Consequently, all drug policies need to be weighed as to whether they increase harm and make addiction more likely—or whether they decrease negative outcomes. One policy change that is completely supported by existing data is decriminalizing low-level possession and personal use of all drugs. Arresting and jailing users makes no more sense and is no more effective for heroin and cocaine users than it is for marijuana users. There may be some drugs for which prohibitions of sales remain sensible—but this shouldn't be the unquestioned default because our policies were not originally made rationally.

People who use drugs also need access to accurate harm reduction information—and this includes high school and college kids, who are at the age when use is most common. They need to know, for example, that mixing depressant drugs like alcohol with pain relievers can be deadly, but that there is no way to overdose on marijuana. Telling kids that all drug use is bad is not effective: teens are exquisitely sensitive to hypocrisy and this makes them dismiss fear tactics that don't jibe with their own experience. Honest, thoughtful information that can withstand critical thinking is essential—but it needs to be targeted and provided in the appropriate context. Above all, science and research need to guide policy—and ways to get research into practice need to be better understood and then implemented.

Celebrating Diversity—Including "Neurodiversity"—Is Critical to Reform

One of the sad ironies of our current drug policy is that the same treatment providers who have been cheerleaders for the war on drugs and who advocate the ongoing criminalization of drug use also claim to want to destigmatize "the disease of addiction." As they call for spirituality, prayer, confession, and restitution as treatment for a "brain disease," they fail to recognize that this moralistic language, the use of coercive and humiliating treatments, and the idea that the criminal justice system can address addiction are by their very nature antithetical to the goal of reducing stigma.

It is impossible to simultaneously criminalize and destigmatize a behavior: one of the key points of criminalization is, in fact, to deliberately create stigma in order to deter lawbreaking. How can you destigmatize a condition as you argue that your patients only respond to the brute force of the law? If punishment is part of the treatment for addiction—and for no other mental or physical illness—it will remain the most stigmatized disorder in medicine. Claiming that "force is the best medicine" for a disease means that you really do not believe it is a disease.

To get out of this loop, we need to see addiction as a developmental disorder—one that is not inherently associated with dishonesty, cruelty, or damage to others but may be connected with such problems in a system that assumes these connections, projects them on everyone who uses drugs, and fails to recognize preexisting personality disorders. We also need to directly confront the racism that drives many aspects of drug policy—and ensure that selective enforcement becomes a thing of the past. Drug policy needs to be about cutting drug-related harm, not sending political messages.

However, reducing stigma will also require more than this. To genuinely increase acceptance of people with addictions and other developmental disorders, we can't see them as subhuman or view them only through the lens of the difficulties associated with their lives. Instead, we need to recognize and accept the wide variety of differences people can have in brain wiring, emphasizing that these are associated with strengths and talents, not just disabilities.

A movement aimed at celebrating this "neurodiversity" has begun to take off among disability rights advocates who are fighting for the civil rights of autistic people. The idea of neurodiversity is that wiring differences—whether they produce autism, ADHD, or any other diagnosable "condition"—are just as deserving of respect as other types of human variety.

And in the same way that accommodations may be necessary for people with physical disabilities, so too should society work to adapt to the needs of people whose brains work in atypical ways. Neurodiversity advocates like Ari Ne'eman of the Autistic Self Advocacy Network are pushing for these kinds of measures, while also highlighting benefits associated with conditions like autism, rather than automatically looking at differences as "problems" or "disorders." In the autism community where the idea developed, these activists view autism itself not as a disability but as a way of being.

Indeed, when autistic people are allowed to act in the ways that make them feel safe and comfortable—even if this seems weird to people who are more neurotypical—disabilities can often be significantly reduced. For example, autistic children may need to learn in quiet and not overly lit settings, or they may need to engage in some repetitive behaviors or sensory soothing activities. When sensory difficulties are mitigated in the way that best suits particular autistic people, they can better focus and pay attention, allowing their skills and talents to grow. Tolerance of neurodiversity can then actually reduce the disabilities that previously seemed inherent to some types of it.

Also, in the same way that accommodating physical disabilities with wheelchair ramps and accessible vehicles can make the difference between a life of difficulty and isolation and one of ease for affected people, so too can accommodating people with brain wiring differences. As addicted folks well know, the way society deals with your condition can make the difference between being able to live a full and productive life and being effectively disabled and marginalized by criminalization and stigma. When we view addiction as a developmental disorder, we recognize the wiring differences that contribute to the condition. People aren't to blame for the way they are wired. We can either accept neurodiversity and create a society where many types flourish or ignore it and make associated disabilities worse.

Moreover, just as autistic people have strengths associated with their wiring, so, too, do people with addictions. The autistic talent of being able to focus intently on systems and ideas often leads to superior skills in music, math, computer programming, and even, sometimes, writing. The same brain that is vulnerable to overload in the wrong conditions can be superior to ordinary brains in the right ones.

Similarly, the ability to persist despite negative consequences can power love and success—not just addiction and obsession. In fact, without this ability, few people with any kind of brain wiring, typical or atypical, would ever succeed as activists, artists, or entrepreneurs. Nor would anyone be able to sustain relationships and raise children. Setbacks, rejections, insults, and obstacles are omnipresent; few goals worth reaching don't include them. The quality of "grit" or sticking with it when the going gets tough is increasingly being recognized as one of the most important aspects of success in all areas of life. And no one can accuse people with addiction of lacking grit when it comes to seeking drugs.

When the compulsion and drive that gets misguided during addiction is channeled in more positive directions, however, the results can be wonderful. We can all think of talented people in nearly every area of achievement who have experienced addictions: from William Stewart Halsted, a founder of Johns Hopkins and surgeon whose techniques are still used today and who was addicted to cocaine and morphine, to writers like Ernest Hemingway, F. Scott Fitzgerald, and Dorothy Parker (all alcoholics), to musicians from Billie Holiday (heroin) to Eric Clapton (heroin) and actors ranging from Judy Garland (multiple drugs) to Robin Williams (multiple drugs) to many of today's superstars. Of course, this doesn't mean that we should only value neurodiversity for the potential talents it may unleash, nor does it mean that we should not try to mitigate disabilities. What it does mean is that we should focus on people's abilities, not their disabilities—and help them find their strengths rather than stigmatize them for their weaknesses.

WHEN I GOT the results of my brain scan, it was hard to stop looking at the images, which were displayed as a movie, zooming in on various regions. It's quite odd to see the structures you have studied and written about so often displayed as they appear in your own head. Although I obviously knew these regions were present in my brain because I'm human, it feels uncanny to actually visualize them. No matter how strongly you feel about the neuroscience, it's still hard to imagine that your self lives in those peculiar folds and wrinkles.

As I'd been warned by Edythe London, professor of addiction studies and psychiatry at UCLA—in whose lab I got scanned—it's almost impossible to draw strong conclusions from any single subject brain scan. Although the neurodiversity movement calls people who aren't autistic and don't have other brain-related diagnoses "neurotypical," there really is no such thing. Every brain is unique, starting even before birth as the initial wiring is laid down. Our genes don't contain nearly enough information to direct the placement of every neuron, every glial cell, and every synapse. Many of the first, primitive connections are random, and many useless. Ruthless pruning takes place during development, killing millions of cells that don't manage to hook up properly or aren't in networks that represent learning. In the brain, it's absolutely literally true that if you don't use particular cells and synapses, you lose them.

So, I didn't turn out to have a giant nucleus accumbens—or a very small one. My caudate and putamen are perfectly normal. I was relieved to see that my ventricles aren't especially large (this occurs in many types of brain disease) and my cortex is appropriately gnarled and twisted. "You have a beautiful brain," London said, adding that she had already ascertained that through our interviews and conversations.

What did stand out, however, was my impulse control. I am apparently excellent at inhibiting responses, which is obviously usually not the case during active addiction. Whether the fact of my recovery built these circuits up or whether the maturation of these circuits actually allowed my recovery is impossible to know. But there is some intriguing evidence that people in long-term recovery do build up extra gray matter in the prefrontal cortex regions responsible for self-monitoring and self-control. One study found that former cocaine and heroin addicts had a greater volume of gray matter in these regions compared not only to active drug users but also to normal controls.

At least in this case, what doesn't kill you can make you stronger—and learning to overcome challenges can leave you better off than if you hadn't faced them in the first place. Of course, this doesn't mean that we should hope for our kids to become addicted so they can triumph over it or wish other types of tragedies on anyone. But it does mean that we should value even those who seem "ill," or "bad," who may at times appear to be devoted only to themselves and only to destruction. Whereas the value of being sensitive and curious and given to intense focus may seem obvious, even qualities that can drive antisocial behavior aren't entirely harmful. The same need for extreme sensation that can produce criminal behavior can also fuel police officers, pilots, and explorers; the same fearlessness that allows psychopathic behavior can also produce heroism.

We all learn to become who we are. And none of us starts from the same place or encounters the same cultural and social contexts in the same way. Our memories and the way our nervous systems react to them make us unique. For those of us with addictions or other differences, we need to fight to be recognized not only for our frailties and faults but for what we have to give.

ACKNOWLEDGMENTS

"You can't do it alone" is one of those recovery clichés that contains an important truth: although many people beat addiction without treatment or even support groups, few do so without strong connections to others. The statement is even more apt when describing book writing: I could not have done it without the help and support of the people and organizations that follow. First, I want to thank the Open Society Foundations for awarding me a 2015 Soros Justice Fellowship, which provided essential financial and institutional support.

Secondly, thanks go to my editor, Nichole Argyres: she understood what this book was meant to be and much improved it with her comments, critiques, suggestions, and enthusiasm. My agent, Andrew Stuart, helped throughout the process and provided much-needed support and guidance. Thanks are due as well to my friends, particularly Alissa Quart and Anne Kornhauser.

I'd also like to acknowledge some of the many people who helped this book in its long journey, which began with writing I did in the late '80s and early '90s: Peter McDermott, Howard Josepher, Trevor Butterworth, Gwen Barrett, Charlie Rose, Bill Moyers; the academics, autodidacts, and AAs of addict-L; Sora Song, Siobhan Reynolds, Walter Armstrong, William Godfrey, Maer Roshan, Bruce D. Perry, Joe Volpicelli, Stanton Peele, Carl Hart, Lisa Rae Coleman, and CLK Transcription. Special thanks to Edythe London and her team at UCLA and Charles Glatt at Cornell and to all of the addiction researchers, autism experts, neuroscientists, psychologists, and psychiatrists I've interviewed over the course of this work, particularly Nora Volkow, Kent Berridge, Jaak Panksepp, Larry Young, Bill Carlezon, Lisa Monteggia, and Eric Hollander. Caffeination from the folks at Perk Café on 37th Street was also particularly appreciated.

Without my attorney, Donald Vogelman, I might never have been able to write this book; extreme gratitude as well to Judge Leslie Crocker Snyder for giving me a second chance. My mom, Nora Staffanell, and my father, of blessed memory, Miklos Szalavitz, also deserve thanks, as do my siblings, Ari, Kira, and Sarah. Penultimate thanks and love to Ted Johnson, my husband, which is an appellation I thought I might never use. Finally, I want to thank all of the drug users and people with addiction who shared their stories with me over the years, who helped me understand and recover. Apologies to everyone I've inevitably and mistakenly excluded, and of course, any additional errors within are mine.

INTRODUCTION

2 *number one cause of accidental death* M. Warner et al., *Drug Poisoning Deaths in the United States, 1980–2008* (Centers for Disease Control, NCHS Data Brief No. 81, December 2011).

2 *1 in 10 American adults* New York State Office of Alcoholism and Substance Abuse Services, "Survey: Ten Percent of American Adults Report Being in Recovery from Substance Abuse or Addiction" (March 16, 2012), accessed August 21, 2015, http://www.oasas.ny.gov/pio/press/20120306Recovery.cfm.

2 *another 23 million currently suffer from some type of substance use disorder* Substance Abuse and Mental Health Services Administration, "The NHSDUH Report: Substance Use and Mental Health Estimates from the 2013 National Survey on Drug Use and Health: Overview of Findings," accessed August 21, 2015, http://www.samhsa.gov/data/sites/default/files/NSDUH-SR200-RecoveryMonth-2014/NSDUH-SR200-RecoveryMonth-2014.htm.

2 *2013 declaration by the American Medical Association that obesity, like addiction, is a disease* A. Pollack, "A.M.A. Recognizes Obesity as a Disease," *New York Times,* June 18, 2013.

3 *people with ADHD often thrive as entrepreneurs or explorers* D. Dobbs, "Restless Genes," *National Geographic,* January 2013.

3 *autistic people can excel at detail-oriented tasks* T. Grandin, *The Autistic Brain* (New York: Houghton Mifflin Harcourt, 2013), 117–35. See also S. Silberman, *Neurotribes: The Legacy of Autism and the Future of Neurodiversity* (New York: Avery, 2015), 273 and generally.

3 *Dyslexia can improve visual processing and pattern finding* M. H. Schneps, "Dyslexia Can Deliver Benefits," *Scientific American,* December 18, 2014.

3 *severely neglected children often develop autistic-like behavior* M. Szalavitz and B. D. Perry, *Born for Love: Why Empathy Is Essential—and Endangered* (New York: William Morrow, 2011), 78.

4 *Addiction is far less common in people who use drugs for the first time after age 25* G. M. Heyman, "Addiction and Choice: Theory and New Data," *Frontiers in Psychiatry* 4 (2013): 31, doi:10.3389/fpsyt.2013.00031, http://www.ncbi.nlm.nih.gov/pmc/articles/PMC3644798/pdf/fpsyt-04-00031.pdf.

4 *only 10–20% of those who try even the most stigmatized drugs* M. Szalavitz, "Genetics: No More Addictive Personality," *Nature* 522, no. 7557 (June 25, 2015): S48–S49, doi:10.1038/522S48a.

4 *the majority of cocaine, alcohol, prescription drug, and cannabis addictions* Heyman, "Addiction and Choice."

4 Most addictions are overcome without treatment: D. A. Dawson et al., "Recovery from DSM-IV Alcohol Dependence: United States, 2001–2002," *Addiction* 100 (2005): 281–92, doi:10.1111/j.1360-0443.2004.00964.x, http://www.ncbi.nlm.nih.gov/pubmed/15733237. See also W. M. Compton et al., "Prevalence, Correlates, Disability, and Comorbidity of DSM-IV Drug Abuse and Dependence in the United States: Results from the National Epidemiologic Survey on Alcohol and Related Conditions," *Archives of General Psychiatry*

64, no. 5 (2007): 566–76, doi:10.1001/archpsyc.64.5.566, http://archpsyc.jamanetwork
.com/article.aspx?articleid=482282.

4 *between one third and one half of children diagnosed with ADHD* M. Shaw et al., "A Systematic
Review and Analysis of Long-Term Outcomes in Attention Deficit Hyperactivity Disorder: Effects of Treatment and Non-Treatment," *BMC Medicine* 10 (2012): 99, doi:10.1186/1741-
7015-10-99, http://www.ncbi.nlm.nih.gov/pmc/articles/PMC3520745/.

4 *treatment doesn't seem to affect whether they outgrow the disorder* L. E. Arnold et al., "Effect of Treatment Modality on Long-Term Outcomes in Attention-Deficit/Hyperactivity Disorder:
A Systematic Review," *PLOS ONE* 10, no. 2 (2015): e0116407, doi:10.1371/journal.pone.0116407,
http://www.ncbi.nlm.nih.gov/pmc/articles/PMC4340791/.

6 *addiction is the psychiatric disorder with the* highest *odds of recovery, not the worst prognosis* Heyman, "Addiction and Choice."

6 *people actually have* increased *odds of recovery as they age* G. M. Heyman, "Quitting Drugs: Quan-
titative and Qualitative Features," *Annual Review of Clinical Psychology* 9 (2013): 29–59, doi:
10.1146/annurev-clinpsy-032511-143041, http://www.ncbi.nlm.nih.gov/pubmed/23330937.

6 *"A person who has been punished is not"* B. F. Skinner, *Beyond Freedom and Dignity* (Middlesex, UK:
Penguin Books, 1971), 83.

6 *Fear and threat also literally shunt energy away* Perry and Szalavitz, *Born for Love*, 48–49.

6 *Changing behavior is far easier if you use social support, empathy, and positive incentives* For instance, see
Tobias Wächter et al., "Differential Effect of Reward and Punishment on Procedural
Learning," *Journal of Neuroscience* 29, no. 2 (January 14, 2009): 436–43, doi:10.1523/JNEUROSCI.
4132-08.2009. For specific effects of reward on addiction, see M. Prendergast et al.,
"Contingency Management for Treatment of Substance Use Disorders: A Meta-Analysis,"
Addiction 101 (2006), 1546–60.

7 *alcoholism and other addictions can and often do remit [with marriage]* D. A. Dawson et al., "Maturing
Out of Alcohol Dependence: The Impact of Transitional Life Events," *Journal of Studies on
Alcohol* 67, no. 2 (March 2006): 195–203.

1. NEEDLE POINT

12 30-day methadone detox shown to be ineffective by 1980s: K. L. Sees et al., "Methadone
Maintenance vs 180-Day Psychosocially Enriched Detoxification for Treatment of Opioid Dependence: A Randomized Controlled Trial," *Journal of the American Medical Association* 283, no. 10 (2000): 1303–10, doi:10.1001/jama.283.10.1303. Note: Study is from 2000 but
it summarizes early data.

15 *At least half of the injection drug users in the city were already infected* J. H. Burack and D. Bangsberg,
"Epidemiology and HIV Transmission in Injection Drug Users," HIV InSite Knowledge
Base Chapter (May 1998), accessed August 19, 2015, http://hivinsite.ucsf.edu/InSite?page
=kb-07-04-01#S3X.

15 *at least one third of all methadone programs still failing to provide an adequate dose* H. A. Pollack and
T. D'Aunno, "Dosage Patterns in Methadone Treatment: Results from a National Sur-
vey, 1988–2005," *Health Services Research* 43, no. 6 (2008): 2143–63, doi:10.1111/j.1475-6773.2008
.00870.x, http://www.ncbi.nlm.nih.gov/pmc/articles/PMC2613988/.

15 *having an intention to do something only predicts* R. Cooke and P. Sheeran, "Moderation of
Cognition-Intention and Cognition-Behaviour Relations: A Meta-Analysis of Properties of Variables from the Theory of Planned Behaviour," *British Journal of Social Psy-*

chology 43, no. 2 (2004): 159–86, doi:10.1348/0144666041501688, cited in G.-J. Y. Peters, R. A. C. Ruiter, and G. Kok, "Threatening Communication: A Critical Re-analysis and a Revised Meta-Analytic Test of Fear Appeal Theory," supplement, *Health Psychology Review* 7, suppl. 1 (2013): S8–S31, doi:10.1080/17437199.2012.703527, http://www.ncbi .nlm.nih.gov/pmc/articles/PMC3678850/.

15 Quantum change: W. R. Miller and J. C'de Baca, *Quantum Change: When Epiphanies and Sudden Insights Transform Ordinary Lives* (New York: Guilford Press, 2001).

17 Many women who exchange drugs for sex or engage in prostitution to obtain drugs were sexually abused: E. Verona, B. Murphy, and S. Javdani, "Gendered Pathways: Violent Childhood Maltreatment, Sex Exchange, and Drug Use," *Psychology of Violence* 2015, a0039126, doi:10.1037/a0039126.

2. A HISTORY OF ADDICTION

21 *Naloxone worsens withdrawal symptoms* D. P. Wermeling, "Review of Naloxone Safety for Opioid Overdose: Practical Considerations for New Technology and Expanded Public Access," *Therapeutic Advances in Drug Safety* 6, no. 1 (2015): 20–31, doi:10.1177/2042098614564776.

22 *One study found lasting psychological damage* M. Lieberman, I. Yalom, and M. Miles, *Encounter Groups: First Facts* (New York: Basic Books, 1973), 174.

22 Species deliberately seek intoxicating plants: G. S. Flory et al., "A Computerized Apparatus Designed to Automatically Dispense, Measure, and Record Alcohol Consumption by Individual Members of a Rhesus Macaque Social Group: Trait-like Drinking across Social- and Single-Cage Conditions," *Methods* 38, no. 3 (March 2006): 178–84, doi:0.1016/j.ymeth. 2005.12.002, http://www.sciencedirect.com/science/article/pii/S1046202305002367. See also R. K. Siegel, *Intoxication: Life in Pursuit of Artificial Paradise* (New York: E. P. Dutton, 1989).

22 Cats like catnip: J. Stromberg, "How Catnip Gets Your Cat High," *Vox,* December 20, 2014, http://www.vox.com/2014/9/12/6136451/catnip-cats-science.

22 Horses will seek out locoweed and take again even if poisoned: J. A. Fister et al., "Grazing of Spotted Locoweed (*Astragalus Lentiginosus*) by Cattle and Horses in Arizona," *Journal of Animal Science* 81 (2003): 2285–93, http://www.ars.usda.gov/research/publications/publi cations.htm?SEQ_NO_115=142862.

22 Goats revealed stimulating properties of coffee: Siegel, *Intoxication,* 39.

22 Civilization started for beer, not bread: S. H. Katz and M. M. Voigt, "Bread and Beer: The Early Use of Cereals in the Human Diet," *Expedition* 28, no. 2, 1986, http://www.penn .museum/documents/publications/expedition/PDFs/28-2/Bread.pdf.

22 There's never been a drug-free human population: The New Zealand government has claimed that the Maori at one time had a drug-free culture, but their ancestors who first settled New Zealand did not and the Maori of today unfortunately have high levels of addiction like other colonized peoples.

22 *the pursuit of altered states of consciousness is a human universal* S. Pinker, *The Blank Slate* (New York: Viking Press, 2002), http://condor.depaul.edu/mfiddler/hyphen/humunivers.htm.

22 Shamans imbibe urine: Siegel, *Intoxication,* 66–67.

22 *around two thirds of Americans over 12 have had at least one drink in the last year, and 1 in 5 are current smokers* Substance Abuse and Mental Health Services Administration, "National Household Survey on Drug Use and Health," 2013, Table 2.1B, http://www.samhsa.gov/data /sites/default/files/NSDUH-DetTabs2013/NSDUH-DetTabs2013.htm#tab2.1b.

22 *In the 1940s and '50s, a whopping 67% of men smoked* Gary A. Giovino, "Epidemiology of To-bacco Use in the United States," *Oncogene* 21 (2002): 7326–40, doi:10.1038/sj.onc.1205808, http://www.nature.com/onc/journal/v21/n48/full/1205808a.html.

22 *Among people ages 21 to 25, 60% have taken an illegal drug at least once* Substance Abuse and Mental Health Services Administration, "National Household Survey on Drug Use and Health," 2012, Table 1.6B, http://www.samhsa.gov/data/NSDUH/2012SummNatFindDetTables /DetTabs/NSDUH-DetTabsSect1peTabs1to46-2012.htm#Tab1.6B.

22 *half of us could suffer from physical withdrawal symptoms if denied our daily coffee* Reuters, "U.S. Coffee Drinking Grinds Lower," *Fortune.com,* March 13, 2015.

23 Drug addiction typically affects 10–20% of users of the most common substances: Sza-lavitz, "Genetics."

23 Cigarettes addict about one third of those who try smoking: US Centers for Disease Control and Prevention, US National Center for Chronic Disease Prevention and Health Promotion, and US Office on Smoking and Health, "How Tobacco Smoke Causes Disease: The Biology and Behavioral Basis for Smoking-Attributable Disease: A Report of the Surgeon General" (Atlanta: Centers for Disease Control and Preven-tion, 2010), 4; "Nicotine Addiction: Past and Present," http://www.ncbi.nlm.nih.gov /books/NBK53018/.

23 *recovery without treatment is the rule rather than the exception* D. S. Hasin et al., "Prevalence, Cor-relates, Disability, and Comorbidity of DSM-IV Alcohol Abuse and Dependence in the United States: Results from the National Epidemiologic Survey on Alcohol and Related Conditions," *Archives of General Psychiatry* 64, no. 7 (2007): 830–42, doi:10.1001/ archpsyc.64.7.830. See also Compton, "Prevalence, Correlates, Disability, and Comor-bidity."

23 Addiction seen as voluntary ("God sends many sore judgments . . ."): S. Danforth, "The Woful Effects of Drunkenness," Boston, 1710, cited in Harry Levine, "The Discov-ery of Addiction: Changing Conceptions of Habitual Drunkenness in America," *Journal of Studies on Alcohol* 15 (1978): 493–506.

24 Quotes from De Quincey: T. De Quincey, "Confessions of an English Opium-Eater," *London Magazine,* September 1821, http://www.gutenberg.org/files/2040/2040-h/2040-h .htm.

24 Benjamin Rush and alcoholism as a disease of the will: W. White, *Slaying the Dragon: The History of Addiction Treatment and Recovery in America* (Bloomington, IL: Chestnut Health Sys-tems/Lighthouse Institute, 1998), 1–4.

24 Rush did not conflate "negritude" as a disease with racist ideas about the character of black people: author interview with Harriet Washington, May 2014, author of *Medical Apartheid: The Dark History of Medical Experimentation on Black Americans from Colonial Times to the Present,* rev. ed. (New York: Anchor, 2008).

25 *"fear of the cocainized black"* David F. Musto, *The American Disease: Origins of Narcotic Control,* rev. ed. (New York: Oxford University Press, 1987), 7.

26 *"white women and Chinamen side by side"* D. F. Duncan et al., "A Brief History of Prohibition and Treatment Solutions for Substance Abusers," *International Journal of Criminology and So-ciology* 3 (2014): 188.

26 *"a New Southern Menace"* E. H. Williams, "Negro Cocaine 'Fiends' Are a New Southern Menace," *New York Times,* February 8, 1914, http://query.nytimes.com/mem/archive-free /pdf?res=9901E5D61F3BE633A2575BC0A9649C946596D6CF.

26 *"Most of the attacks upon white women of the South are the direct result of a cocaine-crazed Negro brain"* Quoted in C. L. Hart, "How the Myth of the 'Negro Cocaine Fiend' Helped Shape American Drug Policy," *The Nation,* February 17, 2014, http://www.thenation.com/article /178158/how-myth-negro-cocaine-fiend-helped-shape-american-drug-policy.

26 Anslinger, cannabis, and the "effect on the degenerate races": Peter Schrag, *Not Fit for Our Society: Immigration and Nativism in America* (Oakland, CA: University of California Press, 2010), https://books.google.com/books?id=sVHx2hXrYQwC&pg=PT214&lpg=PT214&dq =anslinger+degenerate+races&source=bl&ots=cNYlDEu73w&sig=QB1iVwyg9h XXOB5-nS6L1Phislc&hl=en&sa=X&ei=7idvVeXeGKXIsQSTsoKQCA&ved=0CCAQ 6AEwATgU#v=onepage&q=anslinger%20degenerate%20races&f=false.

26 *"reefer makes darkies think they're as good as white men"* and Anslinger quote on "Satanic" music: quoted in M. Gray, *Drug Crazy: How We Got Into This Mess and How We Can Get Out* (New York: Random House, 1998).

27 1906 Pure Food and Drug Act: US Food and Drug Administration, "Promoting Safe and Effective Drugs for 100 Years," *FDA Consumer Magazine,* January-February 2006, The Centennial Edition, http://www.fda.gov/AboutFDA/WhatWeDo/History/CentennialofFDA /CentennialEditionofFDAConsumer/ucm093787.htm.

27 *labeling opiate-containing medicines cut their use* Musto, *The American Disease,* 33.

27 Benjamin Rush's characterization of addiction: White, *Slaying the Dragon,* 5.

27 *"Prohibition represented the single most important bond between Klansmen"* D. J. Hanson, "The Ku Klux Klan (KKK), Alcohol, & Prohibition," Alcohol Problems and Solutions Website, accessed August 6, 2015, http://www2.potsdam.edu/alcohol/Controversies/1107362364 .html#.U9e5KI1dVNM.

27 Attacks on bootleggers and overlapping membership of Klan and Anti-Saloon League: T. R. Pegram, "Hoodwinked: The Anti-Saloon League and the Ku Klux Klan in 1920s Prohibition Enforcement," *Journal of the Gilded Age and Progressive Era* 7, no. 1 (January 2008): 89–119, http://journals.cambridge.org/action/displayAbstract?fromPage=online&aid =7880605.

28 Racial animus against Italian, Irish, German, and Jewish immigrants: M. A. Lerner, "Going Dry: The Coming of Prohibition," *Humanities* 32, no. 5 (September/October 2011), http://www.neh.gov/humanities/2011/septemberoctober/feature/going-dry.

28 Countries without prohibition also saw declines in alcohol-related problems: M. Thornton, "Policy Analysis: Alcohol Prohibition Was a Failure," *Cato Policy Analysis* no. 157 (July 17, 1991), http://www.cato.org/pubs/pas/pa-157.html#30.

28 *insurers estimated that alcoholism rates rose by 300% as Prohibition continued* D. Blum, "The Chemists' War," *Slate,* February 19, 2010, http://www.slate.com/articles/health_and_science/medical _examiner/2010/02/the_chemists_war.html.

28 Homicide rates before and after Prohibition: Schaffer Library of Drug Policy, "Homicide Rates, 1910–1944," accessed August 19, 2015, http://www.druglibrary.org/schaffer/Library /homrate1.htm.

28 Information on government-enforced poisoning of alcohol and death and disability rates: Blum, "The Chemists' War."

28 AA founded in 1935: Alcoholics Anonymous, "Historical Data: The Birth of A.A. and Its Growth in the U.S./Canada," AA Website, accessed August 6, 2015, http://www.aa .org/pages/en_US/historical-data-the-birth-of-aa-and-its-growth-in-the-uscanada.

29 Cocaine comparable to potato chips and peanuts: C. Van Dyke, R. Byck, "Cocaine,"

Scientific American, March 1982, 128–41. To be fair to the authors, rereading the article now, I see that if I'd read it more carefully back then, I would have recognized that they saw only snorting cocaine as nonaddictive—both smoking and injecting they viewed as truly compulsive. But when I read it then, I somehow missed this.

3. THE NATURE OF ADDICTION

32 Clonidine effects in withdrawal: S. L. Walsh, E. C. Strain, and G. E. Bigelow, "Evaluation of the Effects of Lofexidine and Clonidine on Naloxone-Precipitated Withdrawal in Opioid-Dependent Humans," *Addiction* 98 (2003): 427–39, doi:10.1046/j.1360-0443.2003.00372.x.

34 Story about cancer patient from an interview with Gav Pasternak: M. Szalavitz, "The Pleasant Truths about Pain," *Psychology Today,* August 2005, https://www.psychologytoday.com/articles/200508/the-pleasant-truths-about-pain. Prior paragraph also comes from Pasternak.

34 Same brain areas used for psychological and physical pain and are rich in opioids: N. I. Eisenberger, "The Pain of Social Disconnection: Examining the Shared Neural Underpinnings of Physical and Social Pain," *Nature Reviews Neuroscience* 13, no. 6 (May 3, 2012): 421–34, doi:10.1038/nrn3231.

34 *"I know it when I see it"* P. Lattman, "The Origins of Justice Stewart's 'I Know It When I See It,'" *Wall Street Journal Law Blog,* September 27, 2007, http://blogs.wsj.com/law/2007/09/27/the-origins-of-justice-stewarts-i-know-it-when-i-see-it/.

35 No known specific biological basis for DSM diagnoses: A. Frances, *Saving Normal: An Insider's Revolt Against Out-of-Control Psychiatric Diagnosis, DSM 5, Big Pharma and the Medicalization of Ordinary Life* (New York: William Morrow, 2013), 10–34.

35 Genetic profile or unique physiology redefined as "neurological" rather than psychiatric and removed from psychiatry's diagnostic manual: S. DeWeerdt, "Reclassification of Rett Syndrome Diagnosis Spurs Concerns," *SFARI News,* July 4, 2011, accessed August 19, 2015, http://sfari.org/news-and-opinion/news/2011/reclassification-of-rett-syndrome-diagnosis-stirs-concerns.

35 *same gene produces risk for different disorders* Cross-Disorder Group of the Psychiatric Genomics Consortium, "Genetic Relationship Between Five Psychiatric Disorders Estimated from Genome-Wide SNPs," *Nature Genetics* 45, no. 9 (2013): 984–94. doi:10.1038/ng.2711.

35 Missing doses of some blood pressure medications can be deadly: M. M. Reidenberg, "Drug Discontinuation Effects Are Part of the Pharmacology of a Drug," *Journal of Pharmacology and Experimental Therapeutics* 339, no. 2 (2011): 324–28, doi:10.1124/jpet.111.183285, http://www.ncbi.nlm.nih.gov/pmc/articles/PMC3200000/.

36 *some antidepressants can produce a wicked withdrawal syndrome* G. A. Fava et al., "Withdrawal Symptoms after Selective Serotonin Reuptake Inhibitor Discontinuation: A Systematic Review," *Psychotherapy and Psychosomatics* 84, no. 2 (February 21, 2015): 72–81, http://www.ncbi.nlm.nih.gov/pubmed/25721705.

36 *"Addiction is defined as a chronic, relapsing brain disease"* National Institute on Drug Abuse, "Media Guide," National Institute on Drug Abuse website, accessed August 6, 2015, http://www.drugabuse.gov/publications/media-guide/science-drug-abuse-addiction-basics.

36 *"addiction is established in a learning process"* A. R. Lindesmith, *Addiction and Opiates,* revision of 1947 book *Opiate Addiction,* accessed August 6, 2015, http://druglibrary.eu/library/books/adopiates/lindesmith01.pdf.

37 Countless researchers and theorists have made contributions to the conversation: I apologize to all the equally important researchers I have undoubtedly and accidentally omitted.

38 rates of co-occurring disorders with addiction as high as 98%: B. F. Grant et al., "Prevalence and Co-occurrence of Substance Use Disorders and Independent Mood and Anxiety Disorders: Results from the National Epidemiologic Survey on Alcohol and Related Conditions," *Archives of General Psychiatry* 61, no. 8 (2004): 807–816, doi:10.1001/archpsyc.61.8.807; C. Lopez-Quintero et al., "Probability and Predictors of Remission from Lifetime Nicotine, Alcohol, Cannabis, or Cocaine Dependence: Results from the National Epidemiologic Survey on Alcohol and Related Conditions," *Addiction* (Abingdon, England), 106, no. 3 (2011): 657–69, doi:10.1111/j.1360-0443.2010.03194.x; and H. M. Lai et al., "Prevalence of Comorbid Substance Use, Anxiety and Mood Disorders in Epidemiological Surveys, 1990–2014: A Systematic Review and Meta-Analysis," *Drug and Alcohol Dependence* (May 28, 2015), pii:S0376-8716(15)00281-1, doi:10.1016/j.drugalcdep.2015.05.031.

38 *50% of people with one type of addictive behavior also engage in others* Center for Behavioral Health Statistics and Quality, Substance Abuse and Mental Health Services Administration, Treatment Episode Data Set (TEDS), "Table 1.1a, Admissions Aged 12 and Older, by Primary Substance of Abuse: 2002–2012," http://www.samhsa.gov/data/sites/default/files/2002_2012_TEDS_National/2002_2012_Treatment_Episode_Data_Set_National_Tables.htm. See also S. Kedia, M. A. Sell, and G. Relyea, (2007) "Mono- Versus Polydrug Abuse Patterns among Publicly Funded Clients," *Substance Abuse Treatment, Prevention, and Policy* 2 (2007) 2, no. 33., doi:10.1186/1747-597X-2-33.

38 *typical case of depression begins in one's early 30s* National Institute of Mental Health, "What Is Depression?," National Institute on Mental Health website, accessed August 6, 2015, http://www.nimh.nih.gov/health/topics/depression/index.shtml.

38 *odds of alcoholism for those who start drinking at age 14* R. W. Hingson, T. Heeren, and M. R. Winter, "Age at Drinking Onset and Alcohol Dependence: Age at Onset, Duration, and Severity," *Archives of Pediatric and Adolescent Medicine* 160 (2006): 739–46.

38 *risk of rapidly developing addiction to marijuana, cocaine, opioids, and pills* C.-Y. Chen, C. L. Storr, and J. C. Anthony, "Early-Onset Drug Use and Risk for Drug Dependence Problems," *Addictive Behaviors* 34, no. 3 (2009): 319–22, doi:10.1016/j.addbeh.2008.10.021.

4. INTENSE WORLD

42 Background on Asperger's syndrome: National Institute of Neurological Disorders and Stroke, "Asperger Syndrome Factsheet," NINDS website, accessed August 6, 2015, http://www.ninds.nih.gov/disorders/asperger/detail_asperger.htm.

43 *stimulant medication is being prescribed to more than six percent of schoolchildren* Susanna N. Visser et al., "Trends in the Parent-Report of Health Care Provider-Diagnosed and Medicated Attention-Deficit/Hyperactivity Disorder: United States, 2003–2011," *Journal of the American Academy of Child & Adolescent Psychiatry* 53, no. 1 (January 2014): 34–46.e2, http://dx.doi.org/10.1016/j.jaac.2013.09.001, http://www.cdc.gov/ncbddd/adhd/medicated.html.

43 *It would also be many years before Big Pharma* S. A. Mogull, "Chronology of Direct-to-Consumer Advertising Regulation in the United States," *AMWA Journal* 23, no. 3 (2008): 200, https://www.academia.edu/278465/Chronology_of_Direct-to-Consumer_Advertising_Regulation_in_the_United_States.

43 *extreme stress can actually cause physical harm to brain cells* M. Popoli et al., "The Stressed Synapse:

The Impact of Stress and Glucocorticoids on Glutamate Transmission," *Nature Reviews Neuroscience* 13, no. 1 (2011): 22–37, doi:10.1038/nrn3138.

44 Before hearing screening, many deaf people wound up institutionalized: J. Berke, "Deaf History: Deaf, Not Retarded: When Misdiagnoses are Made, Everyone Pays," *About.com Health,* September 29, 2010, http://deafness.about.com/cs/featurearticles/a/retarded _2.htm.

45 Early screening and early signing improve cognitive function in deaf people: D. Corina and J. Singleton, "Developmental Social Cognitive Neuroscience: Insights from Deaf-ness," *Child Development* 80 (2009): 952–67, doi: 10.1111/j.1467-8624.2009.01310.x, http://onlinelibrary.wiley.com/doi/10.1111/j.1467-8624.2009.01310.x/full#b74.

45 *early intervention can significantly reduce associated disability* G. Dawson et al., "Randomized, Controlled Trial of an Intervention for Toddlers with Autism: The Early Start Denver Model," *Pediatrics* 125, no. 1 (January 2010): e17–e23, doi:10.1542/peds.2009-0958. See also A. J. Orinstein et al., "Intervention History of Children and Adolescents with High-Functioning Autism and Optimal Outcomes," *Journal of Developmental and Behavioral Pediatrics* 35, no. 4 (2014): 247–56, doi:10.1097/DBP.0000000000000037.

46 *"sensitive dependence on initial conditions"* and butterfly effect: For a good summary, see J. Gleick, *Chaos: Making a New Science,* reprint edition (New York: Penguin, 2008).

46 Path dependence and keyboards: P. Pierson, "Increasing Returns, Path Dependence and the Study of Politics," *American Political Science Review* 94, no. 2 (June 2000): 254.

48 Suicide in Hungarians and Finns: Andrej Marušič and Anne Farmer, "Genetic Risk Factors As Possible Causes of the Variation in European Suicide Rates," *British Journal of Psychiatry* 179, no. 3 (September 2001): 194–96, doi:10.1192/bjp.179.3.194.

48 Poverty at least doubles risk for depression: V. Lorant et al., "Socioeconomic Inequali-ties in Depression: A Meta-Analysis," *American Journal of Epidemiology* 157, no. 2 (January 15, 2003): 98–112.

50 Definition of hyperlexia: L. Burd, J. Kerbeshian, and W. Fisher, "Inquiry into the Inci-dence of Hyperlexia in a Statewide Population of Children with Pervasive Developmental Disorder," *Psychological Reports* 57 (1985): 236–38. See also X. Wei et al., "Reading and Math Achievement Profiles and Longitudinal Growth Trajectories of Children with an Autism Spectrum Disorder," *Autism* 19, no. 2 (February 2015): 200–10, doi:10.1177/1362361313 516549.

50 Hyperlexia more common in autism: L. Mottron et al., "Veridical Mapping in the De-velopment of Exceptional Autistic Abilities," *Neuroscience & Biobehavioral Reviews* 37, no. 2 (February 2013): 209–28, doi:10.1016/j.neubiorev.2012.11.016.

50 *sensory issues are now recognized as a key diagnostic symptom of autism* Susan L. Hyman, "New DSM-5 Includes Changes to Autism Criteria," *AAP News,* June 4, 2013, doi:10.1542/aapnews. 20130604-1, http://aapnews.aappublications.org/content/early/2013/06/04/aapnews .20130604-1.

52 Pete Townshend quote: Quoted in D. Wholey, *The Courage to Change: Personal Conversations about Alcoholism with Dennis Wholey* (New York: Warner Books, 1984), 26.

53 *"I was always at odds with the entire world"* Quoted in Alcoholics Anonymous, *Alcoholics Anony-mous ("Big Book"),* third edition (New York: A.A. World Services, 1976), 400.

53 *"From the time I was a little girl"* Quoted in Narcotics Anonymous, *Narcotics Anonymous ("N.A. Big Book"),* fourth edition (Van Nuys, CA: World Service Office, 1984), 121.

53 *"[Codeine, an opiate] made me feel right for the first time in my life"* Andrew Weil and Winnifred Rosen, *From Chocolate to Morphine* (Boston / New York: Houghton Mifflin, 1993), 199.

54 ADHD and risk for addiction: S. S. Lee et al., "Prospective Association of Childhood Attention-deficit/hyperactivity Disorder (ADHD) and Substance Use and Abuse/Dependence: A Meta-Analytic Review," *Clinical Psychology Review* 31, no. 3 (2011): 328–41, doi:10.1016/j.cpr.2011.01.006, http://www.ncbi.nlm.nih.gov/pmc/articles/PMC3180912/. See also S. Levy et al., "Childhood ADHD and Risk for Substance Dependence in Adulthood: A Longitudinal, Population-Based Study," *PLOS ONE* 9, no. 8 (2014): e105640, doi:10.1371/journal.pone.0105640, http://www.ncbi.nlm.nih.gov/pmc/articles/PMC4146503/.

54 *research finds few effects of medication, positive or negative, on the later risk* K. L. Humphreys, T. Eng, and S. S. Lee, "Stimulant Medication and Substance Use Outcomes: A Meta-analysis," *JAMA Psychiatry* 70, no. 7 (2013): 740–49, doi:10.1001/jamapsychiatry.2013.1273, http://www.ncbi.nlm.nih.gov/pubmed/23754458.

54 Therapeutic use of potentially addictive drugs in childhood: Opioid drugs in child cancer patients don't typically lead to addiction, for example. See K. M. Foley and M. Szalavitz, "Why Not a National Institute on Pain Research?" In *Cerebrum 2007: Emerging Ideas in Brain Science,* ed. Bruce McEwen (New York: Dana Press, 2007).

54 Some studies do suggest a benefit from medicating for ADHD in childhood: Z. Chang et al., "Stimulant ADHD Medication and Risk for Substance Abuse," *Journal of Child Psychology and Psychiatry* 55 (2014): 878–85, doi:10.1111/jcpp.12164.

55 *between 30% and 50% of autistic people have ADHD symptoms* Y. Leitner, "The Co-Occurrence of Autism and Attention Deficit Hyperactivity Disorder in Children—What Do We Know?," *Frontiers in Human Neuroscience* 8 (2014): 268, doi:10.3389/fnhum.2014.00268, http://www.ncbi.nlm.nih.gov/pmc/articles/PMC4010758/.

55 *around 20% of people with drug use disorders have ADHD* G. Van De Glind et al., "The International ADHD in Substance Use Disorders Prevalence (IASP) Study: Background, Methods and Study Population," *International Journal of Methods in Psychiatric Research* 22, no. 3 (2013): 232–44, doi:10.1002/mpr.1397, http://www.ncbi.nlm.nih.gov/pmc/articles/PMC4085151/.

55 As a child I was learning to cope and changing my development in the process: Some parts of this chapter appeared in slightly different form in M. Szalavitz, "The Boy Whose Brain Could Unlock Autism," *Matter,* December 11, 2013, https://medium.com/matter/the-boy-whose-brain-could-unlock-autism-70c3d64ff221.

5. THE MYTH OF THE ADDICTIVE PERSONALITY

58 *black and Hispanic people are not more likely than whites to become addicted* L.-T. Wu et al., "Racial/Ethnic Variations in Substance-Related Disorders among Adolescents in the United States," *Archives of General Psychiatry* 68, no. 11 (2011): 1176–85, doi:10.1001/archgenpsychiatry.2011.120.

58 Quote from George Koob: Szalavitz, "Genetics" S48.

58 *Research finds no universal character traits that are common to* all *addicted people* Ibid.

58 *Only half have more than one addiction* Center for Behavioral Health Statistics and Quality, SAMHSA, TEDS, "Table 1.1a." See also Kedia, Sell, and Relyea, "Mono- Versus Polydrug Abuse Patterns."

59 *Only 18% of addicts, for example, have* R. B. Goldstein et al., "Antisocial Behavioral Syndromes

and DSM-IV Drug Use Disorders in the United States: Results from the National Epidemiologic Survey on Alcohol and Related Conditions," *Drug and Alcohol Dependence* 90,
nos. 2–3 (2007): 145–58, doi:10.1016/j.drugalcdep.2007.02.023, http://www.ncbi.nlm.nih.gov
/pmc/articles/PMC2633099/.

59 *This is more than four times the rate seen in typical people* B. F. Grant et al., "Prevalence, Correlates,
and Disability of Personality Disorders in the United States: Results from the National
Epidemiologic Survey on Alcohol and Related Conditions," *Journal of Clinical Psychiatry* 65,
no. 7 (July 2004): 948–58, http://www.ncbi.nlm.nih.gov/pubmed/15291684.

59 *Giftedness and high IQ, for instance, are linked with higher rates of illegal drug use* James White, "G
David Batty Research Report: Intelligence Across Childhood in Relation to Illegal
Drug Use in Adulthood: 1970 British Cohort Study," *Journal of Epidemiology and Community
Health* 66, no. 9 (September 2012): 767–74, doi: 10.1136/jech-2011-200252.

59 The data on giftedness are less reliable, but there are some: https://books.google.com
/books?id=36B8YL60P_UC&pg=PA229&lpg=PA229&dq=giftedness+drug+use
&source=bl&ots=RV4zHPccAF&sig=-PFS2Y-4Buzl0ceem2I6X0WiXig&hl=en&sa=X
&ved=0CCUQ6AEwAjgKahUKEwjm_Ieo0_vGAhWGdz4KHZBTDPI#v=onepage&q
=giftedness%20drug%20use&f=falseHealth jech-2011 doi:10.1136/jech-2011-200252.

59 Shedler and Block study: J. Shedler and J. Block, "Adolescent Drug Use and Psychological Health: A Longitudinal Inquiry," *American Psychologist* 45, no. 5 (May 1990): 612–30.

60 *nearly two thirds of high school seniors nationally reported at least trying marijuana* R. A. Miech et al.,
"Monitoring the Future National Survey Results on Drug Use, 1975–2014: Volume I,
Secondary School Students" (Ann Arbor: Institute for Social Research, The University
of Michigan, 2015), Monitoringthefuture.org.

60 Similar data on teen drinking: E. Kuntsche et al., "Not Early Drinking but Early Drunkenness Is a Risk Factor for Problem Behaviors Among Adolescents from 38 European and
North American Countries," *Alcoholism, Clinical and Experimental Research* 37, no. 2 (2013): 308–
14, doi:10.1111/j.1530-0277.2012.01895.x, http://www.ncbi.nlm.nih.gov/pubmed/23240610.

60 Moderate drinking beneficial: J. Chick, "Can Light or Moderate Drinking Benefit Mental
Health?," *European Addiction Research* 5, no. 2 (June 1999): 74–81.

60 Impulsivity pathway to addiction: Szalavitz, "Genetics."

61 Depression/anxiety path to addiction: P. H. Kuo et al., "The Temporal Relationship of
the Onsets of Alcohol Dependence and Major Depression: Using a Genetically Infor-
mative Study Design," *Psychological Medicine* 36, no. 8 (August 2006): 1153–62. See also B. F.
Grant et al., "Sociodemographic and Psychopathologic Predictors of First Incidence of
DSM-IV Substance Use, Mood, and Anxiety Disorders: Results from the Wave 2 National
Epidemiologic Survey on Alcohol and Related Conditions," *Molecular Psychiatry* 14, no. 11
(2009): 1051–66, doi:10.1038/mp.2008.41.

61 Quotes from Shedler and Block study: Shedler and Block, "Adolescent Drug Use."

62 *similar brain circuits are involved in both addiction and obsessive-compulsive disorder (OCD)* G. S. Berlin
and E. Hollander, "Compulsivity, Impulsivity, and the DSM-5 Process," *CNS Spectrums*
19, no. 1 (February 2014): 62–68, doi:10.1017/S1092852913000722.

62 Brain regions involved in addiction: G. F. Koob and N. D. Volkow, "Neurocircuitry of
Addiction," *Neuropsychopharmacology* 35, no. 1 (2010): 217–38, doi:10.1038/npp.2009.110.

62 The insula: N. H. Naqvi and A. Bechara, "The Hidden Island of Addiction: The Insula,"
Trends in Neurosciences 32, no. 1 (2009): 56–67, doi:10.1016/j.tins.2008.09.009.

62 The anterior cingulate: N. Ichikawa et al., "Feeling Bad about Screwing Up: Emotion Reg-

ulation and Action Monitoring in the Anterior Cingulate Cortex," *Cognitive, Affective and Behavioral Neuroscience* 11, no. 3 (2011): 354–71, doi:10.3758/s13415-011-0028-z.

63 The amygdala: G. F. Koob, "Brain Stress Systems in the Amygdala and Addiction," *Brain Research* 1293 (October 13, 2009): 61–75. doi: 10.1016/j.brainres.2009.03.038.

63 As a behavior is learned and becomes more automatic, it engages different parts of the striatum: Barry J. Everitt and Trevor W. Robbins, "From the Ventral to the Dorsal Striatum: Devolving Views of Their Roles in Drug Addiction," *Neuroscience & Biobehavioral Reviews* 37, no. 9, part A (November 2013): 1946–54.

63 *Estimates of the genetic heritability of addiction* A. Agrawal et al., "The Genetics of Addiction—A Translational Perspective," *Translational Psychiatry* 2 (2012): e140, doi:10.1038/tp.2012.54, http://www.nature.com/tp/journal/v2/n7/full/tp201254a.html.

65 *At least two thirds of addicted people have suffered at least one* L. Khoury et al., "Substance Use, Childhood Traumatic Experience, and Posttraumatic Stress Disorder in an Urban Civilian Population," *Depression and Anxiety* 27, no. 12 (2010): 1077–86, doi:10.1002/da.20751.

65 *the higher the exposure to trauma, the greater the addiction risk* V. J. Felitti, "The Relationship of Adverse Childhood Experiences to Adult Health: Turning Gold into Lead," *Zeitschrift für Psychosomatische Medizin und Psychotherapie* 48, no. 4 (2002): 359–69.

65 *the typical preaddiction experience as a "shattered childhood"* S. Darke, "Pathways to Heroin Dependence: Time to Re-appraise Self-Medication," *Addiction* 108 (2013): 659–67, doi:10.1111/j.1360-0443.2012.04001.x, http://www.ncbi.nlm.nih.gov/pubmed/23075121.

65 *humans have a remarkable tendency toward resilience* F. J. Stoddard Jr., "Outcomes of Traumatic Exposure," *Child & Adolescent Psychiatric Clinics of North America* 23, no. 2 (April 2014): 243–56, viii, doi:10.1016/j.chc.2014.01.004.

65 One extreme adversity doubles odds of addiction: G. N. Giordano et al., "Unexpected Adverse Childhood Experiences and Subsequent Drug Use Disorder: A Swedish Population Study (1995–2011)," *Addiction* (Abingdon, England), 109, no. 7 (2014): 1119–27, doi:10.1111/add.12537, http://www.ncbi.nlm.nih.gov/pubmed/24612271.

66 *64% of risk for these addictions could be attributed to child trauma* S. R. Dube et al., "Childhood Abuse, Neglect, and Household Dysfunction and the Risk of Illicit Drug Use: The Adverse Childhood Experiences Study," *Pediatrics* 111, no. 3 (March 2003): 564–72, http://www.ncbi.nlm.nih.gov/pubmed/12612237?dopt=Abstract.

66 *risk of heavy smoking is nearly tripled for people with five or more ACEs* R. F. Anda et al., "Adverse Childhood Experiences and Smoking During Adolescence and Adulthood," *JAMA* 282, no. 17 (1999): 1652–58, doi:10.1001/jama.282.17.1652, http://www.ncbi.nlm.nih.gov/pubmed/10553792?dopt=Abstract.

66 *alcoholism risk is increased by a factor of 7 for those with four or more [ACEs]* Vincent J. Felitti, "Relationship of Childhood Abuse and Household Dysfunction to Many of the Leading Causes of Death in Adults: The Adverse Childhood Experiences (ACE) Study," *American Journal of Preventive Medicine* 14, no. 4 (May 1998): 245–58, http://dx.doi.org/10.1016/S0749-3797(98)00017-8, http://www.sciencedirect.com/science/article/pii/S0749379798000178.

66 Studies on the offspring of Holocaust survivors: R. Yehuda, "Influences of Maternal and Paternal PTSD on Epigenetic Regulation of the Glucocorticoid Receptor Gene in Holocaust Survivor Offspring," *American Journal of Psychiatry* 171, no. 8 (2014): 872–80, doi:10.1176/appi.ajp.2014.13121571, http://www.ncbi.nlm.nih.gov/pubmed/24832930. See also Rachel Yehuda and Linda M. Bierer, "Transgenerational Transmission of Cortisol and PTSD Risk," cited in E. Ronald De Kloet, Melly S. Oitzl, and Eric Vermetten,

eds., *Progress in Brain Research*, vol. 167 (Amsterdam, the Netherlands: Elsevier, 2007): 121–35, http://dx.doi.org/10.1016/S0079-6123(07)67009-5, http://www.sciencedirect.com/science/article/pii/S0079612307670095#; and Rachel Yehuda et al., "Maternal, not Paternal, PTSD Is Related to Increased Risk for PTSD in Offspring of Holocaust Survivors," *Journal of Psychiatric Research* 42, no. 13 (October 2008): 1104–11, http://dx.doi.org/10.1016/j.jpsychires.2008.01.002, http://www.sciencedirect.com/science/article/pii/S0022395608000046.

67 Neglectful childrearing in rats passed down through environment: Gustavo Turecki and Michael J. Meaney, "Effects of the Social Environment and Stress on Glucocorticoid Receptor Gene Methylation: A Systematic Review," *Biological Psychiatry,* in press, doi:10.1016/j.biopsych.2014.11.022.

68 *changes that result from being raised in a stressful environment aren't all negative* M. Rutter, "Achievements and Challenges in the Biology of Environmental Effects," supplement, *Proceedings of the National Academy of Sciences of the United States of America* 109, suppl. 2 (2012): 17149–53, doi:10.1073/pnas.1121258109.

68 Trajectory of brain's stress system development affected by early environment: See generally Szalavitz and Perry, *Born for Love.*

68 *Being wired for a future in a threatening world* J. Belsky and M. Pluess, "Beyond Risk, Resilience, and Dysregulation: Phenotypic Plasticity and Human Development," *Development and Psychopathology* 25, no. 4, part 2 (November 2013): 1243–61, doi: 10.1017/S095457941300059X.

69 Death obsession common in Asperger's and OCD: P. H. Thomsen, "Obsessive-Compulsive Symptoms in Children and Adolescents: A Phenomenological Analysis of 61 Danish Cases," *Psychopathology* 24, no. 1 (1991): 12–18, doi: 10.1159/000284691.

69 Quotes from Ernest Becker: E. Becker, *The Denial of Death* (New York: Simon and Schuster, 1973), ix.

70 Mouse study suggested that specific fears can be passed down via changes in the reading instructions of DNA: B. G. Dias and K. J. Ressler, "Parental Olfactory Experience Influences Behavior and Neural Structure in Subsequent Generations," *Nature Neuroscience* 17, no. 1 (2014): 89–96, doi:10.1038/nn.3594.

70 Shame, self-loathing, and compulsive symptoms: Bunmi O. Olatunji, Rebecca Cox, and Eun Ha Kim, "Self-Disgust Mediates the associations Between Shame and Symptoms of Bulimia and Obsessive-Compulsive Disorder," *Journal of Social and Clinical Psychology* 34, no. 3 (2015): 239–58.

6. LABELS

73 Preschoolers fixate on moral ideas: K. Klein Burhans and C. S. Dweck, "Helplessness in Early Childhood: The Role of Contingent Worth," *Child Development* 66 (1995): 1719–38.

76 *One 2007 study of nearly 400 seventh graders* L. S. Blackwell, K. H. Trzesniewski, and C. S. Dweck, "Implicit Theories of Intelligence Predict Achievement across an Adolescent Transition: A Longitudinal Study and an Intervention," *Child Development* 78 (January-February 2007): 246–63.

76 *a whopping 40% of those who believed that intelligence was fixed actually lied* C. S. Dweck, *Mindset: The New Psychology of Success* (New York: Random House, 2006), 73.

76 *An even more fascinating paper* D. S. Yeager et al., "The Far-Reaching Effects of Believing People Can Change: Implicit Theories of Personality Shape Stress, Health and Achievement During Adolescence," *Journal of Personality and Social Psychology* 106, no. 6 (June 2014): 867–84. doi: 10.1037/a0036335, http://www.ncbi.nlm.nih.gov/pubmed/24841093.

78 *"very concerned about whether their performance means that they are 'good' or 'bad'"* Klein Burhans and Dweck, "Helplessness in Early Childhood."

81 *"hospital flu" is actually opioid withdrawal* US National Library of Medicine, *Medical Encyclopedia: Opiate Withdrawal,* accessed August 23, 2015, http://www.nlm.nih.gov/medlineplus/ency /article/000949.htm.

82 *children of alcoholics have a risk of developing alcoholism that is two to four times greater* M. A. Schuckit, *Educating Yourself about Alcohol and Drugs* (New York and London: Plenum Press, 1995), 110–11.

82 Alcohol dependence found in the general population = 12.5%: D. S. Hasin et al., "Prevalence, Correlates, Disability, and Comorbidity of DSM-IV Alcohol Abuse and Dependence in the United States: Results from the National Epidemiologic Survey on Alcohol and Related Conditions," *Archives of General Psychiatry* 64, no. 7 (2007): 830–42, doi:10.1001/ archpsyc.64.7.830, http://www.ncbi.nlm.nih.gov/pubmed/17606817.

82 Fifty percent of children of alcoholics still don't develop severe drinking problems: 12.5% x 4 = 50%.

82 *about one in seven people seriously dislikes opioids* M. S. Angst et al., "Aversive and Reinforcing Opioid Effects: A Pharmacogenomic Twin Study," *Anesthesiology* 117, no. 1 (2012): 22–37, doi:10.1097/ ALN.0b013e31825a2a4e, http://www.ncbi.nlm.nih.gov/pmc/articles/PMC3428265.

82 Dan Rather said heroin gave him "a hell of a headache": T. Nguyen, "Dan Rather Reminisces about That One Time He Did Heroin," *Mediaite,* January 9, 2014, accessed August 23, 2015, http://www.mediaite.com/tv/dan-rather-reminisces-about-that-one -time-he-did-heroin/.

82 *Two normal volunteers who injected heroin for several days* B. K. Alexander, *The Globalization of Addiction* (Oxford: Oxford University Press, 2008), 182.

82 *A more recent study, which gave 228 healthy adult twins* Angst et al., "Aversive and Reinforcing Opioid Effects." Numbers do not add up to 100% due to rounding.

82 *in Japan, when heavy drinking became de rigueur among businessmen* S. E. Luczak, S. J. Glatt, and T. J. Wall. "Meta-analyses of ALDH2 and ADH1B with alcohol dependence in Asians." *Psychological Bulletin* 132, no. 4 (2006): 607–21, doi:10.1037/0033-2909.132.4.607.

7. HELL IS JUNIOR HIGH SCHOOL

85 *youth who learn these thinking patterns are more prone to adult depression* L. C. Michl et al., "Rumination As a Mechanism Linking Stressful Life Events to Symptoms of Depression and Anxiety: Longitudinal Evidence in Early Adolescents and Adults," *Journal of Abnormal Psychology* 122, no. 2 (2013): 339–52, doi:10.1037/a0031994.

85 *kids who can't self-regulate* F. Mishna, "Learning Disabilities and Bullying: Double Jeopardy," *Journal of Learning Disabilities* 36, no. 4 (July-August 2003): 336–47. See also P. R. Sterzing et al., "Prevalence and Correlates of Bullying Involvement among Adolescents with an Autism Spectrum Disorder," *Archives of Pediatrics & Adolescent Medicine* 166, no. 11 (2012): 1058–64, doi:10.1001/archpediatrics.2012.790.

87 *being a victim of chronic bullying can sometimes* Suzet Tanya Lereya et al., "Adult Mental Health Consequences of Peer Bullying and Maltreatment in Childhood: Two Cohorts in Two Countries," *Lancet Psychiatry* 2, no. 6 (June 2015): 524–31, http://dx.doi .org/10.1016/S2215-0366(15)00165-0, http://www.sciencedirect.com/science/article/pii /S2215036615001650.

88 *kids who exhibit pseudomature behavior at 13* J. P. Allen, "What Ever Happened to the 'Cool' Kids? Long-Term Sequelae of Early Adolescent Pseudomature Behavior," *Child Development,* 85,

no. 5 (2014): 1866–80, http://doi.org/10.1111/cdev.12250; http://www.ncbi.nlm.nih.gov /pubmed/24919537.

90 *we basically need other people to keep severe stress at bay* See generally B. D. Perry and M. Szalavitz, *The Boy Who Was Raised As a Dog and Other Stories from a Child Psychiatrist's Notebook* (New York: Basic, 2006); and Szalavitz and Perry, *Born for Love.*

90 *At age 40 to 64, low-ranked clerks were* M. G. Marmot and M. J. Shipley, "Do Socioeconomic Differences in Mortality Persist After Retirement? 25 Year Follow Up of Civil Servants from the First Whitehall Study," *BMJ: British Medical Journal* 313, no. 7066 (1996): 1177–80.

90 *only one third of the gradient between top and bottom* I. Giesinger et al., "Association of Socioeconomic Position with Smoking and Mortality: The Contribution of Early Life Circumstances in the 1946 Birth Cohort," *Journal of Epidemiology and Community Health* 68, no. 3 (2014): 275–79, doi:10.1136/jech-2013-203159.

90 *A second Whitehall study* M. Marmot, *The Status Syndrome* (New York: Holt, 2005), 45; M. Bartleya et al., "Gender Differences in the Relationship of Partner's Social Class to Behavioural Risk Factors and Social Support in the Whitehall II Study," *Social Science & Medicine* 59, no. 9 (November 2004): 1925–36.

90 *lower-ranked animals are at greater risk for all types of early death* R. Sapolsky, *Why Zebras Don't Get Ulcers* (New York: Barnes and Noble Books, 2000), 293–97.

90 High levels of stress hormones can increase risk of depression: R. M. Sapolsky, "Potential Behavioral Modification of Glucocorticoid Damage to the Hippocampus," *Behavioural Brain Research* 57, no. 2 (November 30, 1993): 175–82.

91 Lower socioeconomic status linked with higher addiction risk: Compton et al., "Prevalence, Correlates, Disability, and Comorbidity."

91 *not all stress is bad* V. K. Parihar et al., "Predictable Chronic Mild Stress Improves Mood, Hippocampal Neurogenesis and Memory," *Molecular Psychiatry* 16, no. 2 (2011): 171–83, doi:10.1038/mp.2009.130.

91 Excess glutamate released during chronic stress can damage the hippocampus: H. Uno et al., "Neurotoxicity of Glucocorticoids in the Primate Brain," *Hormones and Behavior* 28, no. 4 (December 1994): 336–48.

91 *all long-lasting effective methods of treating depression* A. E. Autry and L. M. Monteggia, "Brain-Derived Neurotrophic Factor and Neuropsychiatric Disorders," *Pharmacological Reviews* 64, no. 2 (2012): 238–58, doi:10.1124/pr.111.005108.

91 Severe childhood stress more likely to lead to depression than to PTSD: Bessel A. van der Kolk et al., "Proposal to Include a Developmental Trauma Disorder Diagnosis for Children and Adolescents in DSM-V" (unpublished manuscript, February 1, 2009), http://w.traumacenter.org/announcements/DTD_papers_Oct_09.pdf.

91 Evolutionary links between low status and depression: N. B. Allen and P. Badcock, "Darwinian Models of Depression: A Review of Evolutionary Accounts of Mood and Mood Disorders," *Progress in Neuro-Psychopharmacology and Biological Psychiatry*, 30, no. 5 (August 2006): 815–26.

91 Changing serotonin levels affects dominance status: M. J. Raleigh et al., "Serotonergic Mechanisms Promote Dominance Acquisition in Adult Male Vervet Monkeys," *Brain Research* 559, no. 2 (September 20, 1991): 181–90.

92 Lowest socioeconomic status linked with doubled risk of depression: Carles Muntaner et al., "Socioeconomic Position and Major Mental Disorders," *Epidemiological Reviews* 26, no. 1 (2004): 53–62, doi:10.1093/epirev/mxh001.

92 *45% of those who were bullied in fifth grade* Laura M. Bogart et al., "Peer Victimization in Fifth Grade and Health in Tenth Grade,"*Pediatrics* 133, no. 3 (2014): 440–47, doi:10.1542/peds .2013-3510, http://www.ncbi.nlm.nih.gov/pubmed/24534401.

92 *a study of young adults in North Carolina* W. E. Copeland et al., "Adult Psychiatric Outcomes of Bullying and Being Bullied by Peers in Childhood and Adolescence," *JAMA Psychiatry* 70, no. 4 (2013): 419–26, doi:10.1001/jamapsychiatry.2013.504, http://archpsyc.jamanetwork .com/article.aspx?articleid=1654916.

92 *Another study of the same group* D. Wolke et al., "Impact of Bullying in Childhood on Adult Health, Wealth, Crime and Social Outcomes," *Psychological Science* 24, no. 10 (2013): 1958–70, doi:10.1177/0956797613481608, http://www.ncbi.nlm.nih.gov/pubmed/23959952.

93 Bullies target kids who are emotionally dysregulated: Mishna, "Learning Disabilities and Bullying"; and Sterzing et al., "Prevalence and Correlates of Bullying Involvement."

8. TRANSITIVE NIGHTFALL

97 *90% of all addictions begin during adolescence* National Center on Addiction and Substance Abuse, "Adolescent Substance Use: America's #1 Public Health Problem," June 2011, http://www.casacolumbia.org/addiction-research/reports/adolescent-substance-use.

97 Odds of developing alcoholism lower in people who start after age 21: C. E. Sartor et al., "Timing of First Alcohol Use and Alcohol Dependence: Evidence of Common Genetic Influences," *Addiction* (Abingdon, England) 104, no. 9 (2009): 1512–18, doi:10.1111 /j.1360-0443.2009.02648.x, http://www.ncbi.nlm.nih.gov/pmc/articles/PMC2741422/.

97 Brains of five- and six-year olds are 95% of adult size: J. N. Giedd and J. L. Rapoport, "Structural MRI of Pediatric Brain Development: What Have We Learned and Where Are We Going?" *Neuron* 67, no. 5 (2010): 728–34, doi:10.1016/j.neuron.2010.08.040, http:// www.ncbi.nlm.nih.gov/pmc/articles/PMC3285464/.

98 *The pruning process can be extreme* G. Z. Tau and B. S. Peterson, "Normal Development of Brain Circuits," *Neuropsychopharmacology* 35, no. 1 (2010): 147–68, doi:10.1038/npp.2009.115, http://www.ncbi.nlm.nih.gov/pmc/articles/PMC3055433/.

98 Gray matter in prefrontal cortex shrinks during adolescence: Paul M. Thompson et al., "Structural MRI and Brain Development," *International Review of Neurobiology* 67 (2005): 285–323, http://dx.doi.org/10.1016/S0074-7742(05)67009-2, http://www.ncbi.nlm.nih.gov /pubmed/16291026/.

98 Early brain overgrowth found consistently in autism: Eric Courchesne and Karen Pierce, "Brain Overgrowth in Autism During a Critical Time in Development: Implications for Frontal Pyramidal Neuron and Interneuron Development and Connectivity," *International Journal of Developmental Neuroscience* 23, nos. 2–3 (April–May 2005): 153–70, http://dx.doi.org/10.1016/j.ijdevneu.2005.01.003.

98 Some brain regions in autism are "hyperconnected": K. Markram and H. Markram, "The Intense World Theory—A Unifying Theory of the Neurobiology of Autism," *Frontiers in Human Neuroscience* 4 (2010): 224, doi:10.3389/fnhum.2010.00224.

98 Autistic teens show less pruning of brain cells: Guomei Tang et al., "Loss of mTOR-Dependent Macroautophagy Causes Autistic-like Synaptic Pruning Deficits," *Neuron* 83, no. 5 (September 3, 2014): 1131–43, http://www.cell.com/neuron/abstract/S0896-6273 (14)00651-5.

99 *"while the frontal cortex is still trying to make sense of the assembly instructions"* R. Sapolsky, "Dude,

Where's My Frontal Cortex?" *Nautilus*, July 24, 2014, http://nautil.us/issue/15/turbulence /dude-wheres-my-frontal-cortex.

100 New experiences may spur larger dopamine release in teens: D. Wahlstrom, T. White, and M. Luciana, "Neurobehavioral Evidence for Changes in Dopamine System Activity During Adolescence," *Neuroscience and Biobehavioral Reviews* 34, no. 5 (2010): 631–48, doi:10.1016/j. neubiorev.2009.12.007.

100 Large rewards are disproportionately attractive to teens: Adriana Galvan et al., "Earlier Development of the Accumbens Relative to Orbitofrontal Cortex Might Underlie Risk-Taking Behavior in Adolescents," *Journal of Neuroscience* 26, no. 25 (June 21, 2006): 6885–92, doi:10.1523/JNEUROSCI.1062-06.2006.

100 Small rewards can seem like punishments to adolescents: Sapolsky, "Dude, Where's My Frontal Cortex?"

101 Teens overestimate odds of bad outcomes in risky behavior: V. F. Reyna and F. Farley, "Risk and Rationality in Adolescent Decision-Making: Implications for Theory, Practice, and Public Policy," *Psychological Science in the Public Interest* 7 (2006): 1–44.

102 Statistics on drug use prevalence over time: Miech et al., "Monitoring the Future."

102 *one third of the adult population smoked cigarettes* Centers for Disease Control and Prevention, "Trends in Current Cigarette Smoking among High School Students and Adults, United States, 1965–2011," accessed August 6, 2015, http://www.cdc.gov/tobacco/data_statistics /tables/trends/cig_smoking/.

9. ON DOPE AND DOPAMINE

108 Dopamine is found in less than 1 million neurons: P. Cumming, *Imaging Dopamine* (Cambridge, UK: Cambridge University Press, 2009), 12, https://books.google.com/books?id =VlXEkJKR-ZsC&pg=PA12&lpg=PA12&dq=400,000+dopamine+neurons&source=bl &ots=0wDrvBxVDJ&sig=cLllerJieWc5HvDX5ijHauhAGmc&hl=en&sa=X&ei=a_NDV cG6DornsASh44CADQ&ved=0CCsQ6AEwAzgK#v=onepage&q=400%2C000%20dopa mine%20neurons&f=false.

108 There are 86 billion neurons in the human brain: F. A. C. Azevedo, "Equal Numbers of Neuronal and Nonneuronal Cells Make the Human Brain an Isometrically Scaled-up Primate Brain," *Journal of Comparative Neurology* 513 (2009): 532–41, doi: 10.1002/cne.21974, http://www.ncbi.nlm.nih.gov/pubmed/19226510.

108 *rats seem to enjoy the stimulation so much* Matteo Colombo, "Deep and Beautiful: The Reward Prediction Error Hypothesis of Dopamine," *Studies in History and Philosophy of Science Part C: Studies in History and Philosophy of Biological and Biomedical Sciences* 45 (March 2014): 57–67, http: //dx.doi.org/10.1016/j.shpsc.2013.10.006, http://www.sciencedirect.com/science/article/pii /S1369848613001313.

108 *"It's just like a stimulant"* Author interview with Eric Hollander, 2014.

108 Patients implanted incorrectly will self-stimulate: Vaughan Bell, "Erotic Self Stimulation and Brain Implants," Mindhacks website, September 16, 2008, http://mindhacks.com /2008/09/16/erotic-self-stimulation-and-brain-implants/.

109 Roy Wise theory on dopamine as pleasure: R. A. Wise, "Dopamine and Reward: The Anhedonia Hypothesis 30 Years On," *Neurotoxicity Research* 14, nos. 2–3 (2008): 169–83, doi:10.1007/BF03033808.

109 Antipsychotic drugs blunt pleasure: Ibid.

109 Some dopamine neurons fire more in response to distress than reward: M. Matsumoto and O. Hikosaka, "How Do Dopamine Neurons Represent Positive and Negative Motivational Events?" *Nature* 459, no. 7248 (2009): 837–41, doi:10.1038/nature08028.

109 Parkinson's disease classified as a movement disorder: Y. Smith et al., "Parkinson's Disease Therapeutics: New Developments and Challenges Since the Introduction of Levodopa," *Neuropsychopharmacology* 37, no. 1 (2012): 213–46, doi:10.1038/npp.2011.212.

109 Antipsychotic drugs can cause movement disorders: K. Komossa et al., "Risperidone Versus Other Atypical Antipsychotics for Schizophrenia," *Cochrane Database of Systematic Reviews* 1 (2011): CD006626. doi: 10.1002/14651858.CD006626.pub2.

109 Psychological effects of Parkinson's: Y. Smith et al., "Parkinson's Disease Therapeutics."

110 *"The first qualities of Parkinsonism which were ever described"* Oliver Sacks, *Awakenings* (New York: Harper Perennial, 1990), 6.

110 *"[S]ome of them would sit for hours"* Ibid., 9.

111 Case reports on Parkinson's: Ibid., 10, note 14.

111 Donald Klein labels the pleasures of the hunt and pleasures of the feast: P. Kramer, *Listening to Prozac* (New York: Penguin Books, 1997), 231–32.

111 Klein classified pleasures based on observations of people with addictions: Ibid.

112 *"They wouldn't want to eat"* Author interview with Kent Berridge, 2014.; Also: T. E. Robinson and K. C. Berridge, "The Incentive Sensitization Theory of Addiction: Some Current Issues," *Philosophical Transactions of the Royal Society B: Biological Sciences* 363, no. 1507 (2008): 3137–46, doi:10.1098/rstb.2008.0093.

114 George Koob on "pharmacological Calvinism": M. Szalavitz, "A Combo of Existing Drugs Could Treat Cocaine Addiction," *TIME.com*, August 9, 2012, http://healthland.time.com /2012/08/09/combo-of-fda-approved-drugs-could-fight-cocaine-addiction/.

114 *"I can remember many, many times"* T. Wells and W. Triplett, *Drug Wars: An Oral History from the Trenches* (New York: William Morrow, 1992), 141.

114 *"[M]y body's saying no and my mind's saying no"* D. Waldorf et al., *Cocaine Changes: The Experience of Using and Quitting* (Philadelphia: Temple University Press, 1992), 126.

114 *"I used to smoke some [cocaine] that wasn't good"* D. Crosby and C. Gottlieb, *The Autobiography of David Crosby* (New York: Dell, 1988), 318.

116 *"I kept pumping [cocaine] into my vein"* M. Lewis, *Memoirs of an Addicted Brain* (New York: Public Affairs, 2011), 273.

117 Kandel won Nobel for discoveries about sensitization: http://www.nobelprize.org/nobel _prizes/medicine/laureates/2000/kandel-bio.html.

117 Research on slugs and sensitization and tolerance: E. R. Kandel et al., *Principles of Neural Science*, fourth edition (New York: McGraw Hill, 2000), 1249–52.

117 PTSD may be a form of sensitization learning: S. Lissek S and B. van Meurs, "Learning Models of PTSD: Theoretical Accounts and Psychobiological Evidence," *International Journal of Psychophysiology* (November 20, 2014), doi:10.1016/j.ijpsycho.2014.11.006.

119 Autistic children have sensitized startle response: G. F. Madsen, "Increased Prepulse Inhibition and Sensitization of the Startle Reflex in Autistic Children," *Autism Research* 7 (2014): 94–103, doi: 10.1002/aur.1337.

119 Autistic children have less habituation in the amygdala: J. R. Swartz et al., "Amygdala Habituation and Prefrontal Functional Connectivity in Youth with Autism Spectrum Disorders," *Journal of the American Academy of Child and Adolescent Psychiatry* 52, no. 1 (2013): 84–93, doi:10.1016/j.jaac.2012.10.012, http://www.ncbi.nlm.nih.gov/pubmed/23265636.

10. SET AND SETTING

124 *Among people . . . now starting to hit their 50s . . . just under half report having tried cocaine at least once* L. D. Johnston et al., "Monitoring the Future National Survey Results on Drug Use, 1975–2013: Volume II, College Students and Adults Ages 19–55" (Ann Arbor: Institute for Social Research, The University of Michigan, 2014), http://monitoringthefuture.org/pubs/mono graphs/mtf-vol2_2013.pdf.

124 Statistics on college students and cocaine: Ibid, 390–91.

126 Illegal CIA LSD experiments: Troy Hooper, "Operation Midnight Climax: How the CIA Dosed S.F. Citizens with LSD," *S.F. Weekly*, March 14, 2012, http://www.sfweekly.com/2012 -03-14/news/cia-lsd-wayne-ritchie-george-h-white-mk-ultra/.

127 88% of Vietnam veterans who became addicted to heroin did not get readdicted back home: N. E. Zinberg, *Drug, Set and Setting* (New Haven: Yale University Press, 1984), accessed August 6, 2015, http://www.psychedelic-library.org/zinbergp.htm.

127 Vietnam vets who had ongoing heroin problems had drug problems before the war: L. N. Robins, "The Sixth Thomas James Okey Memorial Lecture: Vietnam Veterans' Rapid Recovery from Heroin Addiction: A Fluke or Normal Expectation?," *Addiction* 88, no. 8 (August 1993): 1041–54.

127 Story of Skinner's discovery of intermittent reinforcement: Indiana University, online course for Psychology 101, accessed August 6, 2015, http://www.indiana.edu/~p1013447 /dictionary/sked.htm.

127 Random reinforcement is most effective: M. D. Zeiler, "Fixed-Interval Behavior: Effects of Percentage Reinforcement," *Journal of the Experimental Analysis of Behavior* 17, no. 2 (1972): 177–89, doi:10.1901/jeab.1972.17-177.

128 Music appreciation involves pattern prediction: D. J. Levitin, *This Is Your Brain on Music* (New York: Plume, 2006), 234–46.

128 Role of dopamine in signaling and wanting: Some researchers argue that dopamine release is really an error prediction signal, not a marker of "wanting," but I'm not sure the two ideas are actually incompatible—an assessment that Nora Volkow, director of the National Institute on Drug Abuse and a leading dopamine researcher, shares. Author interview with Nora Volkow, 2014.

129 *pigeons even developed "superstitions"* B. F. Skinner, " 'Superstition' in the Pigeon," *Journal of Experimental Psychology* 38 (1948): 168–72.

130 Methadone maintenance reduces mortality 75%: John R. M. Caplehorn et al., "Methadone Maintenance and Addicts' Risk of Fatal Heroin Overdose," *Substance Use & Misuse* 31, no. 2 (1996): 177–96, http://informahealthcare.com/action/showCitFormats?doi=10 .3109%2F10826089609045806.

131 Stable methadone patients are not impaired and can work and drive: National Institute on Drug Abuse, "NIDA International Program Methadone Research Web Guide: Part B: 20 Questions and Answers Regarding Methadone Maintenance Treatment Research," accessed August 19, 2015, http://www.drugabuse.gov/sites/default/files/pdf/partb.pdf.

131 Context can make an otherwise safe dose lethal: S. Siegel et al., "Heroin 'Overdose' Death: Contribution of Drug-Associated Environmental Cues," *Science* 216, no. 4544 (April 23, 1982): 436–37.

131 Rats can die of overdose when given their usual heroin dose in new setting: Ibid.

132 Stimulant maintenance so far not as good as opioid maintenance: M. E. Mooney et al.,

"Pilot Study of the Effects of Lisdexamfetamine on Cocaine Use: A Randomized, Double-Blind, Placebo-Controlled Trial," *Drug and Alcohol Dependence* 153 (August 1, 2015): 94–103, doi:10.1016/j.drugalcdep.2015.05.042.

132 Solitary confinement rapidly produces psychiatric symptoms: Stuart Grassian, "Psychiatric Effects of Solitary Confinement," *Washington University Journal of Law and Policy* 22, no. 325 (2006): http://openscholarship.wustl.edu/law_journal_law_policy/vol22/iss1/24.

133 Isolated rats took 19 times as much morphine: P. F. Hadaway et al., "The Effect of Housing and Gender on Preference for Morphine-Sucrose Solutions in Rats," *Psychopharmacology* (Berlin) 66, no. 1 (1979): 87–91. Great, carefully researched description of experiments here: Stuart McMillen, Rat Park Cartoon, accessed August 6, 2015, http://www.stuartmcmillen.com/comics_en/rat-park/.

133 Further description of Rat Park data: S. Peele, *The Meaning of Addiction* (Lexington, MA: Lexington Books, 1985), 83.

134 Some experiments did not replicate Rat Park effect: Ibid., 88.

134 Running wheel reduces cocaine use: P. Kelly et al., "Wheel-Running Attenuates Intravenous Cocaine Self-Administration in Rats: Sex Differences," *Pharmacology Biochemistry and Behavior* 73, no. 3 (October 2002): 663–71, http://dx.doi.org/10.1016/S0091-3057(02)00853-5, http://www.sciencedirect.com/science/article/pii/S0091305702008535.

134 Social defeat increases cocaine use: J. W. Tidey and K. A. Miczek, "Acquisition of Cocaine Self-Administration After Social Stress: Role of Accumbens Dopamine," *Psychopharmacology* (Berlin) 130, no. 3 (April 1997): 203–12, http://www.ncbi.nlm.nih.gov/pubmed/9151353.

134 Rats prefer sugar and even artificial sweetener to cocaine: Magalie Lenoir et al., "Intense Sweetness Surpasses Cocaine Reward," *PLOS One* 2, no. 8 (August 1, 2007): e698 doi:10.1371/journal.pone.0000698, http://www.plosone.org/article/info%3Adoi%2F10.1371%2Fjournal.pone.0000698.

135 LSD and psychedelics may be useful in addiction treatment: Michael P. Bogenschutz and Matthew W. Johnson, "Classic Hallucinogens in the Treatment of Addictions," *Progress in Neuro-Psychopharmacology and Biological Psychiatry,* accessed August 14, 2015, http://dx.doi.org/10.1016/j.pnpbp.2015.03.002, http://www.sciencedirect.com/science/article/pii/S0278584615000512.

11. LOVE AND ADDICTION

138 *"madmen, junkies, comedians, ex-cons"* Carlton Arms Hotel, "About Page," accessed August 6, 2015, http://www.carltonarms.com/about_us.htm.

140 *around 1,000 stories about crack and cocaine appeared* C. Reinarman and H. G. Levine, *Crack in America: Demon Drugs and Social Justice* (Berkeley: University of California Press, 1997), 20.

143 *the world's oldest song that survives in writing* L. Krause, "Egyptian Tomb Yields 'World's Oldest Love Song,'" *National Geographic News,* February 13, 2001, accessed August 6, 2015, http://news.nationalgeographic.com/news/2001/02/0213_1stlovesong.html.

143 C. Sue Carter's research on pair bonding: Maia Szalavitz, "'Cuddle Chemical' Could Treat Mental Illness," *New Scientist,* May 14, 2008; and C. Sue Carter, A. Courtney Devries, and Lowell L. Getz, "Physiological Substrates of Mammalian Monogamy: The Prairie Vole Model," *Neuroscience & Biobehavioral Reviews* 19, no. 2 (Summer 1995): 303–14, http://dx.doi.org/10.1016/0149-7634(94)00070-H, http://www.sciencedirect.com/science/article/pii/014976349400070H.

143 Pert and Snyder discover first opioid receptor: C. B. Pert and S. H. Snyder, "Opiate Receptor: Demonstration in Nervous Tissue," *Science* 179, no. 4077 (March 9, 1973): 1011–14.

144 Five percent of mammals are monogamous: Hemanth P. Nair and Larry J. Young, "Vasopressin and Pair-Bond Formation: Genes to Brain to Behavior," *Physiology* 21, no. 2 (April 1, 2006): 146–52, doi:10.1152/physiol.00049.2005, http://physiologyonline.physiology.org/content/21/2/146.

144 Prairie voles, montane voles, and varying receptors' relationship to monogamy: T. R. Insel, "Is Social Attachment an Addictive Disorder?," *Physiology and Behavior* 79 (2003): 351–57.

144 *"It's taking over parts of the nervous system"* Szalavitz, " 'Cuddle Chemical' Could Treat Mental Illness."

145 Gene affects marriage and relationships: Z. R. Donaldson and L. J. Young, "Oxytocin, Vasopressin, and the Neurogenetics of Sociality," *Science* 322, no. 5903 (November 7, 2008): 900–04, doi:10.1126/science.1158668, http://www.ncbi.nlm.nih.gov/pubmed/18988842.

145 Oxytocin can relieve withdrawal symptoms from heroin and alcohol: Iain S. McGregor and Michael T. Bowen, "Breaking the Loop: Oxytocin As a Potential Treatment for Drug Addiction," *Hormones and Behavior* 61, no. 3 (March 2012): 331–39, http://dx.doi.org/10.1016/j.yhbeh.2011.12.001, http://www.sciencedirect.com/science/article/pii/S0018506X11002765.

145 Intranasal oxytocin indistinguishable from placebo: Szalavitz, " 'Cuddle Chemical' Could Treat Mental Illness."

145 *"oxytocin elevates trust"* T. Baumgartner et al., "Oxytocin Shapes the Neural Circuitry of Trust and Trust Adaptation in Humans," *Neuron* 58, no. 4 (May 22, 2008): 639–50, doi: 10.1016/j.neuron.2008.04.009.

145 Oxytocin helps autistic people better detect emotions: A. J. Guastella et al., "Intranasal Oxytocin Improves Emotion Recognition for Youth with Autism Spectrum Disorders," *Biological Psychiatry* 67, no. 7 (April 1, 2010): 692–94, doi: 10.1016/j.biopsych.2009.09.020.

145 Oxytocin increases bias against outsiders: F. Sheng et al., "Oxytocin Modulates the Racial Bias in Neural Responses to Others' Suffering," *Biological Psychology* 92, no. 2 (February 2013): 380–86, doi:10.1016/j.biopsycho.2012.11.018.

146 Feelings of warmth and social connection modulated by opioids: T. K. Inagaki, M. R. Irwin, and N. I. Eisenberger, "Blocking Opioids Attenuates Physical Warmth-Induced Feelings of Social Connection," *Emotion* 15, no. 4 (August 2015): 494–500, doi:10.1037/emo0000088.

146 Dopamine drives desire for partner: T. R. Insel, "Is Social Attachment an Addictive Disorder?"

146 *adult male rats who were stimulated* P. Paredes-Ramos et al., "Juvenile Play Conditions Sexual Partner Preference in Adult Female Rats," *Physiology & Behavior* 104, no. 5 (October 24, 2011): 1016–23, doi: 10.1016/j.physbeh.2011.06.026.

146 Female rats prefer mating with animals with scent familiar from childhood: Ibid.

146 Neglect and trauma influence neurotransmitters and relationships: M. D. De Bellis and A. Zisk., "The Biological Effects of Childhood Trauma," *Child and Adolescent Psychiatric Clinics of North America* 23, no. 2 (2014): 185–222, doi:10.1016/j.chc.2014.01.002.

147 Oxytocin reduces trust in borderline personality disorder: A. Ebert et al., "Modulation of Interpersonal Trust in Borderline Personality Disorder by Intranasal Oxytocin and Childhood Trauma," *Social Neuroscience* 8, no. 4 (2013): 305–13, doi:10.1080/17470919.2013.807301, http://www.ncbi.nlm.nih.gov/pubmed/23802121. See also J. Bartz et al., "Oxytocin Can Hinder Trust and Cooperation in Borderline Personality Disorder," *Social Cognitive*

and Affective Neuroscience 6, no. 5 (2011): 556–63, doi:10.1093/scan/nsq085, http://www.ncbi.nlm.nih.gov/pubmed/21115541.

148 Codependence isn't a diagnosable disorder: E. J. Chiauzzi and S. Liljegren, "Taboo Topics in Addiction Treatment: An Empirical Review of Clinical Folklore," *Journal of Substance Abuse Treatment* 10, no. 3 (May–June 1993): 303–16.

148 Cartoon about normal parents: J. Berman, *Adult Children of Normal Parents* (New York: Pocket Books, 1994).

151 *one in three infants raised by rotating staff in orphanages died* D. Iwaniec, *Children Who Fail to Thrive: A Practice Guide* (New York: Wiley, 2004), 18.

151 Loneliness as dangerous to health as smoking; more dangerous than obesity: J. Holt-Lunstad, T. B. Smith, and J. B. Layton, "Social Relationships and Mortality Risk: A Meta-analytic Review," *PLOS Medicine* 7, no. 7 (2010): e1000316, doi:10.1371/journal.pmed.1000316.

152 Serotonin levels fall in people in love: D. Marazziti et al., "Alteration of the Platelet Serotonin Transporter in Romantic Love," *Psychological Medicine* 29, no. 3 (May 1999): 741–45.

152 Argument about whether to remove "dependence" from DSM and change in the DSM-5: C. P. O'Brien, N. Volkow, and T. K. Li, "What's in a Word? Addiction Versus Dependence in DSM-V," *American Journal of Psychiatry* 163 (2006): 764–65; and W. M. Compton et al., "Crosswalk Between DSM-IV Dependence and DSM-5 Substance Use Disorders for Opioids, Cannabis, Cocaine and Alcohol," *Drug and Alcohol Dependence*, 132 (September 1, 2013): 387–90, http://doi.org/10.1016/j.drugalcdep.2013.02.036.

12. RISKY BUSINESS

157 Dilaudid is twice as potent as heroin: E. Oviedo-Joekes et al., "Potency Ratio of Hydromorphone and Diacetylmorphine in Substitution Treatment for Long-Term Opioid Dependency," *Journal of Opioid Management* 7, no. 5 (September–October 2011): 371–76, http://www.ncbi.nlm.nih.gov/pubmed/22165036.

158 Teens overestimate HIV risk: V. F. Reyna and F. Farley, "Is the Teen Brain Too Rational?," *Scientific American,* May 17, 2007.

159 Emotions are decision-making algorithms: A. Damasio, *Descartes' Error: Emotion, Reason and the Human Brain* (New York: Penguin Reprint, 2005).

159 One fascinating study found it took teens longer to say no to risk: A. A. Baird and J. A. Fugelsang, "The Emergence of Consequential Thought: Evidence from Neuroscience," *Philosophical Transactions of the Royal Society of London, Series B: Biological Sciences* 359 (2004): 1797–1804, cited in E. Wargo, "Adolescents and Risk: Helping Young People Make Better Choices: Research Facts and Findings," a publication of the ACT for Youth Center of Excellence, September 2007, accessed August 6, 2015, http://www.actforyouth.net/resources/rf/rf_risk_0907.cfm.

161 Ainslie's "hyperbolic discounting" economic theory of addiction: G. Ainslie, "A Research-Based Theory of Addictive Motivation," *Law and Philosophy* 19 (2000): 77–115.

161 Children don't resist marshmallow if researchers are flaky: Celeste Kidd, Holly Palmeri, and Richard N. Aslin, "Rational Snacking: Young Children's Decision-Making on the Marshmallow Task Is Moderated by Beliefs about Environmental Reliability," *Cognition* 126, no. 1 (January 2013): 109–14, http://dx.doi.org/10.1016/j.cognition.2012.08.004.

164 Children reduce happiness: L. Belkin, "Does Having Children Make You Unhappy?," *New York Times,* April 1, 2009, http://parenting.blogs.nytimes.com/2009/04/01/why-does-anyone -have-children/?_r=0.

164 *In fact, in the early '70s, when Jaak Panksepp* Interview with Jaak Panksepp, 2008.

165 Ten percent of women with OCD have onset related to pregnancy: V. Guglielmi et al., "Obsessive-Compulsive Disorder and Female Reproductive Cycle Events: Results from the OCD and Reproduction Collaborative Study," *Depression and Anxiety* 31 (2014): 979–87, doi:10.1002/da.22234, http://www.ncbi.nlm.nih.gov/pubmed/24421066.

13. BUSTED

172 *about two thirds of people with substance addictions showed an elevated emotional response to the prospect of monetary gain* Antoine Bechara, Sara Dolan, and Andrea Hindes, "Decision-Making and Addiction (Part II): Myopia for the Future or Hypersensitivity to Reward?" *Neuropsychologia* 40, no. 10 (2002): 1690–1705, http://dx.doi.org/10.1016/S0028-3932(02)00016-7, http://www.sciencedirect.com/science/article/pii/S0028393202000167#.

172 *remaining third of the participants did not respond to punishment* Ibid.

172 *other studies have found reduced brain activation during punishment (typically monetary loss) in people addicted to cocaine and methamphetamine* Ibid.

14. THE PROBLEM WITH BOTTOM

176 A systematic review of the research on criminal recidivism: Martin Killias and Patrice Villetaz, "The Effects of Custodial vs Non-custodial Sanctions on Reoffending: Lessons from a Systematic Review," *Psicothema* 20, no. 1 (2008): 29–34, http://www.psicothema.com /pdf/3425.pdf.

176 *A study of over 1,300 injection drug users* N. Galai et al., "Longitudinal Patterns of Drug Injection Behavior in the ALIVE Study Cohort, 1988–2000: Description and Determinants," *American Journal of Epidemiology* 158, no. 7 (2003): 695–704, doi:10.1093/aje/kwg209, http://aje.oxfordjournals.org/content/158/7/695.long.

176 *Canadian study of 1,600 IV drug users* K. DeBeck et al., "Incarceration and Drug Use Patterns among a Cohort of Injection Drug Users," *Addiction* 104 (2009): 69–76, doi:10.1111/ j.1360-0443.2008.02387.x, http://onlinelibrary.wiley.com/doi/10.1111/j.1360-0443.2008.02387 .x/full.

177 *One study of over 100,000 American children arrested* A. Aizer and J. J. Doyle, "Juvenile Incarceration & Adult Outcomes: Evidence from Randomly-Assigned Judges," *National Bureau of Economic Research,* February 2011, accessed August 6, 2015, http://www.law.yale .edu/documents/pdf/leo/j.doyle.swingjudges_03032011.pdf.

177 *the odds of adult crime were more than 37 times higher if the teen was actually locked up in a reform school or juvenile prison* U. Gatti, R. E. Tremblay, and F. Vitaro, "Iatrogenic effect of juvenile justice," *Journal of Child Psychology and Psychiatry* 50 (2009): 991–98, doi: 10.1111/j.1469-7610.2008.02057.x, http://onlinelibrary.wiley.com/doi/10.1111/j.1469-7610.2008.02057.x/abstract.

177 *the British government assigned experts* UK Home Office, "Drugs: International Comparators," October 2014, accessed August 19, 2015, https://www.gov.uk/government/uploads/system /uploads/attachment_data/file/368489/DrugsInternationalComparators.pdf.

177 *the same conclusion drawn by a 2008 multinational study* L. Degenhardt et al., "Toward a Global View of Alcohol, Tobacco, Cannabis, and Cocaine Use: Findings from the WHO World Mental Health Surveys," *PLOS Medicine* 5, no. 7 (2008): e141, doi:10.1371/journal.pmed.0050141.

178 America tops charts in cocaine and marijuana addiction: Louisa Degenhardt et al., "Global Burden of Disease Attributable to Illicit Drug Use and Dependence: Findings from the Global Burden of Disease Study 2010," *The Lancet* 382, no. 9904 (November 9–15, 2013): 1564–74, http://dx.doi.org/10.1016/S0140-6736(13)61530-5, http://www.sciencedirect.com/science/article/pii/S0140673613615305#.

178 International statistics on heroin and opiate use rates: U.N. World Drug Report, 2014, http://www.unodc.org/wdr2014/. See table of "opiate" use (heroin and opium); opioid tables include prescription drug misuse, which is higher in the United States but is not the target of street-level enforcement that heroin is.

178 *Funding for the drug war* Associated Press, "AP IMPACT: After 40 Years, $1 Trillion, US War on Drugs Has Failed to Meet Any of Its Goals," *Fox News,* May 13, 2010, accessed August 6, 2015, http://www.foxnews.com/world/2010/05/13/ap-impact-years-trillion-war-drugs-failed-meet-goals/.

178 Epidemiological Catchment Area (ECA) study: D. A. Regier et al., "The NIMH Epidemiologic Catchment Area Program: Historical Context, Major Objectives, and Study Population Characteristics," *Archives of General Psychiatry* 41, no. 10 (1984): 934–41, doi:10.1001/archpsyc.1984.01790210016003, http://www.ncbi.nlm.nih.gov/pubmed/6089692.

178 *around 6.1% of citizens in the first half of the '80s had had some type of illegal drug problem* D. A. Regier et al., "Comorbidity of Mental Disorders with Alcohol and Other Drug Abuse: Results from the Epidemiologic Catchment Area (ECA) Study," *JAMA* 264, no. 19 (1990): 2511–18, doi:10.1001/jama.1990.03450190043026, http://www.ncbi.nlm.nih.gov/pubmed/2232018.

178 National Epidemiological Survey on Alcohol and Related Conditions (NESARC): Compton et al., "Prevalence, Correlates, Disability, and Comorbidity."

179 *our trillion-dollar law enforcement spending spree may have actually increased addiction rates* "Jail, Prison, Parole, and Probation Populations in the US, 1980–2013," ProCon.org, accessed August 6, 2015, http://felonvoting.procon.org/view.resource.php?resourceID=004353.

180 12 steps used in at least 80% of American treatment: Anne Fletcher, *Inside Rehab: The Surprising Truth about Addiction Treatment—and How to Get Help That Works* (New York: Viking Press, 2013), 218–19.

180 ASAM views addiction as a disease of powerlessness: American Society of Addiction Medicine, "Definition of Addiction," accessed August 12, 2015, http://www.asam.org/for-the-public/definition-of-addiction.

180 Summary of AA founders' stories: Nan Robertson, *Getting Better: Inside Alcoholics Anonymous* (New York: Fawcett Crest, 1988); and Alcoholics Anonymous, "Alcoholics Anonymous ('Big Book')," third edition (New York: A.A. World Services, 1976), 1–16, 171–81.

181 *"Why all this insistence that every A.A."* Alcoholics Anonymous, "Twelve Steps and Twelve Traditions," accessed August 6, 2015, http://www.aa.org/assets/en_US/en_step1.pdf.

182 *By 2000, 90% of all addiction treatment was 12 step based* Fletcher, *Inside Rehab.*

182 *"still had their health, their families, their jobs, and even two cars in their garage"* Alcoholics Anonymous, "Twelve Steps and Twelve Traditions."

183 *"Without accountability and consequences, drug abusers"* Michael Rothfeld, "Sheen to Lead Fight

Against Rehab Measure," *L.A. Times,* August 28, 2008, http://articles.latimes.com/2008/aug /28/local/me-sheen28.

183 *legally coerced patients typically do not do better in treatment than those who enter voluntarily* A. S. Hampton et al., "Pathways to Treatment Retention for Individuals Legally Coerced to Substance Use Treatment: The Interaction of Hope and Treatment Motivation," *Drug and Alcohol Dependence* 118, nos. 2–3 (2011): 400–07, doi:10.1016/j.drugalcdep.2011.04.022.

183 *significant evidence that empathetic and empowering approaches . . . are far more effective than treatment that relies on confrontation* W. L. White and W. R. Miller, "The Use of Confrontation in Addiction Treatment: History, Science, and Time for a Change," *Counselor* 8 (2007): 12–30.

183 *"Force is the best medicine"* R. Tiger, "Drug Courts and the Logic of Coerced Treatment," *Sociological Forum* 26 (2011): 169–82, doi:10.1111/j.1573-7861.2010.01229.x.

183 Research shows that some people bounce back after relapses: R. H. Moos and B. S. Moos, "Rates and Predictors of Relapse after Natural and Treated Remission from Alcohol Use Disorders," *Addiction* (Abingdon, England), 101, no. 2 (2006): 212–22, doi:10.1111/j.1360-0443.2006.01310.x.

184 Social capital linked to recovery: C. E. Grella and J. A. Stein, "Remission from Substance Dependence: Differences Between Individuals in a General Population Longitudinal Survey Who Do and Do Not Seek Help," *Drug and Alcohol Dependence* 133, no. 1 (2013): 146–53, doi:10.1016/j.drugalcdep.2013.05.019; and Simon J. Adamson, John Douglas Sellman, and Chris M. A. Frampton, "Patient Predictors of Alcohol Treatment Outcome: A Systematic Review," *Journal of Substance Abuse Treatment* 36, no. 1 (January 2009): 75–86, http://dx.doi .org/10.1016/j.jsat.2008.05.007.

184 Examples of abusive treatment: See generally M. Szalavitz, *Help at Any Cost: How the Troubled-Teen Industry Cons Parents and Hurts Kids* (New York: Riverhead, 2006), 26–28; and M. Szalavitz, "Sick Treatment," *City Limits Magazine,* 2003, http://www.citylimits.org/news/articles /2869/the-big-idea-sick-treatment.

185 *everything from psychosurgery to hydrotherapy* White, *Slaying the Dragon.*

185 1941 article quadrupled AA membership within a year: White, *Slaying the Dragon,* 134.

185 *"Alcoholics Anonymous is based on love, we are based on hate"* Paul Morantz, "Carina Ray Story on Children of Synanon," Paul Morantz website, accessed August 7, 2015, http://www .paulmorantz.com/cult/carina-ray/.

186 New Jersey Study and Synanon history: Szalavitz, *Help at Any Cost,* 26–28.

186 Treatment programs becoming cults: See generally Szalavitz, *Help at Any Cost.*

187 virtually every publicly funded inpatient addiction program is based on Synanon: Ibid.

187 *lasting psychological damage in 9% of the normal college students* Lieberman et al., *Encounter Groups,* 174.

188 Story of Terry McGovern: G. McGovern, *Terry: My Daughter's Life and Death Struggle with Alcoholism* (New York: Plume, 1997).

188 Kurt Cobain's suicide and intervention: G. A. Marlatt, "Come As You Are: Cobain, Addiction and Hope," *Seattle Times,* April 24, 2004.

188 *there are methods of intervening gently* R. J. Meyers et al., "A Randomized Trial of Two Methods for Engaging Treatment-Refusing Drug Users Through Concerned Significant Others," *Journal of Consulting and Clinical Psychology* 70, no. 5 (October 2002): 1182–85.

189 Sense of powerlessness fundamental aspect of how trauma causes PTSD: J. L. Herman, *Trauma and Recovery* (New York: Basic Books, 1992), 33.

189 *having PTSD doubles to quadruples the risk of becoming addicted* J. L. McCauley et al., "Posttraumatic Stress Disorder and Co-Occurring Substance Use Disorders: Advances in Assessment and

Treatment," *Clinical Psychology: A Publication of the Division of Clinical Psychology of the American Psychological Association* 19, no. 3 (2012): 229–40, doi:10.1111/cpsp.12006, http://www.ncbi .nlm.nih.gov/pubmed/24179316.

189 *not a single study has supported the confrontational approach* White and Miller, "The Use of Confrontation in Addiction Treatment."

189 Treatment at KIDS: Szalavitz, *Help at Any Cost.*

15. ANTISOCIAL BEHAVIOR

193 Lack of qualifications of addiction counselors: National Center on Addiction and Substance Abuse, "Addiction Medicine: Closing the Gap Between Science and Practice," June 2012, accessed August 6, 2015, http://www.casacolumbia.org/addiction-research/reports /addiction-medicine.

194 *the more a counselor confronts alcoholic patients, the more they drink* W. R. Miller, R. G. Benefield, and J. S. Tonigan, "Enhancing Motivation for Change in Problem Drinking: A Controlled Comparison of Two Therapist Styles," *Journal of Consulting and Clinical Psychology* 61, no. 3 (June 1993): 455–61.

194 Counselor empathy most important factor in addiction counseling success: W. R. Miller and T. B. Moyers, "The Forest and the Trees: Relational and Specific Factors in Addiction Treatment," *Addiction* 110, no. 3 (March 2015): 401–13, e-pub September 12, 2014, doi:10.1111/add.12693.

195 Betty Ford Center costs $53,000 a month: G. Glaser, "The Billion-Dollar Rehab Racket That Drains Family Savings," *The Daily Beast,* May 2, 2015, http://www.thedailybeast.com /articles/2015/05/02/the-million-dollar-rehab-racket-that-drains-family-savings.html.

195 Films and lectures least effective in rehab: R. K. Hester and W. R. Miller, eds., *Handbook of Alcoholism Treatment Approaches: Effective Alternatives,* 3rd ed. (Boston: Allyn & Bacon, 2003), accessed August 7, 2015, http://www.behaviortherapy.com/ResearchDiv/whatworks.aspx.

197 *early psychological and psychiatric literature often discusses addicts as psychopathic* C. A. Haertzen et al., "Measurement of Psychopathy As a State," *Journal of Psychology* 100(2nd half) (November 1978): 201–14.

198 *prevalence of antisocial behavior among children increases by a factor of 10* T. E. Moffitt, "Adolescence-Limited and Life-Course-Persistent Antisocial Behavior: A Developmental Taxonomy," *Psychological Review* 100, no. 4 (October 1993): 674–701, http://www.ncbi.nlm.nih.gov/pubmed /8255953.

199 *CD is "a repetitive and persistent pattern of behavior"* R. Loeber et al., "Oppositional Defiant and Conduct Disorder: A Review of the Past 10 Years, Part I," *Journal of the American Academy of Child and Adolescent Psychiatry* 39, no. 12 (December 2000): 1468–84, http://www.ncbi.nlm.nih .gov/pubmed/11128323.

199 Having both ADHD and CD produces greater addiction risk: Timothy E. Wilens, "Does ADHD Predict Substance-Use Disorders? A 10-Year Follow-up Study of Young Adults with ADHD," *Journal of the American Academy of Child and Adolescent Psychiatry* 50, no. 6 (June 2011): 543–53.

199 between 40% and 75% of those diagnosed don't grow out of CD: H. L. Gelhorn et al., "DSM-IV Conduct Disorder Criteria As Predictors of Antisocial Personality Disorder," *Comprehensive Psychiatry* 48, no. 6 (2007): 529–38, doi:10.1016/j.comppsych.2007.04.009, http:// www.ncbi.nlm.nih.gov/pmc/articles/PMC2764329/.

199 People with addiction no more likely to lie than anyone else when information is confidential: E. J. Chiauzzi and S. Liljegren, "Taboo Topics in Addiction Treatment: An Empirical Review of Clinical Folklore," *Journal of Substance Abuse Treatment* 10, no. 3 (May–June 1993): 303–16.

200 Statistics on prevalence and co-occurrence of ASPD and addiction: R. B. Goldstein et al., "Antisocial Behavioral Syndromes and DSM-IV Drug Use Disorders in the United States: Results from the National Epidemiologic Survey on Alcohol and Related Conditions," *Drug and Alcohol Dependence* 90, nos. 2–3 (2007): 145–58, doi:10.1016/j.drugalcdep .2007.02.023, http://www.ncbi.nlm.nih.gov/pmc/articles/PMC2633099/.

200 ASPD may be as high as 37% in IV opioid users: M. Kidorf et al., "Prevalence of Psychiatric and Substance Use Disorders in Opioid Abusers in a Community Syringe Exchange Program," *Drug and Alcohol Dependence* 74, no. 2 (May 10, 2004): 115–22, http://www.ncbi.nlm .nih.gov/pubmed/15099655.

200 the majority of people with ASPD don't have addictions: W. M. Compton et al., "Prevalence, Correlates, and Comorbidity of DSM-IV Antisocial Personality Syndromes and Alcohol and Specific Drug Use Disorders in the United States: Results from the National Epidemiologic Survey on Alcohol and Related Conditions," *Journal of Clinical Psychiatry* 66, no. 6 (June 2005): 677–85, http://www.ncbi.nlm.nih.gov/pubmed/15960559.

200 *only 16% of people with alcohol or other drug addictions have [borderline personality disorder]* B. F. Grant et al., "Prevalence, Correlates, Disability, and Comorbidity of DSM-IV Borderline Personality Disorder: Results from the Wave 2 National Epidemiologic Survey on Alcohol and Related Conditions," *Journal of Clinical Psychiatry* 69, no. 4 (2008): 533–45, http://www .ncbi.nlm.nih.gov/pmc/articles/PMC2676679/.

203 *creating a coherent narrative out of your experience may help recovery from trauma* R. O'Kearney and K. Perrott, "Trauma Narratives in Posttraumatic Stress Disorder: A Review," *Journal of Traumatic Stress* 19 (2006): 81–93, doi:10.1002/jts.20099.

203 Same genes increase risk for many psychiatric disorders: Cross-Disorder Group of the Psychiatric Genomics Consortium, "Genetic Relationship Between Five Psychiatric Disorders Estimated from Genome-wide SNPs," *Nature Genetics* 45, no. 9 (2013): 984–94, doi:10.1038/ng.2711.

204 Family therapies best treatment for teen addictions: A. Hogue et al., "Evidence Base on Outpatient Behavioral Treatments for Adolescent Substance Use: Updates and Recommendations 2007–2013," *Journal of Clinical Child & Adolescent Psychology* 43, no. 5 (2014): 695–720, doi:10.1080/15374416.2014.915550.

204 *most important "active ingredient" is a good rapport [with therapists]* Miller and Moyers, "The Forest and the Trees."

16. THE 12-STEP CONUNDRUM

210 Connections among the Oxford Group, Nazis, and AA: Robertson, *Getting Better,* 17–56.

211 *whether a counselor has his or her own addiction history does not affect outcomes* L. S. Aiken et al., "Paraprofessional Versus Professional Drug Counselors: The Progress of Clients in Treatment," *International Journal of Addiction* 19, no. 4 (July 1984): 383–401; and B. Brown and R. Thompson, "The Effectiveness of Formerly Addicted and Nonaddicted Counsellors on Client Functioning," *Drug Forum* 5 (1975–76): 123–28.

211 *Social support is the single most important factor in mitigating severe stress and trauma* Perry and Szalavitz, *The Boy Who Was Raised as a Dog,* 80.

212 CBT is one of most effective therapies for addiction: R. K. McHugh, B. A. Hearon, and M. W. Otto, "Cognitive-Behavioral Therapy for Substance Use Disorders," *Psychiatric Clinics of North America* 33, no. 3 (2010): 511–25, doi:10.1016/j.psc.2010.04.012.

217 2006 Cochrane Review of AA outcomes research: M. Ferri, L. Amato, and M. Davoli, "Alcoholics Anonymous and Other 12-Step Programmes for Alcohol Dependence," *Cochrane Database of Systematic Reviews* no. 3 (2006), art. no. CD005032, doi:10.1002/14651858. CD005032.pub2, http://www.ncbi.nlm.nih.gov/pubmed/16856072.

217 Selection bias: *Bandolier Journal,* Oxford University, accessed August 24, 2015, http://www.medicine.ox.ac.uk/bandolier/booth/glossary/selectbi.html. The issue is also discussed in the article cited below, which unfortunately claims that a study was able to remove the influence of selection bias in AA research. The problem is that the researchers studied only the group that complied with treatment, which is an example of selection bias itself: A. Frakt, "Alcoholics Anonymous and the Challenge of Evidence-Based Medicine," *New York Times,* April 6, 2015.

217 *when you force people into AA, they do no better* R. J. Kownacki and W. R. Shadish, "Does Alcoholics Anonymous Work? The Results from a Meta-analysis of Controlled Experiments," *Substance Use & Misuse* 34, no. 13 (November 1999): 1897–916, http://www.ncbi.nlm.nih.gov/pubmed/10540977.

217 *over 200 workers mandated by their employers to get help* Diana Chapman Walsh et al., "A Randomized Trial of Treatment Options for Alcohol-Abusing Workers," *New England Journal of Medicine* 325 (September 12, 1991): 775–82, doi:10.1056/NEJM199109123251105, http://www.nejm.org/doi/full/10.1056/NEJM199109123251105.

218 *one study followed 628 alcoholics* R. H. Moos and B. S. Moos, "Participation in Treatment and Alcoholics Anonymous: A 16-Year Follow-Up of Initially Untreated Individuals," *Journal of Clinical Psychology* 62, no. 6 (2006): 735–50, doi:10.1002/jclp.20259, http://www.ncbi.nlm.nih.gov/pmc/articles/PMC2220012/.

218 *On average, 70% of those referred to AA drop out* R. F. Forman, K. Humphreys, and J. S. Tonigan, "Response: The Marriage of Drug Abuse Treatment and 12-Step Strategies," *Science & Practice Perspectives* 2, no. 1 (2003): 52–54, http://www.ncbi.nlm.nih.gov/pmc/articles/PMC2851049/.

218 *the minority who do find these programs amenable are more likely to recover* S. E. Zemore, M. Subbaraman, and J. S. Tonigan, "Involvement in 12-step Activities and Treatment Outcomes," *Substance Abuse: Official Publication of the Association for Medical Education and Research in Substance Abuse* 34, no. 1 (2013): 60–69, doi:10.1080/08897077.2012.691452.

219 *large organizations that have inflicted traumatic stress* Szalavitz, *Help at Any Cost.*

219 *two thirds of Americans support employment discrimination against people with addictions* L. Colleen et al., "Stigma, Discrimination, Treatment Effectiveness, and Policy: Public Views about Drug Addiction and Mental Illness," *Psychiatric Services* 65, no. 10 (2014): 1269–72, http://psychiatryonline.org/doi/abs/10.1176/appi.ps.201400140?url_ver=Z39.88-2003&rfr_id=ori%3Arid%3Acrossref.org&rfr_dat=cr_pub%3Dpubmed&.

219 *"bio-psycho-socio-spiritual manifestations"* American Society of Addiction Medicine, "Definition of Addiction."

219 *the more someone believes in the idea that addiction is a disease over which he is powerless* W. R. Miller et al., "What Predicts Relapse? Prospective Testing of Antecedent Models," supplement, *Addiction* 91 (December 1996): S155–72.

220 50% of women in AA have "13th step" experience: Cathy J. Bogart and Carol E. Pearce, "'13th-Stepping': Why Alcoholics Anonymous Is Not Always a Safe Place for Women,"

Journal of Addictions Nursing 14, no. 1 (2003): 43–47, doi:10.1080/10884600305373, http://informahealthcare.com/doi/abs/10.1080/10884600305373.

222 All court decisions in the United States find mandatory AA violates constitution: Case law can be found on the website of SMART Recovery, accessed August 14, 2015, http://www.smartrecovery.org/courts/court-mandated-attendance.htm.

222 *"forever nonprofessional . . . our usual 12ᵗʰ-step work"* Alcoholics Anonymous, "The Twelve Traditions (The Long Form)," accessed August 7, 2015, http://www.aa.org/assets/en_US/en_tradition_longform.pdf.

17. HARM REDUCTION

226 *whites use drugs at the same rates as African Americans and actually sell more* Jonathan Rothwell, "How the War on Drugs Damages Black Social Mobility," Brookings Institution, September 30, 2014, http://www.brookings.edu/blogs/social-mobility-memos/posts/2014/09/30-war-on-drugs-black-social-mobility-rothwell.

226 *the lifetime odds of going to prison for a black man in America are 1 in 3* Christopher J. Lyons and Becky Pettit, "Compounded Disadvantage: Race, Incarceration, and Wage Growth," *Social Problems* 58, no. 2 (May 2011): 257–80, http://www.jstor.org/discover/10.1525/sp.2011.58.2.257?uid=3739832&uid=2&uid=4&uid=3739256&sid=21105640068683.

226 *a figure that has doubled since Richard Nixon declared war on drugs in the 1970s* Rothwell, "How the War on Drugs Damages Black Social Mobility."

226 White men have six percent chance of going to prison: The Sentencing Project, "Facts about Prisons and People in Prison," accessed August 24, 2015, http://sentencingproject.org/doc/publications/inc_Facts%20About%20Prisons.pdf.

226 *Between 1980 and 2011, the annual number of drug arrests of black men* Rothwell, "How the War on Drugs Damages Black Social Mobility"; and author interview with Rothwell, 2015. Rothwell cites H. Snyder and J. Mulako-Wangota, Arrest Data Analysis Tool, September 23, 2014, at www.bjs.gov. Bureau of Justice Statistics, Washington, D.C.

227 *black people are ten times more likely to get arrested for drug crimes [than whites]* Human Rights Watch, "Decades of Disparity: Drug Arrests and Race in the United States," March 2009, http://www.hrw.org/sites/default/files/reports/us0309web_1.pdf.

227 *African Americans are also ten times more likely to be sent to prison for these offenses* Jamie Fellner, "Race, Drugs, and Law Enforcement in the United States," *Stanford Law & Policy Review* 20 (2009): 257–92, http://www.hrw.org/sites/default/files/related_material/8%20Fellner_FINAL.pdf.

227 *on average blacks serve nearly as long for drug offenses as whites* National Association for the Advancement of Colored People, "Criminal Justice Fact Sheet," accessed August 7, 2015, http://www.naacp.org/pages/criminal-justice-fact-sheet.

227 *82% of people sentenced under the harsh federal mandatory minimums for crack cocaine are black* The Sentencing Project, "Crack Cocaine Sentencing Key to Racial Disparity in Federal Prison," accessed August 7, 2015, http://www.sentencingproject.org/doc/advocacy/dp_crackandracialdisparitybriefingsheet.pdf.

227 *"I am obliged to enforce the law however stupid and irrational and barbarous it be"* Report by the Correctional Association of New York, "Stupid and Irrational and Barbarous: New York Judges Speak Against the Rockefeller Drug Laws," December 2001, http://www.correctionalassociation.org/wp-content/uploads/2012/05/judgesreport.pdf.

227 More than 80% of those sentenced had no history of violence: Madison Gray, "A Brief

History of New York's Rockefeller Drug Laws," *Time*, April 2, 2009, http://content.time
.com/time/printout/0,8816,1888864,00.html.

228 *By 2008, more than 90% of New York State's drug prisoners were minorities Testimony Before the New
York State Assembly Committees on Codes, Judiciary, Correction, Health, Alcoholism and Drug Abuse, and
Social Services regarding the Rockefeller Drug Laws* (statement of Robert A. Perry, Legislative Director of the New York Civil Liberties Union), accessed August 7, 2015, http://www.nyclu
.org/content/rockefeller-drug-laws-cause-racial-disparities-huge-taxpayer-burden.

228 *It took until 2009 for significant reform to be enacted* Leslie Kellam and Leigh Bates, "2009 Drug
Law Changes: 2014 Update," New York State Division of Criminal Justice Services, Drug
Law Series, report no. 5 (May 2014), accessed August 20, 2015, http://criminaljustice.ny
.gov/drug-law-reform/documents/dlr-update-report-may-2014.pdf.

228 *the number of New York City residents sent to state prison annually . . . quintupled* James Austin, Michael P.
Jacobson, and Inimai M. Chettiar, "How New York City Reduced Mass Incarceration: A
Model for Change?," Brennan Center Report, January 2013, http://www.brennancenter.org
/sites/default/files/publications/How_NYC_Reduced_Mass_Incarceration.pdf.

228 *the American incarceration rate rose over 400% . . . 200,000 in 1973 to 1.574 million in 2013* Sentencing
Project, "US State and Federal Prison Population, 1925–2013," Bureau of Justice Statistics Prisoners Series, accessed August 6, 2015, http://www.sentencingproject.org
/template/page.cfm?id=107. See also Jeremy Travis, Bruce Western, and Steve Redburn,
eds., *The Growth of Incarceration in the United States: Exploring Causes and Consequences* (Washington, DC: National Academy Press, 2014), http://www.nap.edu/catalog/18613/the-growth
-of-incarceration-in-the-united-states-exploring-causes.

228 *the total correctional population is close to 7 million* Lauren E. Glaze and Danielle Kaeble, "Correctional Populations in the United States, 2013," Bureau of Justice Statistics, NCJ 248479,
December 19, 2014, http://www.bjs.gov/index.cfm?ty=pbdetail&iid=5177.

228 The United States is the world's largest jailer: American Civil Liberties Union, "The Prison
Crisis," accessed August 7, 2015, https://www.aclu.org/safe-communities-fair-sentences
/prison-crisis.

229 Media statistics related to crack: Reinarman and Levine, *Crack in America*.

229 Since the 1970s, most heroin addicts have been white and blacks are underrepresented
since 2010: T. J. Cicero et al., "The Changing Face of Heroin Use in the United States: A
Retrospective Analysis of the Past 50 Years," *JAMA Psychiatry* 71, no. 7 (2014): 821–26,
doi:10.1001/jamapsychiatry.2014.366.

229 *Rates of heroin addiction in people making less than $20,000* Centers for Disease Control, Vital Signs,
Today's Heroin Epidemic Infographics. Data from National Household Survey on Drug
Use and Health, 2011–2013. http://www.cdc.gov/vitalsigns/heroin/infographic.html

230 *"It is clear that the defendant has made extraordinary strides . . . benefit from her work."* From sealed records of New York State Supreme Court: Part 72, July 17, 1992.

232 *a group of IV drug users in the Netherlands* New York State Department of Health, AIDS Institute, "Comprehensive Harm Reduction Reverses the Trend in New HIV Infections," Albany, NY, March 2014, http://www.health.ny.gov/diseases/aids/providers/reports/docs
/sep_report.pdf.

232 *"the spread of HIV is a greater danger to individual and public health than drug misuse"* Advisory
Council on the Misuse of Drugs, "AIDS and Drug Misuse, Part 1" (London: HMSO,
1988): 17.

233 *the percentage of deaths associated with drunk driving has fallen* National Highway Traffic Safety

Administration, "Statistical Analysis of Alcohol-Related Driving Trends, 1982–2005," May 2008, http://www-nrd.nhtsa.dot.gov/Pubs/810942.pdf.

233 Drunk driving fatality statistics: Mothers Against Drunk Driving, "History of Drunk Driving: Drunk Driving Deaths 1982–2013," accessed August 7, 2015, http://www.madd.org/drunk-driving/about/history.html.

234 Leshner calls for ban on use of term "harm reduction": A. Leshner, "By Now, 'Harm Reduction' Harms Both Science and the Public Health," *Clinical Pharmacology & Therapeutics* 83 (2008): 513–14, doi:10.1038/sj.clpt.6100478.

234 Jon Parker and legal decision on needle exchange in New York: New York State Department of Health, AIDS Institute, "Comprehensive Harm Reduction Reverses the Trend in New HIV Infections," Albany, NY, March 2014, http://www.health.ny.gov/diseases/aids/providers/reports/docs/sep_report.pdf.

234 *"the one intervention which could be described as the gold standard of HIV prevention"* Ibid.

236 Erin's story and quote: M. Szalavitz, "AIDS: Words from the Front," *Spin,* July 1991, https://books.google.com/books?id=wbkGT02oWs4C&pg=PA67&lpg=PA67&dq=szalavitz+%22jon+parker%22&source=bl&ots=j0i-mfygbI&sig=pQsv26aGn0YWTRlq5-Hw0R83ttI&hl=en&sa=X&ei=vOEjVYfWHMaLsAXspoHIBg&ved=0CEEQ6AEwBg#v=onepage&q=szalavitz%20%22jon%20parker%22&f=false.

236 *Dan Bigg . . . defines recovery as "any positive change"* M. Szalavitz, "It's Time to Reclaim the Word 'Recovery,'" *Pacific Standard,* December 12, 2014, http://www.psmag.com/health-and-behavior/its-time-to-reclaim-the-word-recovery-addiction-treatment-12-step-programs-96037.

237 *"pain [is] the touchstone of all spiritual progress"* Alcoholics Anonymous, "Twelve Steps and Twelve Traditions," 93.

237 My antidepressant story and facts within: Parts of the preceding few paragraphs appeared in somewhat different form in a 2005 article I wrote on antidepressants for *Reason*: M. Szalavitz, "In Defense of Happy Pills," *Reason,* October 2005, http://reason.com/archives/2005/10/01/in-defense-of-happy-pills.

237 Most serotonin in gut, not brain: J. C. Bornstein, "Serotonin in the Gut: What Does It Do?," *Frontiers in Neuroscience* 6, no. 16 (2012): doi:10.3389/fnins.2012.00016.

241 *students do better in schools where they feel welcomed and safe* Centers for Disease Control and Prevention, "School Connectedness: Strategies for Increasing Protective Factors among Youth" (Atlanta, GA: US Department of Health and Human Services, 2009).

241 *People learn best in environments where they feel* Ibid.

241 *"Harm reduction does not try to remove a person's primary coping . . . in place"* G. Alan Marlatt, "Harm Reduction: Come As You Are," *Addictive Behaviors* 21, no. 6 (November–December 1996): 779–88, http://dx.doi.org/10.1016/0306-4603(96)00042-1, http://www.sciencedirect.com/science/article/pii/0306460396000421.

18. THE KIWI APPROACH

243 Portions of this chapter first appeared in slightly different form in *Pacific Standard* magazine, Szalavitz, "The Drug Lord with a Social Mission." Any unsourced quotes are from interviews I did for this article.

245 Tobacco cuts life expectancy by ten years: US Department of Health and Human Services, "The Health Consequences of Smoking—50 Years of Progress: A Report of the Surgeon General," (Atlanta, GA: Centers for Disease Control and Prevention, National

Center for Chronic Disease Prevention and Health Promotion, Office on Smoking and Health, 2014).

245 Marijuana not associated with increased mortality: L. Degenhardt et al., "The Global Epidemiology and Contribution of Cannabis Use and Dependence to the Global Burden of Disease: Results from the GBD 2010 Study," *PLOS ONE* 8, no. 10 (2013): e76635, doi:10.1371/journal.pone.0076635.

246 Data on making opioids from yeast: Kenneth A. Oye et al., "Drugs: Regulate 'Home-Brew' Opiates," *Nature* 521, no. 281 (May 21, 2015).

247 *heart problems caused by the now-discontinued painkiller Vioxx* R. Knox, "Merck Pulls Arthritis Drug Vioxx from Market," All Things Considered, National Public Radio, September 30, 2004.

249 *"I've banned 33 separate substances"* Andrew Laxon, "Get Your (Ultra Mild) Legal Highs Here," *New Zealand Herald,* May 25, 2013, http://www.nzherald.co.nz/nz/news/article.cfm?c_id=1&objectid=10885984.

250 Supreme Court decision on analog law: S. Hananel, "Supreme Court Overturns Conviction in Synthetic Drug Case,"*Associated Press,* June 18, 2015, accessed August 20, 2015, http://www.huffingtonpost.com/2015/06/18/supreme-court-synthetic-drugs_n_7612622.html.

250 *Testing of these health store products frequently finds new psychoactive substances* P. A. Cohen et al., "An Amphetamine Isomer Whose Efficacy and Safety in Humans Has Never Been Studied, β-methylphenylethylamine (BMPEA), Is Found in Multiple Dietary Supplements," *Drug Testing and Analysis,* in press (2015), doi:10.1002/dta.1793.

251 *booze causes six percent of deaths worldwide* National Institute on Alcohol Abuse and Alcoholism, "Alcohol Facts and Statistics," accessed August 17, 2015, http://www.niaaa.nih.gov/alcohol-health/overview-alcohol-consumption/alcohol-facts-and-statistics.

251 Statistics on death rates of various activities: M. Szalavitz, "The Drug Lord with a Social Mission," *Pacific Standard,* March 2015.

252 *[The FDA] is no longer permitted to weigh* Reuters, "U.S. to Roll Back 'Lost Pleasure' Approach on Health Rules," *New York Times,* March 18, 2015.

252 LSD not linked to mental health problems or overdose death: P. Ø. Johansen and T. S. Krebs, "Psychedelics Not Linked to Mental Health Problems or Suicidal Behavior: A Population Study," *Journal of Psychopharmacology* 29, no. 3 (March 2015): 270–79, doi:10.1177/0269881114568039.

256 United States says we're not violating UN conventions, claims they are flexible: William R. Brownfield, "Trends in Global Drug Policy," US Department of State Press Conference, New York, NY, October 9, 2014, accessed August 17, 2015, http://fpc.state.gov/232813.htm.

256 Federal officials and harm reduction: W. Godfrey, "US Drug Czar Opens National Harm Reduction Conference in Unprecedented Move," *Substance.com,* October 23, 2014, http://www.substance.com/us-drug-czar-opens-national-harm-reduction-conference-in-unprecedented-move/14526/.

256 *a U.S. government agency has spent decades* G. Lopez, "Federal Spending on Medical Marijuana Research Is Pathetic," *Vox.com,* August 20, 2015, http://www.vox.com/2015/8/20/9180047/marijuana-research-federal-government.

256 Synthetic cannabinoid users prefer marijuana: Adam R. Winstock and Monica J. Barratt, "Synthetic Cannabis: A Comparison of Patterns of Use and Effect Profile with Natural Cannabis in a Large Global Sample," *Drug and Alcohol Dependence* 131, nos. 1–2 (July 1,

2013): 106–11, http://dx.doi.org/10.1016/j.drugalcdep.2012.12.011, http://www.sciencedirect.com/science/article/pii/S0376871612004875.

256 Reasons for taking legal highs: Ibid.

19. TEACHING RECOVERY

259 *"In 'treatment,' the focus is on individual pathology"* M. Szalavitz, "Viewpoint: Teaching Recovery, Rather Than Treating Addiction," *TIME.com,* September 30, 2011, http://healthland.time.com/2011/09/30/viewpoint-teaching-recovery-rather-than-treating-addiction/.

261 Evaluation of ARRIVE: H. K. Wexler et al., "ARRIVE: An AIDS Education/Relapse Prevention Model for High-Risk Parolees," *International Journal of the Addictions* 29, no. 3 (February 1994): 361–86.

261 Elsa Gonzalez story and quote: Author interview with Elsa Gonzalez.

262 Mary Gordon quote and Roots of Empathy information: Szalavitz and Perry, *Born for Love,* 4.

264 *"He comes back and he tells me"* "Overcriminalized #2: Substance Abuse," YouTube video, 9:07, by Brave New Films, October 9, 2014, https://www.youtube.com/watch?v=_66uT64YzbY&list=PLQ9B-p5Q-YOMyNkARjA4UDVJIWnb7Xybu&index=5.

264 Misti Barrickman background information: Author interview with Misti Barrickman, 2015; and "Overcriminilazied #2: Substance Abuse."

264 *"you'll come back and say, 'I want to deal with my addiction'"* Saki Knafo, "Change of Habit: How Seattle Cops Fought an Addiction to Locking Up Drug Users," *Huffington Post,* August 28, 2014, http://www.huffingtonpost.com/2014/08/28/seattle-lead-program_n_5697660.html.

264 Statistics on LEAD and rearrest: Susan E. Collins, Heather S. Lonczak, and Seema L. Clifasefi, "LEAD Program Evaluation: Recidivism Report," Harm Reduction Research and Treatment Lab University of Washington—Harborview Medical Center, Seattle, March 27, 2015, http://static1.1.sqspcdn.com/static/f/1185392/26121870/1428513375150/LEAD_EVALUATION_4-7-15.pdf.

265 *Over one third of patients treated for substance addictions* Substance Abuse and Mental Health Services Administration, Treatment Episode Data Set (TEDS) 2002–2012. National Admissions to Substance Abuse Treatment Services, BHSIS series S-71, HHS publication no. (SMA) 14-4850, Rockville, MD (2014), Table 2.6: "Admissions Aged 12 and Older, by Treatment Referral Source and Detailed Criminal Justice Referral According to Primary Substance of Abuse: 2011 Percent Distribution," http://www.samhsa.gov/data/sites/default/files/TEDS2012N_Web.pdf.

265 *in some programs, the proportion is closer to 80% or 90%, according to Kerwin Kaye* Author interview with Kerwin Kaye, 2015.

266 Statistics on drug courts and maintenance treatments: Harlan Matusow et al., "Medication Assisted Treatment in US Drug Courts: Results from a Nationwide Survey of Availability, Barriers and Attitudes," *Journal of Substance Abuse Treatment* 44, no. 5 (May–June 2013): 473–80, http://dx.doi.org/10.1016/j.jsat.2012.10.004, http://www.sciencedirect.com/science/article/pii/S0740547212004205.

266 Additional material on drug courts in this chapter: M. Szalavitz, "How America Overdosed on Drug Courts," *Pacific Standard,* May 18, 2015.

267 Victor Maes quote and story: Knafo, "Change of Habit."

267 Gloucester, MA, offers amnesty to people with heroin addictions; other communities

looking at similar approach: T. McCoy, "The 'Only Town' in America Where Cops Grant Amnesty to Drug Addicts Seeking Help," *Washington Post,* August 17, 2015.

267 Portugal statistics: C. E. Hughes and A. Stevens, "What Can We Learn from the Portuguese Decriminalization of Illicit Drugs?," *British Journal of Criminology* 50, no. 6 (2010): 999–1022, doi:10.1093/bjc/azq038. See also M. Szalavitz, "Drugs in Portugal: Did Decriminalization Work?," *TIME.com,* April 26, 2009.

268 WHO supports decriminalization: World Health Organization, "Policy Brief: Consolidated Guidelines on HIV Prevention, Diagnosis, Treatment and Care for Key Populations," July 2014, accessed August 18, 2015, http://www.who.int/hiv/pub/guidelines/briefs_pwid_2014.pdf?ua=1.

268 Research on "venture" programs: P. J. Conrod et al., "Effectiveness of a Selective, Personality-Targeted Prevention Program for Adolescent Alcohol Use and Misuse: A Cluster Randomized Controlled Trial," *JAMA Psychiatry* 70, no. 3 (2013): 334–42, doi:10.1001/jamapsychiatry.2013.651; and Maeve O'Leary-Barrett et al., "Two-Year Impact of Personality-Targeted, Teacher-Delivered Interventions on Youth Internalizing and Externalizing Problems: A Cluster-Randomized Trial," *Journal of the American Academy of Child & Adolescent Psychiatry* 52, no. 9 (2013): 911–20.

20. NEURODIVERSITY AND THE FUTURE OF ADDICTION

272 UN plans for drug-free world by 2008: J. Buxton, *The Politics of Narcotic Drugs: A Survey* (London: Routledge, 2011), 177.

273 Drug czar speaks at Harm Reduction Coalition: W. Godfrey, "US Drug Czar Opens National Harm Reduction Conference in Unprecedented Move," *Substance.com,* October 23, 2014, http://www.substance.com/us-drug-czar-opens-national-harm-reduction-conference-in-unprecedented-move/14526/.

273 John Walsh quote: Editorial, "Latin America Rethinks Drug Policies," *New York Times,* May 26, 2015, http://www.nytimes.com/2015/05/26/opinion/latin-america-rethinks-drug-policies.html.

275 Smoking statistics: Calculated from charts on smoking trends in Johnston et al., "Monitoring the Future," 394–97.

275 Smokers more likely to be poor and/or have mental illness: Centers for Disease Control and Prevention, "Vital Signs: Current Cigarette Smoking Among Adults Aged ≥18 Years with Mental Illness—United States, 2009–2011," *Morbidity and Mortality Weekly Report* 62, no. 5 (February 8, 2013): 81–87.

277 *the more autonomy and control people feel they have* M. Muraven, M. Gagné, and H. Rosman, "Helpful Self-Control: Autonomy Support, Vitality, and Depletion," *Journal of Experimental Social Psychology* 44, no. 3 (2008): 573–85, doi:10.1016/j.jesp.2007.10.008, http://www.ncbi.nlm.nih.gov/pmc/articles/PMC2390997/.

278 Inpatient treatment no more effective than intensive outpatient: Dennis McCarty et al., "Substance Abuse Intensive Outpatient Programs: Assessing the Evidence," *Psychiatric Services* 65, no. 6 (2014): 718–26.

279 Ineffectiveness of fear-based prevention: Prevention First, "Ineffectiveness of Fear Appeals in Youth Alcohol, Tobacco and Other Drug (ATOD) Prevention" (Springfield, IL: Prevention First, 2008), accessed August 21, 2015, https://www.prevention.org/resources/sapp/documents/ineffectivenessoffearappealsinyouthatodprevention-final.pdf.

280 Programs to improve self-regulation also reduce drug problems: S. G. Kellam et al., "The Good Behavior Game and the Future of Prevention and Treatment," *Addiction Science & Clinical Practice* 6, no. 1 (2011): 73–84.

283 Grit important in success: A. L. Duckworth and J. J. Gross, "Self-control and Grit: Related but Separable Determinants of Success," *Current Directions in Psychological Science* 23, no. 5 (2014): 319–25.

285 *former cocaine and heroin addicts had a greater volume of gray matter in these regions* C. G. Connolly et al., "Dissociated Grey Matter Changes with Prolonged Addiction and Extended Abstinence in Cocaine Users," *PLOS ONE* 8, no. 3 (2013): e59645, doi:10.1371/journal.pone.0059645, cited in M. Lewis, *The Biology of Desire: Why Addiction Is Not a Disease* (New York: Public Affairs, 2015), 151.

INDEX

OCD. *See* obsessive-compulsive disorder
Olds, James, brain's "pleasure center"
  discovery of, 108
opioid maintenance, 132
opium
  California banning of, 26
  physical dependence produced by, 30
orbitofrontal cortex, 62
"overlearning," 39, 98–99
Oxford Group, Christian revival movement,
  210
Oxycontin, 263
oxytocin, 153, 163
  in addiction, 145, 147
  in autism, 145
  as critical to social lives of mammals,
    144
  human bonds relevance in, 145
  during infancy, 147
  as "love hormone," 145
  social connections regarding, 146–47
  wiring of, 147

pain, 33
  Pasternak example of, 34
"pair bonding," 143–45, 163
Palmer, Chen, 250
Panksepp, Jaak, 164–65
parent (parenting)
  compared to OCD and addiction, 165–66
  Szalavitz on love of, for child, 163–64
"parentified child," 206
parenting. *See* parent
Parker, Dorothy, 284
Parker, Jon, 234
Parkinson's disease, 109–11
Pasternak, Gavril, 34
"path dependence," 46–47
Peele, Stanton, 36, 37, 143, 149, 150
personality disorders, 43
Pert, Candace, 143
Pessl, Marisha, 174
PFC. *See* prefrontal cortex
"pharmacological Calvinism," 114
The Player, 214
"pleasures of the feast," 111
"pleasures of the hunt," 111
PLOS Medicine, 177
Poe, Edgar Allan, 20
Point Counter Point (Huxley), 121

Poisons Information Center, Freiburg,
  Germany, 253
Portugal
  "dissuasion committee" in, 267
  drug use results in, 268
  LEAD strategy in, 267–68
posttraumatic stress disorder (PTSD), 67, 91,
  117, 119, 189
Power, Mike, 243
powerlessness, 219–20
prefrontal cortex (PFC), 62, 63, 98, 285
present focused study, 161
"Preventure," 269
prisoner analogy, 162
Prohibition, 2, 24, 172
  failure of, 28–29
  immigrant groups focus of, 28
  industrial alcohol regarding, 28
  Ku Klux Klan as supporter of, 27–28
  murder and alcoholism rates in, 28
pruning process, 98
"pseudomature" behavior, 88
Psychoactive Substances Act, 250, 251
  Parliament regarding, 254–55
Psychoactive Substances Regulatory
  Authority
  defining risk levels in, 245
  interim regulations of, 253–54
  issues in regulation of, 252–53
PTSD. *See* posttraumatic stress disorder
punishment
  addiction regarding, 171–72, 179
  drug courts replacement of, with
    treatment, 182–83
  incarceration failure in, 182–83
  problem of resistance to, 276
Pure Food and Drug Act (1906), 27
"put gratitude in your attitude," 212

Quackenbush, Mrs., 57
"quantum change," 15

racism, 58, 226
  addiction origins of, 25–26, 34–35, 179
  in drug laws, 25, 27, 172–73, 179, 227
  in prohibition of marijuana, 26
Radio City Music Hall, 169
"Rat Park" experiments, 132–34, 274
Rather, Dan, 82
Reagan, Ronald, 103, 172